Optimal and Robust Control

Optimal and Robust Control

Advanced Topics with MATLAB®

Second Edition

Luigi Fortuna
Mattia Frasca
Arturo Buscarino

CRC Press
Taylor & Francis Group
Boca Raton London New York

CRC Press is an imprint of the
Taylor & Francis Group, an **informa** business

Second edition published 2022
by CRC Press
6000 Broken Sound Parkway NW, Suite 300, Boca Raton, FL 33487-2742

and by CRC Press
2 Park Square, Milton Park, Abingdon, Oxon, OX14 4RN

First edition published by CRC Press 2012

CRC Press is an imprint of Taylor & Francis Group, LLC

Library of Congress Cataloging-in-Publication Data

ISBN: 978-1-032-05300-4 (hbk)
ISBN: 978-1-032-05301-1 (pbk)
ISBN: 978-1-003-19692-1 (ebk)
ISBN: 978-1-032-15155-7 (ebk+)

DOI: 10.1201/9781003196921

Publisher's note: This book has been prepared from camera-ready copy provided by the authors.

Access the Support Material: https://www.routledge.com/Optimal-and-Robust-Control-Advanced-Topics-with-MATLAB/Fortuna-Frasca/p/book/9781032053004

eResources are available for this title at: https://www.crcpress.com/9781032053004

Dedicated to our wives

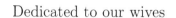

Contents

Preface

The main aim of this book is to provide for undergraduate and graduate students, as well as researchers, who already possess the main concepts of automatic control and system analysis, a self-contained resource collecting advanced techniques for linear system theory and control design. Selected theoretical backgrounds are also presented in the book, together with many numerical exercises and MATLAB® examples.

We intend to offer a complete and easy-to-read handbook of advanced topics in automatic control, including techniques such as the Linear Quadratic Regulator (LQR) and H_∞ control. Large emphasis is also given to Linear Matrix Inequalities (LMIs) with the purpose of demonstrating their use as a unifying tool for system analysis and control design.

In presenting the different approaches to control design, the books explicitly takes into account the problem of the robustness of the obtained closed-loop control. Robustness, in fact, represents the capability of a control system to guarantee the stability in the presence of uncertainty, due to the model itself or to the use of approximated models, and as such is deemed as particularly important in view of the practical implementation of the control techniques.

Many books on LQR control and H_∞ control have been proposed since 1980. The LMI technique has become well-known in the control community, and MATLAB® toolboxes to solve advanced control problems have been developed. However, these subjects are often presented for a specialist audience in materials that are excellent resources for researchers and PhD students. This book, on the contrary, is oriented to illustrate these topics in an easy and concise way, using a language suitable for students, yet maintaining the necessary mathematical rigor.

This book is, therefore, a compendium of many ordered subjects. For specific proofs, the reader is often referred to the proposed literature. Many examples and MATLAB® based exercises are included here to assist the reader in understanding the proposed methods. The book can be considered as a palimpsest of advanced modern topics in automatic control, including an advanced set of analytical examples and MATLAB® exercises. The topics included in the book are mainly illustrated with reference to continuous-time linear systems, even if some results for discrete-time systems are briefly recalled.

The book is organized into chapters structured as follows. The first chapter is an introduction to advanced control, the second discusses some fundamental concepts on stability and provides the tools for studying uncertain systems.

The third presents the Kalman decomposition. The fourth chapter is on singular value decomposition of a matrix, given the importance of numerical techniques for systems analysis. The fifth and sixth chapters are on open-loop balanced realization and reduced order models. The seventh chapter presents the essential aspects of variational calculus and optimal control and the eighth illustrates closed-loop balancing. The properties of positive-real, bounded-real and negative-imaginary systems are the subject of the ninth and tenth chapter. In the eleventh and twelfth chapter, the essential aspects of H_∞ control and LMI techniques commonly used in control systems design are dealt with. The thirteenth chapter is devoted to discuss the class of stabilizing controllers. The fourteenth chapter reviews some of the problems already discussed in the book by introducing an approach based on the steady-state solution of a nonlinear dynamical system. Finally, the fifteenth chapter briefly discusses some fundamental aspects of time-delay systems. The book also includes numerous examples and exercises, considered indispensable for learning the methodology of the topics dealt with, and a list of essential references.

This book is targeted at electrical, electronic, computer science, space and automation engineers interested in advanced topics on automatic control. Mechanical engineers as well as engineers from other fields may also be interested in the topics of the book. The contents of the book can be learned autonomously by the reader in less than a semester.

Symbol List

Symbol	Description
$\mathbb{R}$	the set of real numbers
$\mathbb{C}$	the set of complex numbers
$\mathbb{R}^n$	real vectors of n components
$\mathbb{C}^n$	complex vectors of n components
$\mathbb{R}^{m \times n}$	real matrices of dimensions $m \times n$
$\mathbb{C}^{m \times n}$	complex matrices of dimensions $m \times n$
$\Re e\ (x)$	real part a of a complex number $x = a + jb$
$\Im m\ (x)$	imaginary part b of a complex number $x = a + jb$
$\lvert x \rvert$	absolute value of a real number or the modulus of a complex number
$\lVert x \rVert$	norm of vector x
a_{ij}	coefficient of line i and column j of matrix A
sup	superior extreme of a set
inf	inferior extreme of a set
I	identity matrix of opportune dimensions
A^T	transpose of a matrix
A^*	conjugate transpose of $\mathrm{A} \in \mathbb{C}^{n \times n}$
$\mathrm{A} \otimes \mathrm{B}$	Kronecker product
$\mathrm{A} * \mathrm{B}$	Hadamard product (component by component product of two square matrices, i.e., $\mathrm{A} * \mathrm{B} = \{a_{ij} b_{ij}\}$)
det (A)	determinant of matrix A
$\mathtt{trace}(\mathrm{A})$	trace of matrix A
$\mathrm{P} > 0$	(semi-defined) defined positive matrix
$\mathrm{P} \geq 0$	semi-defined positive matrix
$\mathrm{P} < 0$	(semi-defined) defined negative matrix
$\mathrm{P} \leq 0$	semi-defined negative matrix
$\lambda_i(\mathrm{A})$	i-th eigenvalue of matrix A
$\sigma_i(\mathrm{A})$	i-th singular value of matrix A
$\rho(\mathrm{A})$	spectral radius (maximum eigenvalue) of matrix A
$S(\mathrm{A}, \mathrm{B}, \mathrm{C}, \mathrm{D})$	linear dynamical system $\dot{\mathbf{x}} = \mathrm{A}\mathbf{x} + \mathrm{B}\mathbf{u}$, $\mathbf{y} = \mathrm{C}\mathbf{x} + \mathrm{D}\mathbf{u}$
M_c	controllability matrix of a system
M_o	observability matrix of a system
σ_i	i-th singular value of a system
μ_i	i-th characteristic value of a system
$\mathtt{RIC}(\mathrm{H})$	Riccati equation solution associated with Hamiltonian matrix H

SISO	single input single output system	MIMO	multi input multi output system

1

Modelling of Uncertain Systems and the Robust Control Problem

CONTENTS

In this chapter the main concepts related to robust control are introduced. In particular, the key notions of robustness and uncertainty are presented. In the formalization of uncertainty, both the structured and unstructured cases are dealt with. In this preliminary part, the representation of a linear time invariant system in term of the realization matrix is also given. Some examples of uncertainty generated with MATLAB examples are reported. The essential chronology of the recent history of robust control is also outlined.

1.1 Uncertainty and Robust Control

In control system design, the primary requirement is the asymptotic stability of controlled systems. In automatic control, this is guaranteed through the design of an appropriate controller based on the nominal model of the process. However, in reality there may be several sources of uncertainty which make the nominal model inaccurate. *Robust control* deals explicitly with system uncertainties which account for the differences between the real model and nominal model. Robust control guarantees controlled system performance (primarily, asymptotic stability) when there are uncertainties.

The robust control issue can be summarized as: given a nominal process with acceptable interval values for perturbances, a controller should provide satisfactory performance in a closed-loop system for all the processes with the "acceptable" perturbances.

As regards stability, the requisites of a robust control system should ensure:

1. Closed-loop stability under nominal conditions;
2. Closed-loop stability although there are uncertainties in the model.

DOI: 10.1201/9781003196921-1

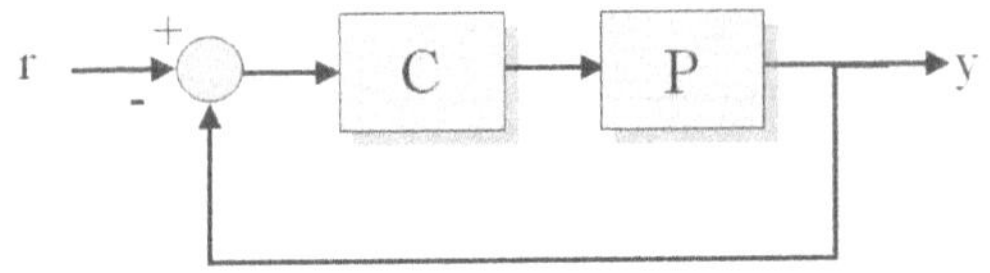

FIGURE 1.1
Unitary feedback control scheme.

As clarified below, uncertainties come in various forms, but, notwithstanding, an acceptable value range for possible uncertainties must be hypothesized within which interval control system performance is guaranteed. This can also provide a way to establish if one controller is more robust than another: the greater the range of acceptable values for uncertainties, the more the controller can be considered robust.

The automatic control theory refers to the feedback control model represented in Figure 1.1 (unitary feedback) or in Figure 1.2 (non-unitary feedback). P represents the process and generally no assumptions are made about its linearity, assuming only it is a time-invariant system. C indicates the controller to be designed. The main issue is to design C, so the closed-loop system is asymptotically stable. Furthermore, in addition to this basic specification, the control system can be required to respond to other criteria. In the case of robust control, another important factor must be accounted for: the process model may have uncertainties.

For example, consider a linear process, described by transfer function $P(s) = \frac{1}{s^2(s+1)}$. Who can say for sure that the pole of this system is exactly -1? And, in the same way, on what basis could the system gain be considered perfectly unitary? For this reason, in the theory of robust control, the system is indicated by $P(s) = \frac{g_1}{s^2(s+p_1)}$, inputting parameters which account for system uncertainties. $g_1 = 1$ and $p_1 = 1$ should be considered the nominal values of the parameters that, however, in practice may eventually have different values, with a variation range that is usually known. This type of uncertainty concerns parameter values which characterize the system and so is defined *parametric uncertainty* or *structured uncertainty*. Examples of parametric uncertainty can be seen in daily life. Think of any size measurement (e.g., the length of a table): the measurement depends on the accuracy of the measuring instrument.

There are also other causes of parametric uncertainty. Parameter values may also vary depending on the operating conditions of the system. Think of a resistor heated by the Joule effect: electrical resistivity increases, modifying the value of that resistance. Or think of an airplane, as the fuel is consumed, the plane total mass decreases.

Obviously, if parameter values change (even if in a predictable range), control system performance drops off compared to a nominal one (in fact,

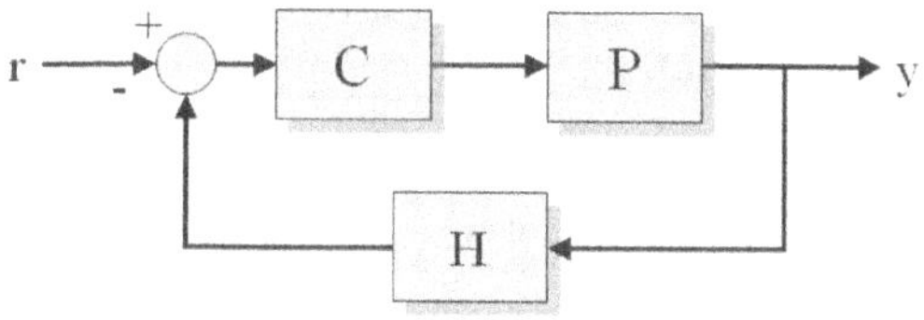

FIGURE 1.2
Feedback control scheme.

the design was related to nominal parameter values). The first goal of robust control is to ensure that, even with changing parameter values, the system asymptotic stability is guaranteed.

Above, there was an example of parametric uncertainty about the coefficients of the transfer function of a linear time-invariant system. Now consider a nonlinear time-invariant system (with one input and output, for example), described by the equations:

$$\begin{aligned} \dot{\mathbf{x}} &= \mathrm{f}(\mathbf{x}) + \mathrm{g}(\mathbf{x})u \\ y &= \mathrm{h}(\mathbf{x})u \end{aligned} \tag{1.1}$$

with $\mathbf{x} \in \mathbb{R}^n$ (state variables), $u \in \mathbb{R}$ (system input) and $y \in \mathbb{R}$ (system output), $\mathrm{g} : \mathbb{R}^n \rightarrow \mathbb{R}^n$, $\mathrm{f} : \mathbb{R}^n \rightarrow \mathbb{R}^n$ and $\mathrm{h} : \mathbb{R}^n \rightarrow \mathbb{R}^n$. With robust control, parametric uncertainty in the model is highlighted by using the parameters α, β and γ:

$$\begin{aligned} \dot{\mathbf{x}} &= \mathrm{f}(\mathbf{x}, \alpha) + \mathrm{g}(\mathbf{x}, \beta)u \\ y &= \mathrm{h}(\mathbf{x}, \gamma)u \end{aligned} \tag{1.2}$$

In this case, $\bar{\alpha}$, $\bar{\beta}$ and $\bar{\gamma}$ indicate the nominal values of the parameters α, β and γ. Robust control for system (1.2) means designing C to guarantee closed-loop asymptotic stability, given the variation spans of parameters α, β and γ (generally, vectors of arbitrary size).

Similarly, one can consider linear systems in state-space form or other nonlinear models. Especially, for a linear time-invariant system in state-space form:

$$\begin{aligned} \dot{\mathbf{x}} &= \mathrm{A}\mathbf{x} + \mathrm{B}\mathbf{u} \\ \mathbf{y} &= \mathrm{C}\mathbf{x} + \mathrm{D}\mathbf{u} \end{aligned} \tag{1.3}$$

with $\mathbf{x} \in \mathbb{R}^n$, $\mathbf{u} \in \mathbb{R}^m$, $\mathbf{y} \in \mathbb{R}^p$, $\mathrm{A} \in \mathbb{R}^{n \times n}$, $\mathrm{B} \in \mathbb{R}^{n \times m}$, $\mathrm{C} \in \mathbb{R}^{p \times n}$, $\mathrm{D} \in \mathbb{R}^{p \times m}$, coefficients of A, B, C and D vary in certain intervals. What is required for robust control, once the controller is designed for nominal parameter values, is that asymptotic stability is guaranteed even when the parameters are not nominal.

The linear time-invariant systems can also be described in a more compact form with the *realization matrix*:

$$\mathrm{R} = \begin{bmatrix} \mathrm{A} & \mathrm{B} \\ \mathrm{C} & \mathrm{D} \end{bmatrix} \tag{1.4}$$

If the number of inputs equals the number of system outputs ($p = m$) then matrix R is square ($\mathrm{R} \in \mathbb{R}^{(n+m)\times(n+m)}$). The eigenvalues of this matrix can easily be shown not to depend on the reference system. In fact, a state transformation leads to:

$$\tilde{\mathrm{R}} = \mathrm{T}^{-1}\mathrm{R}\mathrm{T} \tag{1.5}$$

which is clearly a similarity relation.

To verify if a system is minimal, it is possible to calculate the eigenvalues of matrix R which are represented by $\bar{\lambda}_1, \bar{\lambda}_2, \ldots, \bar{\lambda}_{n+m}$, whereas the eigenvalues of A by $\lambda_1, \lambda_2, \ldots, \lambda_n$. These quantities are system invariants, that is, they do not depend on the state-space representation. The system is minimal if none of eigenvalues of A coincide with those of R.

There is another type of uncertainty which is more difficult to deal with than parametric uncertainty. Called *structural uncertainty* or *unstructured uncertainty*, it concerns the model structure. Consider again the example of the process described by $P = \frac{1}{s^2(s+1)}$, structural uncertainty takes into account the possibility that the modelling did not account perhaps for an additional pole and an additional zero in the process transfer function:

$$P = \frac{(\frac{s}{\alpha_0} + 1)}{s^2(s+1)(\frac{s}{\alpha} + 1)}$$

So, structural uncertainty derives from incorrect modelling, which possibly overlooked some dynamics of the real process. Neglecting some dynamics in modelling is actually very common and can have consequences for closed-loop stability.

Example 1.1

Consider system $P(s) = \frac{1}{(s+1)^2}$. The transfer function of the closed-loop system (Fig. 1.1) is given by $M(s) = \frac{Y(s)}{R(s)} = \frac{C(s)P(s)}{1+C(s)P(s)}$. Using a simple proportional controller $C(s) = k$, you obtain $M(s) = \frac{k}{(s+1)^2+k}$ and the closed-loop system is asymptotically stable $\forall k > 0$.

For example, if $C(s) = 100$, notice that this controller is robust to parametric variations of position of the pole of $P(s)$ with double multiplicity. In fact, if instead of $p = -1$, it was $p = -\alpha$, the controller $C(s) = 100$ would continue to guarantee closed-loop asymptotic stability to a large value of the parameter ($\alpha > 0$).

In the case of structural uncertainty, the scenario is different. Suppose that there is an uncertainty, say on system order, so that $P(s) = \frac{1}{(s+1)^3}$. The characteristic closed-loop equation is given by $(s+1)^3 + k = 0$, i.e., $s^3 + 3s^2 + 3s + 1 + k = 0$. Applying the Routh criterion, we note that in this case the system is asymptotically stable if $0 < k < 8$.

Controller $C(s) = 100$, designed for the nominal system $P(s) = \frac{1}{(s+1)^2}$, no longer guarantees asymptotic stability because of the structural uncertainty of an additional pole at -1. The conclusion is that the consequences of uncertainty on system order, having overlooked some dynamics (which could also be a parasite dynamic which triggers in certain conditions), can be very important.

Defining parametric uncertainty is done by expressing a parameter variation range. For example, given a linear time-invariant system described by its transfer function

$$P(s) = \frac{N(s)}{D(s)} = \frac{b_0 s^m + b_1 s^{m-1} + \ldots + b_{m-1}s + b_m}{s^n + a_1 s^{n-1} + \ldots + a_{n-1}s + a_n}$$

once the minimum and maximum values of the various coefficients are assigned, parametric uncertainty is completely characterized:

$$\begin{array}{c} a_1^m \leq a_1 \leq a_1^M \\ \ldots \\ a_n^m \leq a_n \leq a_n^M \\ b_0^m \leq b_0 \leq b_0^M \\ \ldots \\ b_m^m \leq b_m \leq b_m^M \end{array} \tag{1.6}$$

If there is structural uncertainty we assume $P(s) = P_0(s) + \Delta P(s)$, where $P_0(s)$ represents the nominal model and $\Delta P(s)$ represents uncertainty. This structural uncertainty is *additive*. Structural uncertainty can also be *multiplicative*: $P(s) = P_0(s)\Delta P(s)$. Structural uncertainty is measured by a norm of the transfer function or of the transfer matrix. We will see how to define such a norm in Chapter 9.

The greater the required robustness, the greater the precision required to model the process. A particularly important role in robust control is testing. The popularity of personal computers now makes it possible to perform tests through numerical simulations even when the robust control problem has no analytical solutions. Numerical simulations help verify operation and controller performance in the variation interval indicated by the uncertainty.

The most critical issue is structural uncertainty. Suppose we consider a horizontal beam fixed at one end and subject to a stimulus at the other. This system has distributed parameters and could be modelled by an infinite series of mass-spring systems. The free-end deflection has infinite modes which can be described with a transfer function of type $G(s) = \sum_{i=0}^{\infty} G_i(s)$, where $G_i(s)$ is the transfer function describing the mode associated with the n-th section. However, this approach is not so easy to carry out and so an approximation $G(s) = G_0(s) + \Delta G(s)$ is used which takes into account the most important modes (modelled by $G_0(s)$). Approximation is necessary for the design of a non-distributed controller (note too that additive uncertainty helps model the high frequency dynamic neglected by the approximation). In these cases only the dominant part is considered, but it is essential that the controller is robust enough for any uncertainty in the dynamics not described in the model. This introduces another very important issue treated below: how to figure out reduced order models which can allow to deal with the control problem in a simpler way. It seems clear that this operation introduces an error which can be evaluated and taken into account in the robust controller design.

Now, let us turn to the case of the linear time-invariant systems described by equations (1.3). In the next example, we will discuss parametric and structural uncertainty in the state matrix A and illustrate how the position of the eigenvalues of this matrix is affected by this uncertainty.

MATLAB® Exercise 1.1

Let us consider a linear time-invariant system with state matrix

$$A = \begin{bmatrix} 0 & 1 & 0 \\ 0 & 0 & 1 \\ -1+k_1 & -3+k2 & -3 \end{bmatrix} \tag{1.7}$$

It contains two parameters k_1 and k_2 which are uncertain. This system is therefore an example of a system subject to parametric uncertainty. Note that for $k_1 = k_2 = 0$ the system is asymptotically stable, being all eigenvalues of A equal and negative, namely, $\lambda_1 = \lambda_2 = \lambda_3 = -1$.

Now consider the case that k_1 and k_2 are random numbers drawn from a normal distribution and calculate the eigenvalues repeating the computation for different instances of the random process. To this aim, let us the following MATLAB commands:

```
for i=1:1000
    k1=randn;
    k2=randn;
    A=[0 1 0;0 0 1;-1+k1 -3+k2 -3];
    figure(1); plot(k1,k2,'k.'); hold on;
    figure(2); plot(real(eig(A)),imag(eig(A)),'k.'); hold on;
end
```

The result is illustrated in Figure 1.3, which shows that in most of the cases the stability is preserved, but in some cases eigenvalues with positive real part appear, yielding an unstable system.

The exercise can be repeated considering a smaller uncertainty, for instance, using `k1=0.1*randn;` and `k2=0.1*randn;` to obtain a system that with high probability remains stable.

The effect of other distributions for k_1 and k_2 can also be studied. For instance, considering an uniform distribution in $[-0.5, 0.5]$ with the commands `k1=rand-0.5` and `k2=rand-0.5`, we obtain the result shown in Figure 1.4. In this case, the system is always stable.

The characteristic polynomial of A is given by $p(\lambda) = \lambda^3 + 3\lambda^2 + (3 - k_2)\lambda + 1 - k_1$ as one can also verify using symbolic calculus in MATLAB:

```
>> syms k1 k2 real positive
>> A=[0 1 0;0 0 1;-1+k1 -3+k2 -3];
>> p=charpoly(A)
```

Applying the Routh criterion for stability, one derives that the following conditions need to be satisfied: $k_1 < 1$, $k_2 < 3$ and $8 - 3k_2 + k_1 > 0$. This explains why, when k_1 and k_2 are drawn from an uniform distribution in $[-0.5, 0.5]$, one always obtains a stable polynomial.

Let us now discuss an example of unstructured uncertainty, where the uncertainty is no more located on some specific parameters of the state matrix A. In particular, let us consider the following matrix:

$$A = \begin{bmatrix} 0 & 1 & 0 \\ 0 & 0 & 1 \\ -1 & -3 & -3 \end{bmatrix} + \Delta A \tag{1.8}$$

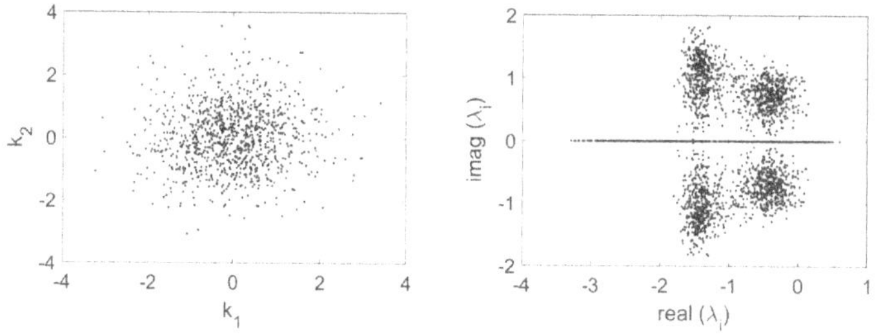

(a) (b)

FIGURE 1.3
Stability of linear systems in presence of parametric uncertainty: (a) distribution of the parameters k_1 and k_2; (b) eigenvalues of the state matrix A (1.7) when k_1 and k_2 are drawn from a normal distribution.

where ΔA is the uncertainty. Note that, in absence of perturbation, i.e., $\Delta A = 0$, the same state matrix previously considered is recovered.
Suppose now that the coefficients of ΔA assume random values drawn from a normal distribution with zero mean and unitary variance. The following MATLAB command can be used to numerically study the problem:

```
A=[0 1 0; 0 0 1;-1 -3 -3]
for i=1:1000
    DeltaA=1*randn(3);
    A=A+DeltaA;
    plot(real(eig(A)),imag(eig(A)),'.k')
    hold on
end
```

Figure 1.5 shows an example of the location of the eigenvalues of A.

For linear time-invariant systems described by equations (1.3), under opportune hypotheses, a controller can be designed using a control law which operates on state variables (the linear state regulator $\mathbf{u} = -\mathbf{kx}$). Remember, should not all the state variables be accessible, they would have to be reconstructed through an asymptotic state observer. This common technique usually includes the design of a linear regulator and an asymptotic observer which together constitute the compensator.

When designing a linear state regulator, the system must be completely controllable (i.e., matrix $M_c = \begin{bmatrix} B & AB & A^2B & \dots & A^{n-1}B \end{bmatrix}$ has maximum rank), whereas designing an asymptotic observer needs the system to be completely observable (i.e., matrix $M_o = \begin{bmatrix} C & CA & CA^2 & \dots & CA^{n-1} \end{bmatrix}^T$ has maximum rank).

Furthermore, note that, since the observer is a dynamical system (linear and time-invariant) of order n, then the compensator obtained is a dynamical system of order n. High-order systems require the design of high-order

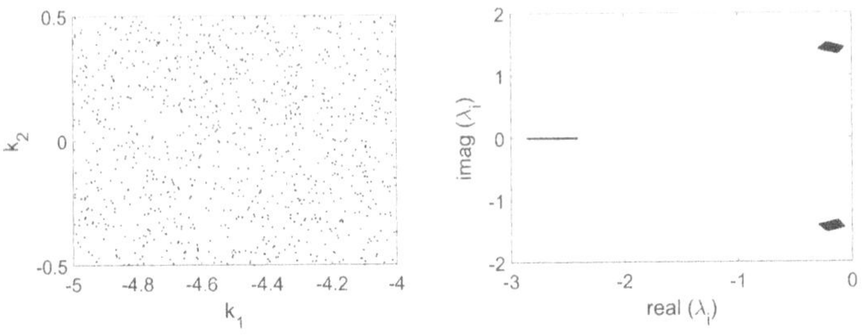

(a) (b)

FIGURE 1.4
Stability of linear systems in presence of parametric uncertainty: (a) distribution of the parameters k_1 and k_2; (b) eigenvalues of the state matrix A (1.7) when k_1 and k_2 are drawn from an uniform distribution in $[-0.5, 0.5]$.

compensators. So, now, the problem of formulating a lower-order model also includes the design of a compensator. To do this, there are two different techniques. The first is to design a compensator and then build the lower-order model. The second is to design the compensator directly on the lower-order model (e.g., considering $P_0(s)$ and neglecting $\Delta P(s)$).

It is important to remark that the uncertainties regard the process P and not the controller C. This is so, as it is supposed that the controller is the result of a design, so any uncertainties about its parameters are minimal. More recently, it has been discovered that compensators designed with the theory of robust control are fragile, that is, that compared to the compensator parameters, robustness is poor, so even minimal uncertainties about compensator coefficients can lead to system de-stabilization. At the heart of the compensator issue is that when it is designed to be robust against the uncertainties in the model of process P, it remains fragile with respect to the uncertainties in the model of controller C. Fragility means that a small perturbation in the controller parameters can even destabilize the closed-loop system.

This chapter has briefly described the main robust control issues to be dealt with in subsequent chapters. Before that, certain subjects will be briefly presented which are the pillars of systems theory (e.g., stability). This approach was decided on for two reasons: first, is the importance of these subjects to robust control, and, second, is that undergraduate courses are organized on two levels. The course which this book is part of is a Master course for students who have likely dealt with automation previously. So, this makes a brief discourse on systems theory necessary.

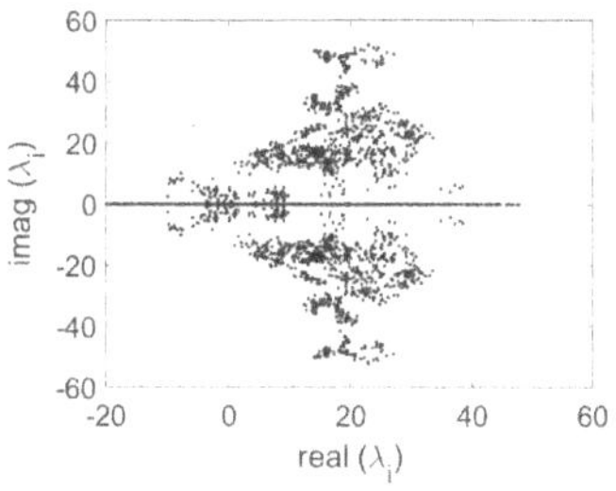

FIGURE 1.5
Stability of linear systems in presence of unstructured uncertainty as in equation (1.8): eigenvalues of the state matrix A.

1.2 The Essential Chronology of Major Findings in Robust Control

Parameterization of stabilizing controllers (Youla, 1976).
Poor robustness of LQG controllers (Doyle, 1978).
Formulating H_∞ problems for SISO systems (Zames, 1981).
Balanced realizations (Moore, 1981).
Definition of μ (Doyle, 1982).
Definition of multivariable stability margins (Safonov, 1982).
H_∞ synthesis algorithms for systems in state-space form: high order control (Doyle, 1984).
First book on robust control (Francis, 1987).
H_∞ synthesis algorithms for systems in state-space form: low order control (Doyle, Glover, Khargonekar, Francis, 1988).

2

Fundamentals of Stability

CONTENTS

This chapter deals with the problem of stability, the main requirement of a control system. In going over some basics in system analysis, the emphasis is on some advanced mathematical tools (e.g., Lyapunov equations), which will be useful below. Particular attention is given to Lyapunov theory for the case of linear time-invariant systems. The Lyapunov linear matrix equation is introduced in detail and the concept of positive definite matrix is also discussed. Some criteria to verify this property are reported and the solution of Lyapunov equations via vectorization is presented with numerical examples. The Kharitonov criterion to test the stability of uncertain system is also introduced. At the end of the chapter, several worked examples are included.

2.1 Lyapunov Criteria

The equilibrium points of an autonomous dynamical system $\dot{\mathbf{x}} = f(\mathbf{x})$ with $\mathbf{x} = \begin{bmatrix} x_1 & x_2 & \dots & x_n \end{bmatrix}^T$ and $f(\mathbf{x}) = \begin{bmatrix} f_1(\mathbf{x}) & f_2(\mathbf{x}) & \dots & f_n(\mathbf{x}) \end{bmatrix}^T$ can be calculated by $\dot{\mathbf{x}} = 0$ (i.e., solving the systems of equations $f(\mathbf{x}) = 0$). Generally, the equations are nonlinear and may have one or more solutions, whereas for autonomous linear systems, the equilibrium points can be calculated by $A\mathbf{x} = 0$, which has only one solution ($\mathbf{x} = 0$) if $\det(A) \neq 0$.

DOI: 10.1201/9781003196921-2

So, nonlinear systems may have more than one equilibrium point. In addition, each of these equilibrium points has its own stability characteristics. For example, think of a pendulum. It has two equilibrium points as shown in Figure 2.1. Only the second of the two equilibrium points is stable. If the pendulum were to start from a position near equilibrium point (a), it would not return to its equilibrium position, differently from what happens at point (b). For this reason, in nonlinear systems, stability is a property of equilibrium points, and not of the whole system. The stability of an equilibrium point can be studied through the criteria introduced by the Russian mathematician Aleksandr Mikhailovich Lyapunov (1857–1918).

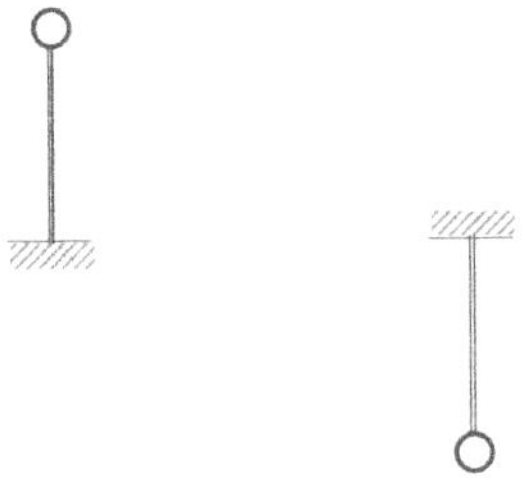

(a) (b)

FIGURE 2.1
Equilibrium points of a pendulum.

The first of Lyapunov's criteria defines stability of an equilibrium point. It says that if a system starts from initial conditions very close to a stable equilibrium point, the state evolution is confined to the neighborhood of that equilibrium point. If the initial conditions $\mathbf{x}(0) = \mathbf{x}_0$ are close to equilibrium, the response obtained by $\mathbf{x}_0$, i.e., the *perturbed response*, is sufficiently close to equilibrium point $\bar{\mathbf{x}}$ (the *nominal response*, i.e., the motion obtained starting exactly from $\bar{\mathbf{x}}$, is in fact constant over time and equal to the equilibrium point). To measure the distance between perturbed and nominal motion in finite space, any vector norm can be used. The criterion is formally expressed by:

Definition 1 (Lyapunov I criterion) *The equilibrium point $\bar{\mathbf{x}}$ of the dynamical system $\dot{\mathbf{x}} = f(\mathbf{x})$ ($f(\bar{\mathbf{x}}) = 0$) is defined as stable when:*

$$\forall \varepsilon > 0, \exists \delta > 0 \text{ such that } \forall \mathbf{x}_0 \text{ with } \| \mathbf{x}_0 - \bar{\mathbf{x}} \| < \delta \text{ one has}$$
$$\| \mathbf{x}(t) - \bar{\mathbf{x}} \| < \varepsilon \text{ for } \forall t > 0$$

Stability is said to be asymptotic, if this additional condition applies:

$$\lim_{t \to +\infty} \mathbf{x}(t) = \bar{\mathbf{x}} \tag{2.1}$$

In this case, the perturbed motion not only stays in the neighborhood of the equilibrium point, but asymptotically tends to it.

Definition 2 (Asymptotically stable equilibrium point) *The equilibrium point* $\bar{\mathbf{x}}$ *of the dynamical system* $\dot{\mathbf{x}} = f(\mathbf{x})$ $(f(\bar{\mathbf{x}}) = 0)$ *is defined as asymptotically stable if it is stable and furthermore condition (2.1) holds.*

Before discussing Lyapunov's second criterion, some preliminary notions need to be discussed.

Note first that, without loss of generality, it is possible to consider $\bar{\mathbf{x}} = 0$, since it is always possible to shift the generic equilibrium point $\bar{\mathbf{x}}$ back to 0 by variable translation.

Now, the concept of positive definite functions needs to be introduced. Let us consider a neighborhood $\Omega \subseteq \mathbb{R}^n$ of point $\bar{\mathbf{x}} = 0$.

Definition 3 (Positive definite function) *Function* $V(\mathbf{x}) : \mathbb{R}^n \rightarrow \mathbb{R}$ *is positive definite in* $\bar{\mathbf{x}} = 0$ *if the following hold:*

1. $V(0) = 0$
2. $V(\mathbf{x}) > 0 \; \forall \mathbf{x} \in \Omega, \; \mathbf{x} \neq 0$

A function is positive semi-definite if $V(0) = 0$ and if $V(\mathbf{x}) \geq 0 \; \forall \mathbf{x} \in \Omega$, $\mathbf{x} \neq 0$. A function is negative definite if $V(0) = 0$ and if $V(\mathbf{x}) < 0 \; \forall \mathbf{x} \in \Omega$, $\mathbf{x} \neq 0$. Finally, a function is negative semi-definite if $V(0) = 0$ and if $V(\mathbf{x}) \leq 0$ $\forall \mathbf{x} \in \Omega$, $\mathbf{x} \neq 0$.

Theorem 1 (Lyapunov II criterion) *The equilibrium* $\bar{\mathbf{x}} = 0$ *of the dynamical system* $\dot{\mathbf{x}} = f(\mathbf{x})$ *is stable if there is a scalar function of the state, the so-called Lyapunov function, which has the following properties:*

1. $V(\mathbf{x})$ *is positive definite for* $\bar{\mathbf{x}} = 0$
2. $\dot{V}(\mathbf{x}) = \frac{dV(\mathbf{x})}{dt}$ *is negative semi-definite for* $\bar{\mathbf{x}} = 0$

The stability is asymptotic if $\dot{V}(\mathbf{x})$ *is negative definite.*

From a physical point of view, the Lyapunov function represents system energy. If the energy grows over time, it means that the state variables grow (thus diverge). Conversely, if the energy tends to zero, the system is dissipating energy.

Note that many Lyapunov functions satisfy the first condition of Lyapunov II criterion; the difficulty lies in verifying the second condition.

2.2 Positive Definite Matrices

For linear systems, a Lyapunov function allowing to assess system stability is easily found. In fact, for nonlinear systems, there is no general procedure for finding Lyapunov functions, while for linear systems there is. Effectively, Lyapunov second criterion for linear systems can be simplified. Before introducing the criterion, firstly, positive definite matrices are defined.

Given matrix $\mathrm{P} \in \mathbb{R}^{n\times n}$, a quadratic form can be associated to it: $V(\mathbf{x}) = \mathbf{x}^T\mathrm{P}\mathbf{x}$, which is clearly a scalar function of vector $\mathbf{x} \in \mathbb{R}^n$.

Definition 4 *A matrix* P *is positive definite if the quadratic form associated to it is positive definite.*

Likewise, it is possible to define negative definite, positive semi-definite and negative semi-definite matrices.

Now, we discuss how to determine if a matrix is positive definite. It turns out that this depends only on the so-called *symmetric part* of the matrix. In fact, matrix P can always be written as the sum of two matrices, one symmetric and one asymmetric, named *symmetric part* P_s and *antisymmetric part* P_{as} of matrix:

$$\mathrm{P} = \mathrm{P}_s + \mathrm{P}_{as}$$

where the symmetric part is

$$\mathrm{P}_s = \frac{1}{2}(\mathrm{P} + \mathrm{P}^T)$$

and the antisymmetric part is

$$\mathrm{P}_{as} = \frac{1}{2}(\mathrm{P} - \mathrm{P}^T).$$

Clearly, only the symmetric part of a matrix determines if a matrix is positive definite as:

$$V(\mathbf{x}) = \mathbf{x}^T\mathrm{P}\mathbf{x} = \mathbf{x}^T\mathrm{P}_s\mathbf{x}$$

In fact, the asymmetric part does not contribute. This can be seen considering that:

$$V(\mathbf{x}) = \mathbf{x}^T\mathrm{P}\mathbf{x} = p_{11}x_1^2 + \ldots + p_{nn}x_n^2 + (p_{12} + p_{21})x_1x_2 + (p_{13} + p_{31})x_1x_3 + \ldots$$
$$\ldots + (p_{n-1,n} + p_{n,n-1})x_{n-1}x_n$$

On the other hand, from the definition of P_s one has:

$$\mathrm{P}_s = \begin{bmatrix} p_{11} & \frac{p_{21}+p_{12}}{2} & \ldots & \frac{p_{n1}+p_{1n}}{2} \\ \frac{p_{21}+p_{12}}{2} & p_{22} & \ldots & \frac{p_{n2}+p_{2n}}{2} \\ \vdots & \vdots & & \vdots \\ \frac{p_{1n}+p_{n1}}{2} & \frac{p_{n2}+p_{2n}}{2} & \ldots & p_{nn} \end{bmatrix}$$

from which it follows that $V(\mathbf{x}) = \mathbf{x}^T\mathrm{P}_s\mathbf{x}$.

Let us recall some properties of symmetric matrices that will be useful below.

- The eigenvalues of a symmetric matrix are real.
- A symmetric matrix is always diagonalizable. This means that there is always a set of linear independent and orthogonal eigenvectors, such that $V_j^T V_i = 0$ if $i \neq j$. For symmetric matrices it is always possible to find a set of orthonormal eigenvectors such that if
  $\mathrm{V} = \begin{bmatrix} V_1 & V_2 & \dots & V_n \end{bmatrix}$ then:

$$\mathrm{V}^T\mathrm{V} = \mathrm{I}$$

$$\mathrm{V}^T = \mathrm{V}^{-1}$$

and

$$P = \mathrm{V}\Lambda\mathrm{V}^T$$

From this, we obtain a significant finding for positive definite matrices. Consider the quadratic form $V(\mathbf{x})$ of P, which equals:

$$V(\mathbf{x}) = \mathbf{x}^T\mathrm{P}_s\mathbf{x} = \mathbf{x}^T\mathrm{V}\Lambda\mathrm{V}^T\mathbf{x}$$

Consider the state transformation $\tilde{\mathbf{x}} = \mathrm{V}^T\mathbf{x}$, then:

$$V(\tilde{\mathbf{x}}) = \tilde{\mathbf{x}}^T\Lambda\tilde{\mathbf{x}} \tag{2.2}$$

Clearly, $V(\mathbf{x})$ and $V(\tilde{\mathbf{x}})$ have the same properties, but understanding if $V(\tilde{\mathbf{x}})$ is positive definite is much simpler. In fact, if all the eigenvalues of P_s are positive, then $V(\tilde{\mathbf{x}})$ is positive definite. This result can be summarized in the theorem below.

Theorem 2 *A symmetric matrix* P *is positive definite if all its eigenvalues are positive.*

The matrix is positive semi-definite if all its eigenvalues are non-negative and there is at least one null eigenvalue. Moreover, P is negative definite if all its eigenvalues are negative and negative semi-definite if all its eigenvalues are non-positive and at least one of them is null. If the matrix is not symmetric, the same conditions apply except that the eigenvalues are those of the symmetric part of the matrix.

The following theorem is useful for determining if a symmetric matrix is positive or not without calculating the eigenvalues.

Theorem 3 (Sylvester test) *A symmetric matrix* P *is positive definite if and only if all of its* n *principal minors* $\mathrm{D}_1, \mathrm{D}_2, \dots, \mathrm{D}_n$ *are positive, i.e.:*

$$\mathrm{D}_1 = p_{11} > 0; \mathrm{D}_2 = \det\begin{bmatrix} p_{11} & p_{12} \\ p_{21} & p_{22} \end{bmatrix} > 0; \dots; \mathrm{D}_n = \det\mathrm{P} > 0 \tag{2.3}$$

MATLAB® Exercise 2.1

Consider $P = \begin{bmatrix} 1 & 2 & 5 \\ 2 & 5 & -1 \\ 5 & -1 & 0 \end{bmatrix}$. Since the matrix is symmetric, the Sylvester test can be applied to determine if it is positive definite or not.

Let us first define matrix P in MATLAB with the command:

```
>> P=[1 2 5; 2 5 -1; 5 -1 0]
```

and then compute D_1, D_2 and D_3 as follows:

```
>> D1=det(P(1,1))
>> D2=det(P(1:2,1:2))
>> D3=det(P)
```

One obtains: $D_1 = 1$, $D_2 = 1$ and $D_3 = -146$. Since $D_3 < 0$, P is not positive definite.

2.3 Lyapunov Theory for Linear Time-Invariant Systems

The Lyapunov second criterion for linear time-invariant systems concerns the *system stability*. In fact, in linear systems, since all the equilibrium points have the same stability properties, it is possible to refer to the stability of the system. The criterion is expressed by the following theorem.

Theorem 4 (Lyapunov II criterion for linear time-invariant systems) *A linear time-invariant system* $\dot{\mathbf{x}} = \mathrm{A}\mathbf{x}$ *is asymptotically stable if and only if for any positive definite matrix* Q *there exists a unique positive definite matrix* P, *which satisfies the following equation (the so-called Lyapunov equation):*

$$\mathrm{A}^{\mathrm{T}}\mathrm{P} + \mathrm{PA} = -\mathrm{Q} \tag{2.4}$$

Equation (2.4) is a *linear* matrix equation, because the unknown P appears with a maximum degree equal to 1.

For simplicity, usually symmetric Q is chosen. Here also P is symmetric:

$$(\mathrm{A}^{\mathrm{T}}\mathrm{P} + \mathrm{PA})^T = -\mathrm{Q}^T \Rightarrow \mathrm{P}^T\mathrm{A} + \mathrm{A}^T\mathrm{P}^T = -\mathrm{Q}^T$$
$$\Rightarrow \mathrm{P}^T\mathrm{A} + \mathrm{A}^T\mathrm{P}^T = -\mathrm{Q} \Rightarrow \mathrm{P}^T = \mathrm{P}$$

One way of solving a Lyapunov equation is by *vectorization*.

Example 2.1

Consider an example with $n = 2$:

$$\mathrm{A} = \begin{bmatrix} -2 & 0 \\ 0 & -5 \end{bmatrix}$$

Since the Lyapunov criterion states that equation (2.4) should hold for any positive definite Q, then let us choose Q = I (indeed, it suffices to prove it for one Q matrix, to show that it holds for any).

Consider matrix P made up of three unknowns p_{11}, p_{12} and p_{22}:

$$P = \begin{bmatrix} p_{11} & p_{12} \\ p_{12} & p_{22} \end{bmatrix}.$$

Solving the Lyapunov equation means solving a system of three equations with three unknowns. By substituting in equation (2.4), we get:

$$\begin{bmatrix} -2 & 0 \\ 0 & -5 \end{bmatrix}\begin{bmatrix} p_{11} & p_{12} \\ p_{12} & p_{22} \end{bmatrix} + \begin{bmatrix} p_{11} & p_{12} \\ p_{12} & p_{22} \end{bmatrix}\begin{bmatrix} -2 & 0 \\ 0 & -5 \end{bmatrix} = \begin{bmatrix} -1 & 0 \\ 0 & -1 \end{bmatrix}$$

$$\Rightarrow \begin{cases} 4p_{11} = 1 \\ 7p_{12} = 0 \\ 10p_{22} = 1 \end{cases}$$

The three equations with three unknowns can be rewritten in matrix form:

$$\begin{bmatrix} 4 & 0 & 0 \\ 0 & 7 & 0 \\ 0 & 0 & 10 \end{bmatrix}\begin{bmatrix} p_{11} \\ p_{12} \\ p_{22} \end{bmatrix} = \begin{bmatrix} 1 \\ 0 \\ 1 \end{bmatrix}$$

This leads to: $P = \begin{bmatrix} \frac{1}{4} & 0 \\ 0 & \frac{1}{10} \end{bmatrix}$. Matrix P is positive definite, so the system is asymptotically stable.

Notice that, since in Example 2.1 matrix A is diagonal, concluding that the system is stable is immediate from the inspection of the eigenvalues of A which are all in the open left half plane. Generally, testing system stability via the Lyapunov second criterion is not numerically very efficient, since the first Lyapunov equation has to be solved and then the positive definiteness of the solution has to be verified. Tests on the eigenvalues of A or criteria such as the Routh criterion are more direct and efficient. If A is known, the Lyapunov second criterion is not the most efficient method to test the stability of the linear system. However, it turns out that the Lyapunov second criterion is a powerful theoretical tool for proving system stability.

Generally, if Q is not symmetric the vectorization is:

$$M \begin{bmatrix} p_{11} \\ p_{21} \\ p_{31} \\ \vdots \\ p_{nn} \end{bmatrix} = - \begin{bmatrix} q_{11} \\ q_{21} \\ q_{31} \\ \vdots \\ q_{nn} \end{bmatrix}$$

where $M \in \mathbb{R}^{n^2 \times n^2}$.

Proof *(Proof of sufficiency of the second criterion of Lyapunov for linear time-invariant systems). This proves the sufficient part of the Lyapunov criterion for linear time-invariant systems, so if $\forall$ Q that is positive definite, there exists a matrix P satisfying Lyapunov equation (2.4), then the system is asymptotically stable.*

Consider the Lyapunov quadratic function $V(\mathbf{x}) = \mathbf{x}^T P \mathbf{x}$. Since P is a positive definite matrix, $V(\mathbf{x})$ is a positive definite function. The derivative against time of this function is:

$$\dot{V}(\mathbf{x}) = \dot{\mathbf{x}}^T \mathrm{P}\mathbf{x} + \mathbf{x}^T \mathrm{P}\dot{\mathbf{x}}$$

Since $\dot{\mathbf{x}} = \mathrm{A}\mathbf{x}$, *then:*

$$\dot{V}(\mathbf{x}) = \mathbf{x}^T \mathrm{A}^T \mathrm{P}\mathbf{x} + \mathbf{x}^T \mathrm{PA}\mathbf{x} =$$

$$= \mathbf{x}^T (\mathrm{A}^T \mathrm{P} + \mathrm{PA})\mathbf{x} = -\mathbf{x}^T \mathrm{Q}\mathbf{x}$$

Since Q *is a positive definite matrix,* $\dot{V}(\mathbf{x})$ *is a negative definite function. Function* $V(\mathbf{x}) = \mathbf{x}^T \mathrm{P}\mathbf{x}$ *satisfies the hypotheses of Lyapunov II criterion for dynamical systems and thus the equilibrium point (and so the system) is asymptotically stable.*

Before proving the necessary part of the theorem, an important property of positive definite matrices should be highlighted.

Theorem 5 *If* $\mathrm{Q} \in \mathbb{R}^{n\times n}$ *is a positive definite matrix and* $\mathrm{M} \in \mathbb{R}^{n\times n}$ *is a full rank matrix, then also* $\mathrm{M}^T\mathrm{QM}$ *is a positive definite matrix.*

Proof *The quadratic form associated to* $\mathrm{M}^T\mathrm{QM}$ *is*

$$V(\mathbf{x}) = \mathbf{x}^T \mathrm{M}^T \mathrm{QM}\mathbf{x} = \mathbf{x}^T \mathrm{M}^T \mathrm{V}\Lambda \mathrm{V}^T \mathrm{M}\mathbf{x}$$

where V *is the orthonormal vector matrix which diagonalizes matrix* Q, *i.e.,* $\mathrm{Q} = \mathrm{V}\Lambda\mathrm{V}^T$.

Given the state transformation $\tilde{\mathbf{x}} = \mathrm{V}^T\mathrm{M}\mathbf{x}$, *then*

$$V(\tilde{\mathbf{x}}) = \tilde{\mathbf{x}}^T \Lambda \tilde{\mathbf{x}}$$

which shows that $\mathrm{M}^T\mathrm{QM}$ *is positive definite.*

Generalizations of Theorem 5. Notice that with a similar procedure, if instead matrix M is not a full rank, matrix $\mathrm{M}^T\mathrm{QM}$ is positive semi-definite. Finally, if matrix M is not square ($\mathrm{M} \in \mathbb{R}^{n\times m}$) it can be shown that matrix $\mathrm{M}^T\mathrm{QM}$ is positive semi-definite. In both cases, in fact, the transformation $\tilde{\mathbf{x}} = \mathrm{V}^T\mathrm{M}\mathbf{x}$ is no longer an invertible transformation (also if $\mathrm{M} \in \mathbb{R}^{n\times m}$, then $\mathbf{x} \in \mathbb{R}^m$), so it can only be concluded that the product matrix is positive semi-definite.

Proof *(Proof of the necessary part of Lyapunov's II criterion for linear systems). The proof of the necessary part of the theorem, that is if the system is asymptotically stable, then for every positive definite matrix* Q *there is a positive definite matrix* P *which is the solution of the Lyapunov equation (2.4), is a constructive demonstration introducing the integral solution of the Lyapunov equation.*

Remember that, if a linear time-invariant system is asymptotically stable, then the zero-input response tends to zero for any initial condition

$$\lim_{t\to+\infty} \mathbf{x}(t) = 0 \quad \forall \mathbf{x}_0$$

and, since $\mathbf{x}(t) = e^{\mathrm{A}t}\mathbf{x}(0)$, *also matrix* $e^{\mathrm{A}t}$ *tends to zero:*

$$\lim_{t\to+\infty} e^{\mathrm{A}t} = 0$$

Remember also that

$$[e^{\mathrm{A}t}]^T = e^{\mathrm{A}^T t}$$

This property can be easily proved by considering the definition of an exponential matrix:

$$e^{\mathrm{A}t} = \mathrm{I} + \mathrm{A}t + \frac{\mathrm{A}^2t^2}{2!} + \frac{\mathrm{A}^3t^3}{3!} + \cdots \Rightarrow$$

$$(e^{\mathrm{A}t})^T = \mathrm{I} + \mathrm{A}^T t + \frac{(\mathrm{A}^2)^T t^2}{2!} + \frac{(\mathrm{A}^3)^T t^3}{3!} + \cdots =$$

$$= \mathrm{I} + \mathrm{A}^T t + \frac{(\mathrm{A}^T)^2 t^2}{2!} + \frac{(\mathrm{A}^T)^3 t^3}{3!} + \cdots = e^{\mathrm{A}^T t}$$

The improper integral $\mathrm{P} = \int_0^\infty e^{\mathrm{A}^T t}\mathrm{Q}e^{\mathrm{A}t}dt$ *is finite, since each term tends to zero. Matrix* P, *because of the properties of Theorem 5, is the integral of a positive definite matrix for each t and therefore is positive definite. Note that matrix* P *is also symmetric (if* Q *is symmetric).*

Given that $\mathrm{P} = \int_0^\infty e^{\mathrm{A}^T t}\mathrm{Q}e^{\mathrm{A}t}dt$ *is a positive definite symmetric matrix, the solution of Lyapunov equation (2.4) can be verified.*

By substituting P *into the Lyapunov equation we obtain:*

$$\mathrm{A}^T \int_0^\infty e^{\mathrm{A}^T t}\mathrm{Q}e^{\mathrm{A}t}dt + \int_0^\infty e^{\mathrm{A}^T t}\mathrm{Q}e^{\mathrm{A}t}dt\mathrm{A} =$$

$$= \int_0^\infty (\mathrm{A}^T e^{\mathrm{A}^T t}\mathrm{Q}e^{\mathrm{A}t} + e^{\mathrm{A}^T t}\mathrm{Q}e^{\mathrm{A}t}\mathrm{A})dt =$$

$$= \int_0^\infty \frac{d}{dt}(e^{\mathrm{A}^T t}\mathrm{Q}e^{\mathrm{A}t})dt = \left[e^{\mathrm{A}^T t}\mathrm{Q}e^{\mathrm{A}t}\right]_0^\infty = -Q$$

where we used the fact that

$$\lim_{t\to+\infty} e^{\mathrm{A}t} = 0$$

and that $e^{\mathrm{A}0} = I$.

Note 1. The solution to Lyapunov equation $\mathrm{AP} + \mathrm{PA}^T = -\mathrm{Q}$ is $\mathrm{P} = \int_0^\infty e^{\mathrm{A}t}\mathrm{Q}e^{\mathrm{A}^Tt}dt$.

Note 2. To study the asymptotic stability of a linear time-invariant system from its characteristic polynomial, besides the well-known Routh criterion, the Hurwitz criterion can be used. It facilitates knowing when a given polynomial is a Hurwitz polynomial (i.e., when it has only roots with negative real part).

Theorem 6 (Hurwitz criterion) *Polynomial*

$$a(\lambda) = a_n\lambda^n + a_{n-1}\lambda^{n-1} + \ldots + a_1\lambda + a_0$$

with $a_n > 0$ *is Hurwitz if all the principal minors of the so-called Hurwitz matrix are greater than zero:*

$$\mathrm{H} = \begin{bmatrix} a_{n-1} & a_{n-3} & a_{n-5} & \cdots & 0 & 0 \\ a_n & a_{n-2} & a_{n-4} & \cdots & 0 & 0 \\ 0 & a_{n-1} & a_{n-3} & \cdots & 0 & 0 \\ 0 & a_n & a_{n-2} & \cdots & 0 & 0 \\ \vdots & \vdots & \vdots & \vdots & \vdots & \vdots \\ 0 & 0 & \cdots & \cdots & a_1 & 0 \\ 0 & 0 & \cdots & \cdots & a_2 & a_0 \end{bmatrix}$$

MATLAB® Exercise 2.2

Consider the following polynomial which is obviously not a Hurwitz polynomial:

$$a(\lambda) = \lambda^4 + 2\lambda^3 + 5\lambda^2 - 10\lambda + 0.7$$

The Hurwitz matrix is given by:

$$\mathrm{H} = \begin{bmatrix} 2 & -10 & 0 & 0 \\ 1 & 5 & 0.7 & 0 \\ 0 & 2 & -10 & 0 \\ 0 & 1 & 5 & 0.7 \end{bmatrix}$$

Calculating the principal minors produces: $H_1 = 2$; $H_2 = 20$; $H_3 = -202.8$ and $H_4 = -141.96$. As expected at least one of the minors is not positive ($H_3 < 0$ and $H_4 < 0$).

2.4 Lyapunov Equations

In the previous section, we saw that, if the linear time-invariant system is asymptotically stable, the Lyapunov equation (2.4) has only one solution.

Let us consider now a generic Lyapunov equation:

$$\mathrm{AX} + \mathrm{XB} = -\mathrm{C} \tag{2.5}$$

where A, B and C are matrices of $\mathbb{R}^{n\times n}$ (in the most general case, they are non-symmetric), and X is a $n \times n$ unknown matrix.

Even here, the equation can be vectorized:

$$\mathrm{M}\begin{bmatrix} x_{11} \\ x_{21} \\ x_{31} \\ \vdots \\ x_{nn} \end{bmatrix} = -\begin{bmatrix} c_{11} \\ c_{21} \\ c_{31} \\ \vdots \\ c_{nn} \end{bmatrix}$$

where M is a suitable matrix in $\mathbb{R}^{n^2\times n^2}$ (if $\mathrm{B} = \mathrm{A}^T$ and C is a symmetric matrix, the number of variables reduces to $k = \frac{n(n+1)}{2}$ and $\mathrm{M} \in \mathbb{R}^{k\times k}$).

The existence of a solution is related to the invertibility of matrix M. This matrix can be obtained with Kronecker algebra. Let us recall therefore some of the definitions and properties of Kronecker algebra.

The Kronecker product of two matrices A and B of $\mathbb{R}^{n\times n}$ is a matrix of $\mathbb{R}^{n^2\times n^2}$ defined by:

$$\mathrm{A} \otimes \mathrm{B} = \begin{bmatrix} a_{11}\mathrm{B} & a_{12}\mathrm{B} & \dots & a_{1n}\mathrm{B} \\ a_{21}\mathrm{B} & a_{22}\mathrm{B} & \dots & a_{2n}\mathrm{B} \\ \vdots & \vdots & & \vdots \\ a_{n1}\mathrm{B} & a_{n2}\mathrm{B} & \dots & a_{nn}\mathrm{B} \end{bmatrix}$$

The matrix eigenvalues can be calculated from those of matrix A and B. Let us indicate with $\lambda_1, \lambda_2, \dots, \lambda_n$ the eigenvalues of A and with $\mu_1, \mu_2, \dots, \mu_n$ the eigenvalues of B. Then, the n^2 eigenvalues of $\mathrm{A} \otimes \mathrm{B}$ are:

$$\lambda_1\mu_1, \lambda_1\mu_2, \dots, \lambda_1\mu_n, \lambda_2\mu_1, \dots, \lambda_n\mu_n$$

It can be shown that the vectorization matrix is given by:

$$\mathrm{M} = \mathrm{I} \otimes \mathrm{A} + \mathrm{B}^T \otimes \mathrm{I}$$

with $\mathbf{I} \in \mathbb{R}^{n\times n}$.

Also the eigenvalues of matrix M are simply linked to those of matrices A and B. The eigenvalues of $\mathrm{M} = \mathrm{I} \otimes \mathrm{A} + \mathrm{B}^T \otimes \mathrm{I}$ are indeed given by all the possible n^2 combinations of sums between an eigenvalue from A and one from B:

$$\lambda_1 + \mu_1, \lambda_1 + \mu_2, \dots, \lambda_1 + \mu_n, \lambda_2 + \mu_1, \dots, \lambda_n + \mu_n$$

M is invertible when there are no null eigenvalues:

$$\lambda_i + \mu_j \neq 0, \quad \forall i, j$$

This leads to answer the question whether a Lyapunov equation has a solution or not. Formally, the condition above can be expressed with the following theorem.

Theorem 7 *The Lyapunov equation*

$$\mathrm{AX} + \mathrm{XB} = -\mathrm{C}$$

with $\mathrm{A} \in \mathbb{R}^{n\times n}$, $\mathrm{B} \in \mathbb{R}^{n\times n}$ *and* $\mathrm{C} \in \mathbb{R}^{n\times n}$ *has only one solution if and only if*

$$\lambda_i + \mu_j \neq 0, \quad \forall i, j$$

where $\lambda_1, \ldots, \lambda_n$ *are the eigenvalues of* A *and* $\mu_1, \ldots, \mu_n$ *are those of* B.

An immediate consequence of this theorem is that the Lyapunov equation $\mathrm{A}^T\mathrm{P} + \mathrm{PA} = -\mathrm{Q}$ has only one solution if A has no eigenvalues on the imaginary axis and if there are no pairs of real eigenvalues with equal module and opposite sign (for example, $\lambda_1 = -1$ and $\lambda_2 = +1$), all of which conditions are verified by the hypothesis of asymptotic stability.

We have seen that vectorizing the Lyapunov equation requires inverting a matrix of size $n^2 \times n^2$ (or a matrix of size $\frac{n(n+1)}{2} \times \frac{n(n+1)}{2}$). Even for a system of order $n = 10$ this yields a matrix of ten thousand coefficients. This method is more laborious to calculate than other iterative methods used routinely for solving Lyapunov equations. Below is the sketch of an algorithm based on a Schur decomposition for solving Lyapunov equations of type $\mathrm{A}^T\mathrm{P}+\mathrm{PA} = -\mathrm{Q}$. A Schur decomposition able to obtain $\mathrm{A} = \mathrm{U}\bar{\mathrm{A}}\mathrm{U}^T$ with $\mathrm{U}^T\mathrm{U} = \mathrm{I}$ and higher triangular matrix $\bar{\mathrm{A}}$ is considered. At this point, all the Lyapunov equation terms are multiplied by U^T left and U right. Thus, $\bar{\mathrm{A}}^T\bar{\mathrm{P}} + \bar{\mathrm{P}}\bar{\mathrm{A}} = -\bar{\mathrm{Q}}$ is obtained with $\bar{\mathrm{Q}} = \mathrm{U}^T\mathrm{QU}$ and $\bar{\mathrm{P}} = \mathrm{U}^T\mathrm{PU}$. Here, triangular matrix A can be used to iteratively solve the equation and then to find the solution P from $\mathrm{P} = \mathrm{U}\bar{\mathrm{P}}\mathrm{U}^T$.

MATLAB® Exercise 2.3

The aim of this MATLAB exercise is to familiarize with some MATLAB commands for resolving control issues. It is always advisable to refer to the "help" command to learn about how each function works.
Let us consider a linear time-invariant system with state matrix

$$\mathrm{A} = \begin{bmatrix} -1 & 0 & 1 & 0 \\ 2 & 0 & 3 & 5 \\ 1 & 10 & 0.5 & 0 \\ -2 & 3 & -5 & -10 \end{bmatrix}$$

and then calculate matrix P of Lyapunov equation (2.4) with $\mathrm{Q} = \mathrm{I}$ by vectorization. To do this, matrix $\mathrm{M} = \mathrm{I} \otimes \mathrm{A}^T + \mathrm{A}^T \otimes \mathrm{I}$ with $\mathrm{C} = \mathrm{I}$ is constructed.

1. Define matrix A:
```
>> A=[-1 0 1 0; 2 0 3 5; 1 10 0.5 0; -2 3 -5 -10];
```
and identity matrix:
```
>> MatriceI=eye(4);
```
2. Construct the vector $\begin{bmatrix} c_{11} \\ c_{21} \\ c_{31} \\ \vdots \\ c_{nn} \end{bmatrix}$

```
>> c=MatriceI(:)
```

3. Construct matrix M

```
>> M=kron(eye(4),A')+kron(A',eye(4))
```

4. Find the solution vector by inverting matrix M

```
>> Psol=-inv(M)*c;
```

and obtain matrix P with an appropriate size (4×4)

```
>> P=reshape(Psol,4,4)
```

One obtains $P = \begin{bmatrix} 0.4727 & 0.2383 & -0.1661 & 0.1690 \\ 0.2383 & 1.7516 & -0.3188 & 0.8961 \\ -0.1661 & -0.3188 & -0.0466 & -0.1292 \\ 0.1690 & 0.8961 & -0.1292 & 0.4980 \end{bmatrix}$.

5. Verify that matrix P solves the problem

```
>> A'*P+P*A
```

MATLAB can solve Lyapunov equations more efficiently with the `lyap` command. In this case one can use:

```
>> P=lyap(A',eye(4))
```

In conclusion, using command `eig(P)` to calculate the eigenvalues of P, it can be verified that P is not a positive definite matrix and that the system is not asymptotically stable which is the case after calculating the eigenvalues of A ($\lambda_1 = -13.0222$, $\lambda_2 = 4.7681$, $\lambda_3 = -1.1229 + 0.8902i$, $\lambda_4 = -1.1229 - 0.8902i$).

2.5 Stability with Uncertainty

Here, we illustrate a brief digression on stability with uncertainty. We will limit the discussion to presenting one key finding for the stability of systems with parametric uncertainty.

Let $D(s)$ be the characteristic polynomial of a linear time-invariant system. Suppose the uncertainty of the system can be expressed in terms of variation ranges of coefficients of polynomial $D(s) = a_n s^n + a_{n-1}s^{n-1} + \ldots + a_1 s + a_0$:

$$\begin{gathered} a_0^m \leq a_0 \leq a_0^M \\ \ldots \\ a_n^m \leq a_n \leq a_n^M \end{gathered} \tag{2.6}$$

So, the hypothesis is that parametric uncertainty can be characterized by assigning maximum and minimum values that the various coefficients of $D(s)$ might assume. Under this hypothesis, stability with uncertainty can be studied very effectively via the Kharitonov criterion.

Theorem 8 (Kharitonov criterion) *Let* $D(s) = a_n s^n + a_{n-1}s^{n-1} + a_{n-2}s^{n-2} + a_{n-3}s^{n-3} + \ldots$ *be the characteristic polynomial of a linear time-invariant system with coefficients such that* $a_i^m \leq a_i \leq a_i^M$ *for* $i = 0, 1, \ldots, n$, *if and only if the four polynomials*

$$
\begin{array}{l}
D_1(s) = a_n^m s^n + a_{n-1}^m s^{n-1} + a_{n-2}^M s^{n-2} + a_{n-3}^M s^{n-3} + \dots \\
D_2(s) = a_n^M s^n + a_{n-1}^M s^{n-1} + a_{n-2}^m s^{n-2} + a_{n-3}^m s^{n-3} + \dots \\
D_3(s) = a_n^m s^n + a_{n-1}^M s^{n-1} + a_{n-2}^M s^{n-2} + a_{n-3}^m s^{n-3} + \dots \\
D_4(s) = a_n^M s^n + a_{n-1}^m s^{n-1} + a_{n-2}^m s^{n-2} + a_{n-3}^M s^{n-3} + \dots
\end{array}
\tag{2.7}
$$

are stable, then the polynomial $D(s)$ is stable for any parameter in the given ranges.

The Kharitonov criterion facilitates the study of stability in a parametric polynomial (aided for example by the Routh criterion). Quite independently of the order of the system and therefore of the number of parameters, this criterion can determine the stability of an infinite number of systems provided they have bounded polynomial coefficients.

Example 2.2

Let us consider a third-order system having the following characteristic polynomial

$$
D(s) = a_3 s^3 + a_2 s^2 + a_1 s + a_0 \tag{2.8}
$$

with uncertain parameters:

$$
\begin{array}{l}
1 = a_3^m \leq a_3 \leq a_3^M = 3 \\
2 = a_2^m \leq a_2 \leq a_2^M = 4 \\
2 = a_1^m \leq a_1 \leq a_1^M = 5 \\
1 = a_0^m \leq a_0 \leq a_0^M = 2
\end{array}
$$

and let us study the stability by the Kharitonov criterion.
First of all, we recall that, for a third-order system, the direct application of the Routh criterion yields the conclusion that stability requires two conditions: 1) all the coefficients are positive, i.e., $a_0, a_1, a_2, a_3 > 0$; 2) $a_2 a_1 > a_3 a_0$. Therefore, to apply the Kharitonov criterion we need to consider the following polynomials

$$
\begin{array}{l}
D_1(s) = a_3^m s^3 + a_2^m s^2 + a_1^M s + a_0^M \\
D_2(s) = a_3^M s^3 + a_2^M s^2 + a_1^m s + a_0^m \\
D_3(s) = a_3^m s^3 + a_2^M s^2 + a_1^M s + a_0^m \\
D_4(s) = a_3^M s^3 + a_2^m s^2 + a_1^m s + a_0^M
\end{array}
\tag{2.9}
$$

and ultimately check the conditions

$$
\begin{array}{l}
a_2^m a_1^M > a_3^m a_0^M \\
a_2^M a_1^m > a_3^M a_0^m \\
a_2^M a_1^M > a_3^m a_0^m \\
a_2^m a_1^m > a_3^M a_0^M
\end{array}
\tag{2.10}
$$

Since $a_2^m a_1^m < a_3^M a_0^M$, we conclude that the system is not stable in the presence of the considered parametric uncertainty.
Similarly, one can conclude that the system is stable for the following parametric uncertainty

$$
\begin{array}{l}
1 = a_3^m \leq a_3 \leq a_3^M = 3 \\
2 = a_2^m \leq a_2 \leq a_2^M = 4 \\
4 = a_1^m \leq a_1 \leq a_1^M = 5 \\
1 = a_0^m \leq a_0 \leq a_0^M = 2
\end{array}
$$

MATLAB® Exercise 2.4

Let us consider now a system with characteristic polynomial

$$D(s) = a_5 s^5 + a_4 s^4 + a_3 s^3 + a_2 s^2 + a_1 s + a_0 \tag{2.11}$$

and uncertain parameters:

$$\begin{array}{c} 1 \leq a_5 \leq 2 \\ 3 \leq a_4 \leq 5 \\ 5 \leq a_3 \leq 7 \\ \frac{4}{3} \leq a_2 \leq \frac{5}{2} \\ \frac{1}{2} \leq a_1 \leq \frac{3}{4} \\ 1 \leq a_0 \leq 2 \end{array}$$

To apply the Kharitonov criterion the following characteristic polynomials have to be considered:

$$\begin{array}{l} D_1(s) = s^5 + 3s^4 + 7s^3 + \frac{5}{2}s^2 + \frac{1}{2}s + 1 \\ D_2(s) = 2s^5 + 5s^4 + 5s^3 + \frac{4}{3}s^2 + \frac{3}{4}s + 2 \\ D_3(s) = s^5 + 5s^4 + 7s^3 + \frac{4}{3}s^2 + \frac{3}{4}s + 2 \\ D_4(s) = 2s^5 + 3s^4 + 5s^3 + \frac{5}{2}s^2 + \frac{3}{4}s + 1 \end{array} \tag{2.12}$$

These can be defined in MATLAB as follows

```
>> D1=[1 3 7 5/2 1/2 1]
>> D2=[2 5 5 4/3 3/4 2]
>> D3=[1 5 7 4/3 1/2 2]
>> D4=[2 3 5 5/2 3/4 1]
```

Their roots are then calculated:

```
>> roots(D1)
>> roots(D2)
>> roots(D3)
>> roots(D4)
```

obtaining the following solutions:

- $D_1(s)$: $s_{1,2} = -1.3073 \pm j2.0671$; $s_3 = -0.7059$; $s_{4,5} = 0.1603 \pm j0.4595$;
- $D_2(s)$: $s_{1,2} = -1.0224 \pm j0.9384$; $s_3 = -1.1541$; $s_{4,5} = 0.3494 \pm j0.5726$;
- $D_3(s)$: $s_{1,2} = -2.3261 \pm j0.4235$; $s_3 = -0.9566$; $s_{4,5} = 0.3044 \pm j0.5304$;
- $D_4(s)$: $s_{1,2} = -0.5107 \pm j1.3009$; $s_3 = -0.8031$; $s_{4,5} = 0.1623 \pm j0.5408$.

From the inspection of the roots of $D_1(s)$, $D_2(s)$, $D_3(s)$, and $D_4(s)$, it can be concluded that the system with characteristic polynomial (2.11) is not stable in the whole range of parameter variations.

MATLAB® Exercise 2.5

Determine the locus of the closed-loop eigenvalues for $k = 10$, $k = 30$ and $k = 120$ for the system $P(s,q) = \frac{1}{s(s^2+(8+q_1)s+(20+q_2))}$ with $-2 \leq q_1 \leq 2$ and $-4 \leq q_2 \leq 4$.

Solution

First, determine the characteristic closed-loop polynomial:

$$p(s,q,k) = s^3 + (8+q_1)s^2 + (20+q_2)s + k$$

This is a parametric polynomial that can be rewritten as:

$$p(s,q,k) = s^3 + As^2 + Bs + k$$

with $A = 8 + q_1$ and $B = 20 + q_2$. The two parameters in the characteristic polynomial vary in $[6, 10]$ and in $[16, 24]$, respectively. At this point, apply the Kharithonov criterion to know if, by fixing k, the asymptotic stability is guaranteed in the parametric uncertainty interval. Indicating $A_{min} = 6$, $A_{max} = 10$, $B_{min} = 16$ and $B_{max} = 24$, we need to study the roots of the four polynomials:

$$p(s, q, k) = s^3 + A_{min}s^2 + B_{max}s + k$$

$$p(s, q, k) = s^3 + A_{max}s^2 + B_{min}s + k$$

$$p(s, q, k) = s^3 + A_{min}s^2 + B_{min}s + k$$

$$p(s, q, k) = s^3 + A_{max}s^2 + B_{max}s + k$$

To do this, the following commands in MATLAB are used:

```
>> k=12;
>> Amin=6; Amax=10; Bmin=16; Bmax=24;
>> A=Amin; B=Bmax; roots([1 A B k])
>> A=Amax; B=Bmin; roots([1 A B k])
>> A=Amin; B=Bmin; roots([1 A B k])
>> A=Amax; B=Bmax; roots([1 A B k])
```

While, for $k = 10$ and $k = 30$, we obtain polynomials having roots with negative real part, for $k = 120$ the third polynomial has two roots with positive real part. After this preliminary analysis, the locus can be built with a Monte Carlo simulation, assigning the parameter values one by one, calculating their roots and plotting them on a graph. By fixing k, we can use the following commands in MATLAB:

```
>> plot(0,0,'m.'); hold on;
>> for A=[Amin:0.1:Amax]
>>     for B=[Bmin:0.1:Bmax]
>>         p=roots([1 A B k]);
>>         plot(real(p), imag(p),'x')
>>     end
>> end
```

Figure 2.2 shows the plots for four different values of k. Notice that for $k = 120$ some roots lie on the right half plane.

2.6 Further Results on the Lyapunov Theory

In this section, some further results on the Lyapunov theory are discussed. They can be skipped in basic courses and dealt with in more advanced ones.

2.6.1 Hystorical Notes

The book that is considered a fundamental reference for the study of the Lyapunov theory is "Nonlinear Dynamical Systems and Control: A Lyapunov-Based Approach" written by Wassim M. Haddad and VijaySekhar Chellaboina. The cover represents a poster that was printed for a meeting celebrating Archimedes and held in April 1961 in Siracusa. The original painting entitled "Archimedes" was realized by Niccoló Barbarino in 1860 and is kept in the Museum Revoltella in Trieste.

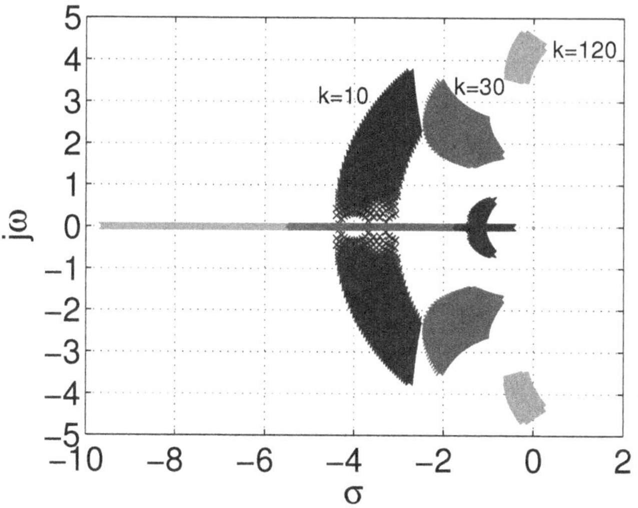

FIGURE 2.2
Locus of the closed-loop eigenvalues for Exercise 2.5.

The book reports a list of intriguing sentences regarding dynamical systems and here summarized:

- Anything that has a beginning and an end cannot at the same time be infinite and everlasting (Anassimander);
- Everything is in a state of flux and nothing is stationary (Eraclito);
- Man cannot step into the same river twice, because neither the man nor the river are the same (Eraclito);
- Give me a place to stand and I'll move the Earth (Archimedes).

Alexandr M. Lyapunov was a Russian mathematician who carried out research in different areas such as hydraulics-hydrostatics, ellipsoidal form of rotating fluids, stability, astronomical applications. About Lyapunov, Smirnov wrote the following observation: "Here at first, the research activity of Lyapunov was cut short. It was necessary to work out courses and put together notes for students, which took up much time".

2.6.2 Lyapunov Stability

In order to present a comprehensive discussion on this subject, let us now go back to a few considerations illustrated in Section 2.1 and extend the definitions there presented.

Let us then consider a general nonlinear autonomous dynamical system defined by

$$\dot{\mathbf{x}} = f(\mathbf{x}) \tag{2.13}$$

for $t \in \mathbb{R}^+$ and with initial conditions $\mathbf{x}(0) = \mathbf{x}_0$. Taking into account that $\mathbf{x} = \begin{bmatrix} x_1 & x_2 & \dots & x_n \end{bmatrix}^T$ and $f(\mathbf{x}) = \begin{bmatrix} f_1(\mathbf{x}) & f_2(\mathbf{x}) & \dots & f_n(\mathbf{x}) \end{bmatrix}^T$, these equations may be rewritten as:

$$\begin{aligned} \dot{x}_1 &= f_1(x_1, x_2, \dots, x_n) \\ \dot{x}_2 &= f_2(x_1, x_2, \dots, x_n) \\ &\dots \\ \dot{x}_n &= f_n(x_1, x_2, \dots, x_n) \end{aligned} \tag{2.14}$$

Since in general the solutions $\mathbf{x}(t)$ cannot be obtained in closed form, here our interest is to provide some important information about the time evolution of the state variables.

The equilibrium states are obtained by solving the algebraic nonlinear system $f(\mathbf{x}) = 0$. Suppose to indicate one of such states as $\bar{\mathbf{x}}$. The Lyapunov theory provides a characterization of the behavior of the solution when a small perturbation to the equilibrium state is applied.

Let us, first, recall (see Definition 1) that such an equilibrium is said to be Lyapunov stable if $\forall \varepsilon > 0$ there exists $\delta = \delta(\varepsilon) > 0$ such that $\forall \mathbf{x}_0$ with $\| \mathbf{x}_0 - \bar{\mathbf{x}} \| < \delta$ one has that $\| \mathbf{x}(t) - \bar{\mathbf{x}} \| < \varepsilon$ for $\forall t > 0$.

In addition, the stability is asymptotic if the equilibrium point is $\bar{\mathbf{x}}$ stable and $\lim_{t \to +\infty} \mathbf{x}(t) = \bar{\mathbf{x}}$.

Here it is important to remark that the previous definitions represent *local* properties. Let us now consider the case in which the property of asymptotic stability holds *globally*, that is, it holds for any initial condition $\mathbf{x}_0$, as formally expressed by the following definition.

Definition 5 *The equilibrium point $\bar{\mathbf{x}}$ is said to be globally asymptotically stable if it is Lyapunov stable and if for any $\mathbf{x}_0 \in \mathbb{R}^n$ it holds that*

$$\lim_{t \to +\infty} \mathbf{x}(t) = \bar{\mathbf{x}} \tag{2.15}$$

Another important property of stability is expressed by the following definition.

Definition 6 *An asymptotically locally stable equilibrium point $\bar{\mathbf{x}}$ is said to be exponentially stable if there exist two constants $\alpha > 0$ and $\beta > 0$ such that, if $\| \mathbf{x}_0 - \bar{\mathbf{x}} \| < \delta$, then*

$$\| \mathbf{x}(t) - \bar{\mathbf{x}} \| \le \alpha \| \mathbf{x}_0 - \bar{\mathbf{x}} \| e^{-\beta t}, \forall t \ge 0 \tag{2.16}$$

Finally, globally exponentially stability can also be considered, as formally expressed by the following definition.

Definition 7 *The equilibrium point* $\bar{\mathbf{x}}$ *is said to be globally exponentially stable if there exist two constants* $\alpha > 0$ *and* $\beta > 0$ *such that*

$$\| \mathbf{x}(t) - \bar{\mathbf{x}} \| \leq \alpha \| \mathbf{x}_0 - \bar{\mathbf{x}} \| e^{-\beta t}, \forall t \geq 0, \forall \mathbf{x}_0 \in \mathbb{R}^n \tag{2.17}$$

Although often control engineering aims at providing a theoretical and general approach for the solution of the problems under consideration, the application of numerical methods to study the characteristics of the solution $\mathbf{x}(t)$ is also useful, especially when closed-form solutions cannot be obtained.

The easiest way to obtain a numerical solution of a system of nonlinear differential equations as in (2.13), starting from initial conditions $\mathbf{x}_0$, is the Euler method, which is based on the approximation $\dot{\mathbf{x}} \approx \frac{\mathbf{x}((k+1)T) - \mathbf{x}(kT)}{T}$, where T represents the sampling interval/integration step size and k the discrete time index. Using this approximation, the following recursive difference equation is obtained:

$$\mathbf{x}((k+1)T) = \mathbf{x}(kT) + Tf(\mathbf{x}(kT)) \tag{2.18}$$

Another method, which, when the same integration step size T is used, is, in general, more accurate than the previous one is the fourth-order Runge-Kutta method. Let us consider a nonlinear system defined by $\dot{\mathbf{x}} = f(t, \mathbf{x})$ and use the index k for the samples of $\mathbf{x}(t)$ at time $t = t_k$ and so on, with $t_{k+1} = t_k + T$. Then, according to the fourth-order Runge-Kutta method, the next sample of the solution $\mathbf{x}(t)$ (with initial condition, fixed as usual, in $\mathbf{x}_0$) is calculated as:

$$\mathbf{x}(k+1) = \mathbf{x}(k) + \frac{1}{6}(\mathbf{k}_1 + 2\mathbf{k}_2 + 2\mathbf{k}_3 + \mathbf{k}_4) \tag{2.19}$$

where

$$\begin{aligned}
\mathbf{k}_1 &= Tf(t_k, \mathbf{x}(k)) \\
\mathbf{k}_2 &= Tf(t_k + \tfrac{T}{2}, \mathbf{x}(k) + \tfrac{\mathbf{k}_1}{2}) \\
\mathbf{k}_3 &= Tf(t_k + \tfrac{T}{2}, \mathbf{x}(k) + \tfrac{\mathbf{k}_2}{2}) \\
\mathbf{k}_4 &= Tf(t_k + T, \mathbf{x}(k) + \mathbf{k}_3)
\end{aligned} \tag{2.20}$$

This method can be extended to consider a recursive equation of the type:

$$\mathbf{x}(k+1) = \mathbf{x}(k) + \sum_{i=1}^{m} b_i \mathbf{k}_i \tag{2.21}$$

where the terms $\mathbf{k}_i$ with $i = 1, \ldots, m$ generalize the terms $\mathbf{k}_1, \mathbf{k}_2, \ldots, \mathbf{k}_4$ appearing in (2.20).

MATLAB® Exercise 2.6

Let us consider the following nonlinear system

$$\begin{aligned}\dot{x}_1 &= -x_1 + x_2^3 \\ \dot{x}_2 &= -x_1 - x_2\end{aligned} \tag{2.22}$$

with initial conditions $x_1(0) = 0.1$ and $x_2(0) = 0.2$. The numerical solution $\mathbf{x}(t)$ starting from these initial conditions can be obtained by using the MATLAB command `ode45` as follows:

```
>> [t,x]=ode45('nonlinsys1',[0 10],[0.1 0.2])
```

Here, the vector `[0 10]` contains the initial and final time of integration, whereas the vector `[0.1 0.2]` the initial conditions for the two variables. The function `nonlinsys1` has to be properly defined (in the directory of current use or in one of the MATLAB path) to represent the dynamical equations of the system. For the example under study, we have:

```
function xdot=nonlinsys1(t,x)
        xdot=[-x(1)+x(2)^3; -x(1)-x(2)];
end
```

Example 2.3

Consider the following nonlinear dynamical system representing a rigid spacecraft:

$$\begin{aligned}\dot{x}_1 &= I_{23}x_2x_3 \\ \dot{x}_2 &= I_{31}x_1x_3 \\ \dot{x}_3 &= I_{12}x_1x_2\end{aligned} \tag{2.23}$$

with

$$\begin{aligned}I_{23} &= \frac{I_2 - I_3}{I_1} \\ I_{31} &= \frac{I_3 - I_1}{I_2} \\ I_{12} &= \frac{I_1 - I_2}{I_3}\end{aligned} \tag{2.24}$$

where $I_1 > I_2 > I_3$ are the principal moments of inertia of the spacecraft. For the sake of example we consider $I_1 = 3$, $I_2 = 2$ and $I_3 = 1$, such that $I_{12} = 1$, $I_{23} = 1/3$, $I_{31} = -1$.

Let us now analyse numerically the stability property of the equilibrium point $\bar{\mathbf{x}} = 0$. In particular, let us calculate the effect of a small perturbation of the equilibrium point. To this aim, we use the command:

```
>> [t,x]=ode45('nonlinsys2',[0:0.01:100],[0.1 0.2 0.3])
```

where the function `nonlinsys2` is defined as follows:

```
function xdot=nonlinsys2(t,x)
       I23=1/3;
       I31=-1;
       I12=1;
       xdot=[I23*x(2)*x(3);I31*x(3)*x(1);I12*x(1)*x(2)];
end
```

The solution shows an oscillatory behavior that yields the conclusion that the equilibrium state $\bar{\mathbf{x}} = 0$ is stable, but not asymptotically. This can be appreciated by plotting the obtained waveform:

```
>> figure, plot(t,x)
```

The result is shown in Figure 2.3.

Example 2.4

Consider the following nonlinear dynamical system:

$$\begin{aligned}\dot{x}_1 &= \frac{-6x_1}{(1+x_1^2)^2} + 2x_2 \\ \dot{x}_2 &= -\frac{2(x_1+x_2)}{(1+x_1^2)^2}\end{aligned} \tag{2.25}$$

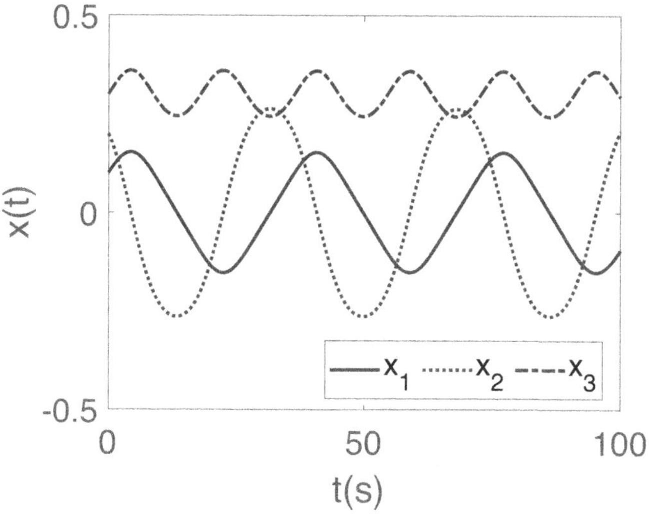

FIGURE 2.3
Time evolution of the state variables $\mathbf{x}(t)$ of system (2.24) in Example 2.3.

The system has an equilibrium state $\bar{\mathbf{x}} = 0$. To numerically study its stability, we integrate in MATLAB the system defined by:

```
function xdot=nonlinsys3(t,x)
        xdot=[-6*x(1)/((1+x(1)^2)^2)+2*x(2);
              -2*(x(1)+x(2))/((1+x(1)^2)*2)];
end
```

We find that trajectories starting from random arbitrary initial conditions converge the equilibrium state $\bar{\mathbf{x}} = 0$. However, it can be noticed that some initial conditions require a long transitory before reaching the equilibrium state.

Example 2.5

Consider now again the nonlinear dynamical system in equation (2.22) discussed in the MATLAB Exercise 2.6.
Repeating the procedure of the previous exampes, we can verify that in this case the equilibrium state $\bar{\mathbf{x}} = 0$ is globally asymptotically stable.

Example 2.6

Consider now the nonlinear dynamical system given by:

$$\begin{aligned} \dot{x}_1 &= x_1(x_1^2 + x_2^2 - 1) - x_2 \\ \dot{x}_2 &= x_1 + x_2(x_1^2 + x_2^2 - 1) \end{aligned} \tag{2.26}$$

The MATLAB function for its numerical study is given by:

```
function xdot=nonlinsys4(t,x)
        xdot=[x(1)*(x(1)^2+x(2)^2-1)-x(2); x(1)+x(2)*(x(1)^2+x(2)^2-1)];
end
```

Integrating again the system for different initial conditions, we notice that, if the initial conditions are selected inside the unit circle, then the trajectory approaches the equilibrium point $\bar{\mathbf{x}} = 0$, otherwise it does not. This supports the conclusion that the equilibrium point $\bar{\mathbf{x}} = 0$ is locally asymptotically stable.

We have already seen that, to determine if an equilibrium point $\bar{\mathbf{x}} = 0$ is asymptotically stable, we can use Theorem 1. Therefore, we can search a function $V(\mathbf{x})$ that is positive definite for $\bar{\mathbf{x}} = 0$ and whose derivative is negative definite in $\bar{\mathbf{x}} = 0$, that is, $\dot{V}(\mathbf{x}) = \frac{dV(\mathbf{x})}{dt} = \left(\frac{\partial V}{\partial \mathbf{x}}\right)^T \dot{\mathbf{x}} < 0$. If, on the contrary, $\dot{V}(\mathbf{x}) = \frac{dV(\mathbf{x})}{dt} = \left(\frac{\partial V}{\partial \mathbf{x}}\right)^T \dot{\mathbf{x}} \leq 0$, then the stability is simple and not asymptotic.

In addition, if there exist four constants α, β, γ and p such that

$$\alpha \parallel \mathbf{x} \parallel^p < V(\mathbf{x}) < \beta \parallel \mathbf{x} \parallel^p \tag{2.27}$$

and

$$\left(\frac{\partial V}{\partial \mathbf{x}}\right)^T f < -\epsilon V \tag{2.28}$$

then the equilibrium state $\bar{\mathbf{x}} = 0$ is exponentially stable.

Example 2.7

Let us consider again system (2.23) and consider the function

$$V(x_1, x_2, x_3) = \frac{1}{2}\alpha_1 x_1^2 + \frac{1}{2}\alpha_2 x_2^2 + \frac{1}{2}\alpha_3 x_3^2 \tag{2.29}$$

with $\alpha_1, \alpha_2, \alpha_3 > 0$.
In this case, one gets:

$$\begin{aligned} \dot{V} &= \left(\frac{\partial V}{\partial \mathbf{x}}\right)^T f = \begin{bmatrix} \alpha_1 x_1 & \alpha_2 x_2 & \alpha_3 x_3 \end{bmatrix} \begin{bmatrix} I_{23} x_2 x_3 \\ I_{31} x_1 x_3 \\ I_{12} x_1 x_2 \end{bmatrix} = \\ &= x_1 x_2 x_3 (\alpha_1 I_{23} + \alpha_2 I_{31} + \alpha_3 I_{12}) \end{aligned} \tag{2.30}$$

Due to the fact that I_{31} is negative, we select: $\alpha_1 = 1$, $\alpha_2 = -3/2$ and $\alpha_3 = 1$, such that $\dot{V} = 0 \ \forall \mathbf{x}$. This shows that the equilibrium state $\bar{\mathbf{x}} = 0$ is stable, as also the numerical simulations performed were indicating.

Notice that the theoretical calculations require the proper choice of a Lyapunov function, which is a non-trivial step. The simulations are an empirical method that often provides an immediate picture of the system behavior, but not a rigorous proof.

Example 2.8

Let us now consider again system (2.25) and study the following Lyapunov function:

$$V(x_1, x_2) = \frac{x_1^2}{1 + x_1^2} + x_2^2 \tag{2.31}$$

In this case, one obtains:

$$\begin{aligned} \dot{V} &= \left(\frac{\partial V}{\partial \mathbf{x}}\right)^T f = \begin{bmatrix} \frac{2x_1}{(1+x_1^2)^2} & 2x_2 \end{bmatrix} \begin{bmatrix} -\frac{6x_1}{(1+x_1^2)^2} + 2x_2 \\ -\frac{2(x_1+x_2)}{(1+x_1^2)^2} \end{bmatrix} = \\ &= -\frac{12x_1^2}{(1+x_1^2)^4} - \frac{4x_2^2}{(1+x_1^2)^2} \end{aligned} \tag{2.32}$$

Since $\dot{V} < 0\ \forall \mathbf{x}$, we conclude that the equilibrium point $\bar{\mathbf{x}} = 0$ is globally asymptotically stable.

2.7 Exercises

1. Apply the vectorization method to solve the Lyapunov equation:

$$\mathrm{A}^T\mathrm{P} + \mathrm{PA} = -\mathrm{I}$$

with

$$\mathrm{A} = \begin{bmatrix} 0 & 1 & -1 \\ 2 & -5 & -1 \\ 3 & 1 & -2 \end{bmatrix}$$

2. Given the nonlinear system $\dot{x} = x^3 - 8x^2 + 17x + u$ calculate the equilibrium points for $u = -10$ and study their stability.

3. Given the nonlinear system

$$\begin{cases} \dot{x}_1 = -x_1 + 2x_2 \\ \dot{x}_2 = -2x_1 - x_2 + x_2^3 \end{cases}$$

analyze numerically the stability of the equilibrium point $\bar{\mathrm{x}} = 0$.

4. Study the stability of system with transfer function

$$G(s) = \frac{s^2 + 2s + 2}{s^4 + a_1s^3 + a_2s^2 + a_3s + a_4}$$

with $1 \leq a_1 \leq 3$, $4 \leq a_2 \leq 7$, $1 \leq a_3 \leq 2$, $0.5 \leq a_4 \leq 2$.

5. Given the polynomial $p(s, a) = s^4 + 5s^3 + 8s^2 + 8s + 3$ with $a = \begin{bmatrix} 3 & 8 & 8 & 5 \end{bmatrix}$, find $p(s, b)$ with $b = \begin{bmatrix} (b_0^-, b_0^+) & \ldots & (b_3^-, b_3^+) \end{bmatrix}$ so that the polynomial class $p(s, b)$ is Hurwitz.

6. Study the stability of the system $G(s) = \frac{s^2+3s+2}{s^4+q_1s^3+5s^2+q_2s+q_3}$ with parameters $q_1 \in [1, 3]$, $q_2 \in [5, 10]$, $q_3 \in [2, 18]$.

3

Kalman Canonical Decomposition

CONTENTS

This chapter describes Kalman canonical decomposition, which highlights the state variables that do not affect the input/output properties of the system, but which nevertheless may be very important. This is a classical subject of system theory, but still fundamental to discuss in order to understand the importance of non-minimal order systems in designing electrical circuits. In addition, this topic is preliminary to the advanced concepts of weak and strong controllability and observability that are faced in the book. Exhaustive examples in MATLAB are reported in order to clarify the key concepts reported in the chapter. The discussion refers to continuous-time systems, so there is no distinction between the controllability properties and the reachability of the system.

3.1 Introduction

Given a continuous-time linear time-invariant system

$$\begin{aligned} \dot{\mathbf{x}} &= \mathrm{A}\mathbf{x} + \mathrm{B}\mathbf{u} \\ \mathbf{y} &= \mathrm{C}\mathbf{x} \end{aligned} \tag{3.1}$$

recall that the transfer function matrix of the system is given by $\mathrm{G}(s) = \mathrm{C}(s\mathrm{I} - \mathrm{A})^{-1}\mathrm{B}$, the controllability matrix $\mathrm{M}_c = \begin{bmatrix} \mathrm{B} & \mathrm{AB} & \ldots & \mathrm{A}^{n-1}\mathrm{B} \end{bmatrix}$, the

DOI: 10.1201/9781003196921-3

observability matrix $M_o = \begin{bmatrix} C \\ CA \\ \vdots \\ CA^{n-1} \end{bmatrix}$ and the Hankel matrix

$$H = M_o M_c = \begin{bmatrix} CB & CAB & \dots & CA^{n-1}B \\ CAB & CA^2B & \dots & \dots \\ \vdots & \vdots & \vdots & \vdots \\ CA^{n-1}B & \dots & \dots & \dots \end{bmatrix}.$$

Remember also that if the controllability matrix has full rank, that is equal to n (i.e., the system order), then the system is completely controllable. If the observability matrix has full rank, that is equal to n, then the system is completely observable. If the Hankel matrix has rank n, then the system is completely controllable and observable.

Now consider an invertible linear transformation $\tilde{\mathbf{x}} = T^{-1}\mathbf{x}$. The state matrices of the new reference system are related to the original state matrices through the relations:

$$\begin{aligned} \tilde{A} &= T^{-1}AT \\ \tilde{B} &= T^{-1}B \\ \tilde{C} &= CT \end{aligned} \tag{3.2}$$

and, as we know, the transfer function matrix is a system invariant: it does not change even as the reference system changes: $\tilde{G}(s) = \tilde{C}(sI - \tilde{A})^{-1}\tilde{B} = C(sI - A)^{-1}B = G(s)$.

The controllability and observability matrices vary as the reference system changes. In particular, they are given by:

$$\begin{aligned} \tilde{M}_c &= T^{-1}M_c \\ \tilde{M}_o &= M_oT \end{aligned} \tag{3.3}$$

Since matrix T is invertible and therefore the rank of $\tilde{M}_c$ and $\tilde{M}_o$ correspond to the rank of M_c and of M_o, from the relations (3.3) we deduce the well-known property that controllability and observability are *structural properties* of the system.

From relations (3.3), note that $\tilde{M}_o\tilde{M}_c = M_oM_c$, so the Hankel matrix is also an invariant of the system. Remember also that a system is minimal when it is completely controllable and observable, that is when the rank of the Hankel matrix equals n.

The number of state variables which characterize a minimal system is the minimum number of state variables needed to express all the input/output relations of the system. A system is minimal when its minimal form has order n, the system order.

For example, the system with transfer function $G(s) = \frac{s+1}{(s+1)(s+2)}$ is not minimal. However, we could say this is a "lucky" case, because the underlying dynamics (i.e., the one which does not influence the input/output relations) is stable. Generally, even hidden state variables are very significant, and it is the stability associated to the hidden dynamics that is the most significant property to be taken into consideration.

Consider also another example, represented by system $G(s) = \frac{s+0.98}{(s+1)(s+2)}$. There is not an exact simplification between the pole and zero, but clearly the pole and zero are very close and in the presence of uncertainty a simplification can occur. Conversely, a seemingly exact simplification is less precise in cases of uncertainty. The issue of determining a system minimal form is thus closely related to structural uncertainty.

In this chapter, we will examine the Kalman decomposition which takes into consideration when the simplifications between the poles and zeros of a system are exact, but in the next chapters we will raise the issue of determining a system approximation that takes into account the most significant dynamic for the input/output relation and that is the most robust against structural uncertainties.

3.2 Controllability Canonical Partition

Consider system (3.1) and suppose it is not completely controllable. This means there are internal state variables whose value, starting from a given initial condition and acting through the inputs, cannot be set to a value arbitrarily fixed in the system state-space. An example of an uncontrollable system is shown in Figure 3.1. Let $x_1(t)$ and $x_2(t)$ indicate the voltage across capacitors C_1 and C_2, respectively. Suppose $C_1 \neq C_2$, since the two capacitors are connected in series, the charge held by them is equal, then $\frac{C_1}{C_2} = \frac{x_1}{x_2}$. Therefore, the system is not controllable, and for example the state $x_1 = x_2 = 5V$ is not attainable by the system.

The canonical decomposition for controllability is a state representation which highlights the division of state variables into controllable state variables and uncontrollable state variables. The canonical decomposition thus emphasizes the existence of a controllable part and an uncontrollable part of the system. $\mathbf{z}$ indicate the state variables of the new reference system. They can be partitioned into two subsets: $\mathbf{z}_r$ are controllable variables and $\mathbf{z}_{nr}$ the uncontrollable variables. The system equations in Kalman canonical form for controllability can be expressed as follows:

$$
\begin{aligned}
\dot{\mathbf{z}}_r &= \mathrm{A}_1\mathbf{z}_r + \mathrm{A}_{12}\mathbf{z}_{nr} + \mathrm{B}_1\mathbf{u} \\
\dot{\mathbf{z}}_{nr} &= \mathrm{A}_2\mathbf{z}_{nr}
\end{aligned}
\tag{3.4}
$$

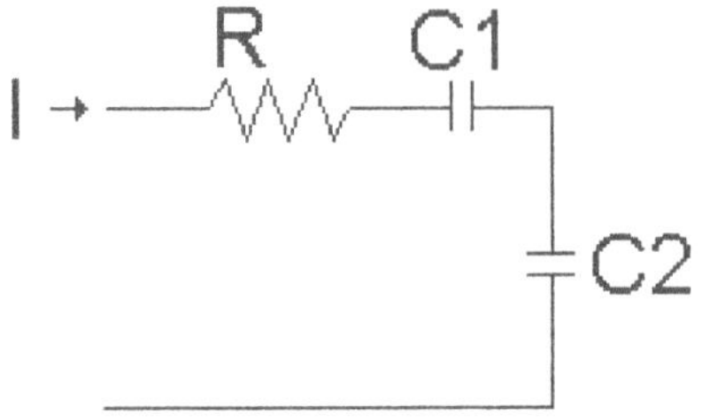

FIGURE 3.1
Example of a non-controllable system.

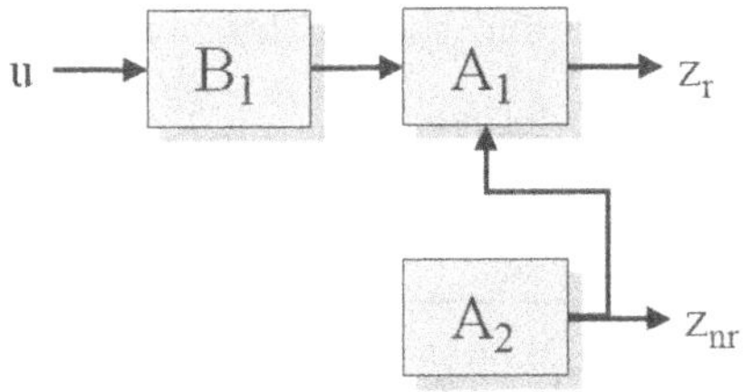

FIGURE 3.2
Canonical decomposition for controllability.

The state matrices are thus block-partitioned matrices: $\tilde{A} = \begin{bmatrix} A_1 & A_{12} \\ 0 & A_2 \end{bmatrix}$ and $\tilde{B} = \begin{bmatrix} B_1 \\ 0 \end{bmatrix}$.

The diagram of this model is shown in Figure 3.2. Note that it is not possible to act through the inputs on the variables $\mathbf{z}_{nr}$ directly given that $B_2 = 0$, nor indirectly through the other state variables $\mathbf{z}_r$ given that $A_{21} = 0$.

Note also that, given the structure of matrix $\tilde{A}$, the system eigenvalues are given by the union of those of A_1 and those of A_2. If A_1 has unstable eigenvalues, those acting through a control law can be moved to the left half plane. Instead, if A_2 has unstable eigenvalues, it is impossible to act on them. This second case is the most serious, because the system cannot be stabilized by feedback, in which case the system is known as non-stabilizable. By contrast, a system for which all unstable eigenvalues belong to the controllable part is known as stabilizable.

Having explained the characteristics of canonical decomposition for controllability, let us see how to calculate matrix T which allows us to switch from the original form to the particular structure of canonical decomposition.

Consider the controllability matrix M_c. It is rank deficient. Note, however, that the linearly independent columns of this matrix define the subspace of controllability of the system. The subspace of uncontrollability can be defined as orthogonal to that of controllability. Matrix T is constructed from the linearly independent columns of M_c. In particular, $T = \begin{bmatrix} T_1 & T_2 \end{bmatrix}$ where T_1 is the matrix formed by the linearly independent columns of M_c and T_2 by the column-vectors orthogonal to those linearly independent of M_c. Applying this transformation matrix to the original system, we obtain:

$$\tilde{A} = T^{-1}AT = \begin{bmatrix} U_1 \\ U_2 \end{bmatrix} A \begin{bmatrix} T_1 & T_2 \end{bmatrix} = \begin{bmatrix} U_1AT_1 & U_1AT_2 \\ U_2AT_1 & U_2AT_2 \end{bmatrix}$$

but, since the columns of U_2 are orthogonal to the linearly independent columns of A, then $U_2AT_1 = 0$ and

$$\tilde{A} = \begin{bmatrix} U_1AT_1 & U_1AT_2 \\ 0 & U_2AT_2 \end{bmatrix}$$

In the same way we obtain that

$$\tilde{B} = T^{-1}B = \begin{bmatrix} U_1B_1 \\ 0 \end{bmatrix}$$

3.3 Observability Canonical Partition

There is a dual decomposition for observability which can identify the observable and unobservable parts of a system. The reference diagram is shown in Figure 3.3, while the system model in this particular state-space representation is expressed by the following equations:

$$\begin{aligned} \dot{\mathbf{z}}_o &= A_1\mathbf{z}_o + B_1\mathbf{u} \\ \dot{\mathbf{z}}_{no} &= A_{21}\mathbf{z}_o + A_2\mathbf{z}_{no} + B_2\mathbf{u} \\ \mathbf{y} &= C_1\mathbf{z}_o \end{aligned} \tag{3.5}$$

Again the state matrices are partitioned in blocks: $\tilde{A} = \begin{bmatrix} A_1 & 0 \\ A_{21} & A_2 \end{bmatrix}$, $\tilde{B} = \begin{bmatrix} B_1 \\ B_2 \end{bmatrix}$ and $\tilde{C} = \begin{bmatrix} C_1 & 0 \end{bmatrix}$.

The model can be derived analogously to the canonical decomposition for controllability. The unobservable variables z_{no} cannot be reconstructed from input/output measurements so there cannot be any direct link between these variables and the output. Because $C_2 = 0$, the unobservable variables do not directly affect system output. Similarly, since the unobservable variables cannot be reconstructed from the output information, even indirectly, they cannot influence the dynamics of the observable variables. So, $A_{12} = 0$.

The transformation matrix from the original reference system to the canonical decomposition for observability can be found with a dual procedure of the previous one.

This time, let us consider the linearly independent rows of observability matrix M_o to construct matrix U_1. Suppose there are n_o linearly independent rows. Consider the subspace of size $n - n_o$ orthogonal to the defined one and determine a subspace basis to construct matrix U_2. So, we have $T^{-1} = \begin{bmatrix} U_1 \\ U_2 \end{bmatrix}$. Applying this transformation matrix to the original system, we have

$$\tilde{A} = \begin{bmatrix} U_1AT_1 & U_1AT_2 \\ U_2AT_1 & U_2AT_2 \end{bmatrix} = \begin{bmatrix} U_1AT_1 & 0 \\ U_2AT_1 & U_2AT_2 \end{bmatrix}$$

and

$$\tilde{C} = \begin{bmatrix} CT_1 & CT_2 \end{bmatrix} = \begin{bmatrix} CT_1 & 0 \end{bmatrix}$$

Again the system eigenvalues are the union of the eigenvalues of A_1 and those of A_2. If the system is unstable and controllable, but not observable, we need to know to which part belong the unstable modes of the system. If they are located in the unobservable part, these modes cannot be observed and so they cannot be controlled. However, if they are located in the observable part of the system, they can be reconstructed and then controlled.

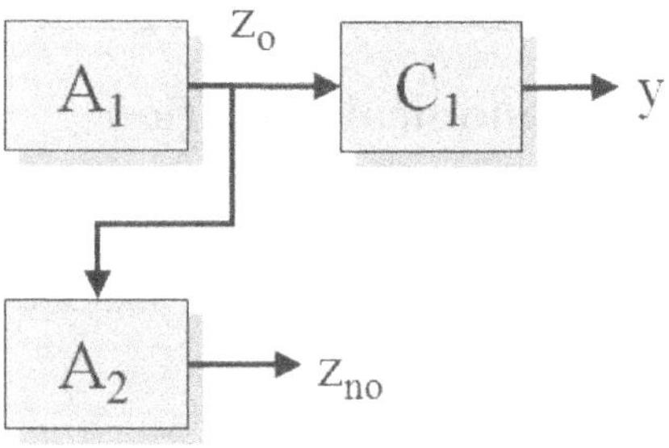

FIGURE 3.3
Observability canonical decomposition.

3.4 General Partition

The two decompositions in the paragraphs above are elementary decompositions which can decouple a system into a controllable part and an uncontrollable part or into an observable part and an unobservable part. There is a more general decomposition which can split a system into four parts: a controllable and unobservable part (part A); a controllable and observable part (part B); an uncontrollable and unobservable part (part C); and an uncontrollable and observable part (part D). Clearly, this decomposition can be applied to the most general case in which the system is not completely controllable nor observable.

The state variables corresponding to part A of the system are $\mathbf{z}_{r,no}$; part B $\mathbf{z}_{r,o}$; part C $\mathbf{z}_{nr,no}$; and to part D $\mathbf{z}_{nr,o}$. The block diagram of the Kalman decomposition is shown in Figure 3.4 and the corresponding state-space equations are characterized by the following matrices:

$$\tilde{\mathrm{A}} = \begin{bmatrix} \mathrm{A}_A & \mathrm{A}_{AB} & \mathrm{A}_{AC} & \mathrm{A}_{AD} \\ 0 & \mathrm{A}_B & 0 & \mathrm{A}_{BD} \\ 0 & 0 & \mathrm{A}_C & \mathrm{A}_{AD} \\ 0 & 0 & 0 & \mathrm{A}_D \end{bmatrix}; \tilde{\mathrm{B}} = \begin{bmatrix} \mathrm{B}_A \\ \mathrm{B}_B \\ 0 \\ 0 \end{bmatrix}; \tilde{\mathrm{C}} = \begin{bmatrix} 0 & \mathrm{C}_B & 0 & \mathrm{C}_D \end{bmatrix}$$

Note that the matrix $\tilde{\mathrm{A}}$ is a triangular-block matrix. The zero blocks reflect no direct link between two parts of the system. For example, since the controllable and the unobservable part of the system cannot influence the controllable and observable part of the system (otherwise, it would lose its property of being unobservable), block A_{BA} is zero. Analogous considerations apply to all the possible connections from any block in Figure 3.4 to any other block below. Any of these links would violate one of the hypotheses underlying the structural properties of the various parts of the system, so they are impossible. Conversely, the upward links between blocks are permitted except any link between the uncontrollable and unobservable part (part C) and the controllable and observable part (part B). In fact, if there was such a link, part C would no longer be unobservable. For this reason $\mathrm{A}_{BC} = 0$.

Now let us analyze how to find transition matrix T. Its construction requires defining four eigenvector groups which form the columns of matrix T and the bases of four subspaces which will be defined below, after having recalled some preliminary notions.

A subspace is A-invariant if, when any subspace vector is multiplied by matrix A, it still belongs to the subspace.

The reachability subspace and unobservability subspace are A-invariant subspaces. The first subspace constructed to define matrix T is given by the intersection of the reachability subspace X_r and the unobservability subspace X_{no}:

$$\mathrm{X}_A = \mathrm{X}_r \bigcap \mathrm{X}_{no} \tag{3.6}$$

This is an A-invariant subspace, since it is intersection of two A-invariant subspaces.

From X_A we then define X_B, so that the direct sum between X_A and X_B is the reachability subspace:

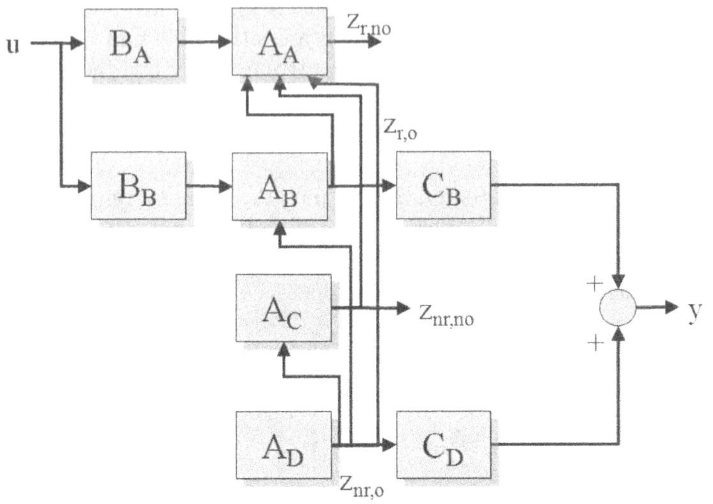

FIGURE 3.4
Canonical decomposition of a linear system.

$$X_r = X_A \oplus X_B \tag{3.7}$$

Similarly, X_C is defined so that the direct sum of X_A and X_C is the unobservability subspace:

$$X_{no} = X_A \oplus X_C \tag{3.8}$$

Finally, we construct X_D so as to obtain the entire state-space as a direct sum of X_r and X_{no} and of X_D.

$$X = (X_r + X_{no}) \oplus X_D \tag{3.9}$$

Since

$$X_A \oplus X_B \oplus X_C = (X_A \oplus X_B) + (X_A \oplus X_C) = X_r + X_{no}$$

then

$$X = X_A \oplus X_B \oplus X_C \oplus X_D$$

It follows that any state vector of the system can be represented uniquely as the sum of four vectors which belong to the spaces defined above. So, the state-space is decomposed into the direct sum of the four subspaces defined as X_A, X_B, X_C and X_D. From these subspaces the columns of matrix T can be generated. However, these subspaces are defined quite arbitrarily which can

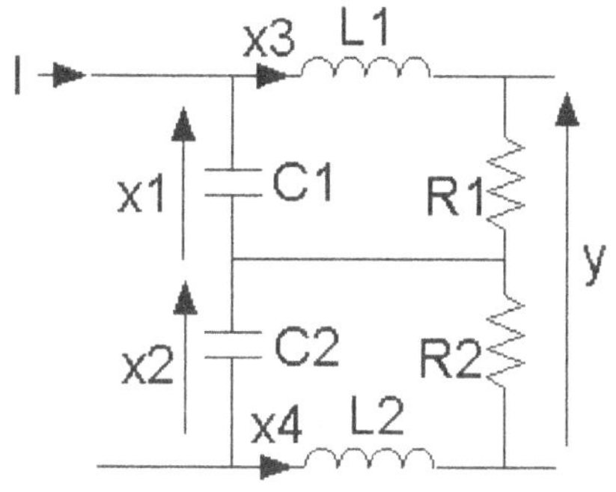

FIGURE 3.5
Example of electrical circuit illustrating Kalman decomposition.

be prevented by imposing additional restrictions to them. In particular, we can make X_B and X_C orthogonal to X_A and that X_D is orthogonal to $(X_r + X_{no})$, thus obtaining new subspaces $\bar{X}_B$, $\bar{X}_C$ and $\bar{X}_D$. With these additional restrictions, subspaces X_A, $\bar{X}_B$, $\bar{X}_C$ and $\bar{X}_D$ can unequivocally be defined:

$$\begin{aligned} X_A &= X_r \bigcap X_{no} \\ \bar{X}_B &= X_r \bigcap (X_{nr} + X_o) \\ \bar{X}_C &= X_{no} \bigcap (X_{nr} + X_o) \\ \bar{X}_D &== X_{nr} \bigcap X_o \end{aligned} \qquad (3.10)$$

Once these subspaces are defined, we can obtain the transition matrix T. The columns of this matrix are provided by the bases of the generated subspaces. The procedure for calculating matrix T is described in detail in the following example.

Example 3.1

Consider the electrical circuit shown in Figure 3.5 and suppose that all the circuit components have normalized values: $L_1 = 1$, $L_2 = 1$, $C_1 = 1$, $C_2 = 1$, $R_1 = 1$ and $R_2 = 1$. The state variables of this system are the two voltages on the capacitors shown in the figure, $x_1(t)$ and $x_2(t)$, and the two currents in the inductors, $x_3(t)$ and $x_4(t)$. Given these state variables, the system is described by the following matrices:

$$A = \begin{bmatrix} 0 & 0 & -1 & 0 \\ 0 & 0 & 0 & 1 \\ 1 & 0 & -1 & 0 \\ 0 & -1 & 0 & -1 \end{bmatrix}; B = \begin{bmatrix} 1 \\ 1 \\ 0 \\ 0 \end{bmatrix}; C = \begin{bmatrix} 0 & 0 & 1 & -1 \end{bmatrix}$$

Let us first calculate the observability and controllability matrices:

$$M_o^T = \begin{bmatrix} 0 & 1 & -1 & 0 \\ 0 & 1 & -1 & 0 \\ 1 & -1 & 0 & 1 \\ -1 & 1 & 0 & 1 \end{bmatrix} \qquad (3.11)$$

$$M_c = \begin{bmatrix} 1 & 0 & -1 & 1 \\ 1 & 0 & -1 & 1 \\ 0 & 1 & -1 & 0 \\ 0 & -1 & 1 & 0 \end{bmatrix} \tag{3.12}$$

Both matrices have rank equal to two. A basis for the observability subspace X_o can be established from the observability matrix, taking for example the first and third columns (linearly independent). The unobservability subspace X_{no} is constructed orthogonal to the observability subspace X_o and is generated by the vectors:

$$V_1 = \begin{bmatrix} 1 \\ -1 \\ 0 \\ 0 \end{bmatrix}; V_2 = \begin{bmatrix} 0 \\ 0 \\ 1 \\ 1 \end{bmatrix}$$

The reachability subspace X_r is identified by the linearly independent columns of M_c. Take for example the first two columns of M_c. The unreachability subspace is determined by considering the subspace orthogonal to that identified by the first two columns of M_c.
Note that in this case $X_o = X_r$. So $X_r = X_{no}^{\perp}$ and $X_r \perp X_{no}$.
As the reachability subspace is orthogonal to the unobservability subspace, their intersection is the empty set. The intersection of these subspaces (see formula (3.6)) is X_A. So $X_A = \{\emptyset\}$.
Moreover, since $X_D = X_{nr} \bigcap X_o$, also $X_D = \{\emptyset\}$.
Finally, applying equation (3.10), $\bar{X}_B$ and $\bar{X}_C$ can be found:

$$\bar{X}_B = X_r \bigcap (X_{nr} + X_r) = X_r$$

$$\bar{X}_C = X_{no} \bigcap (X_{nr} + X_r) = X_{no}$$

From this, it follows that the transition matrix can be defined by taking the first two columns of matrix M_c and the vectors that form the basis of X_{no} (V_1 and V_2):

$$T = \begin{bmatrix} 1 & 0 & 1 & 0 \\ 1 & 0 & -1 & 0 \\ 0 & 1 & 0 & 1 \\ 0 & -1 & 0 & 1 \end{bmatrix} \tag{3.13}$$

At this point, the state matrices $\tilde{A}$, $\tilde{B}$ and $\tilde{C}$ in the state-space representation of the Kalman decomposition can be calculated:

$$\tilde{A} = T^{-1}AT = \begin{bmatrix} 0 & -1 & 0 & 0 \\ 1 & -1 & 0 & 0 \\ 0 & 0 & 0 & -1 \\ 0 & 0 & 1 & -1 \end{bmatrix}$$

$$\tilde{B} = T^{-1}B = \begin{bmatrix} 1 \\ 0 \\ 0 \\ 0 \end{bmatrix}$$

$$\tilde{C} = CT = \begin{bmatrix} 0 & 2 & 0 & 0 \end{bmatrix}$$

As you can see, part B of the system (i.e., the controllable and observable part) is second order and so also part C (the uncontrollable and unobservable part), whereas there is no controllable and unobservable part nor is there an observable and uncontrollable part. The transfer function of the system is therefore second order.

The next MATLAB® exercise shows another example of Kalman decomposition and how to obtain it by using MATLAB®.

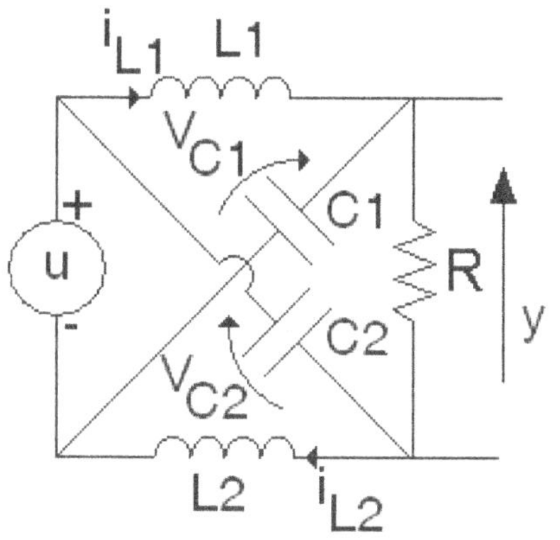

FIGURE 3.6
An *all-pass* electrical circuit.

MATLAB® Exercise 3.1

Consider the circuit shown in Figure 3.6 (the reader is referred to the book of Wiener for more deep discussion on this circuit). Given $L_1 = L_2 = L$, $C_1 = C_2 = C$ and $R = \sqrt{\frac{L}{C}}$, it has the following transfer function:

$$\frac{Y(s)}{U(s)} = \frac{1 - s\sqrt{LC}}{1 + s\sqrt{LC}} \tag{3.14}$$

and it is an *all-pass* system (these systems will be encountered again in Chapter 6 and in Chapter 7). These are systems characterized by a frequency response $G(j\omega)$ with $|G(j\omega)| = 1 \; \forall \omega$, and a flat magnitude Bode diagram. This example is particularly relevant as it shows the use of non-minimal form systems to obtain an all-pass system with passive components.

We now derive the state-space equations and discuss the Kalman decomposition by using MATLAB®.

Taking into account the following state variables

$$\begin{aligned} x_1 &= i_{L1} \\ x_2 &= v_{C2} \\ x_3 &= i_{L2} \\ x_4 &= v_{C1} \end{aligned} \tag{3.15}$$

the state-space representation of the circuit can be derived by applying Kirchhoff's circuit laws. The following equations are obtained:

$$\begin{aligned} \dot{x}_1 &= -\tfrac{x_4}{L1} + \tfrac{u}{L_1} \\ \dot{x}_2 &= -\tfrac{x_2}{C_2 R} + \tfrac{x_3}{C_2} - \tfrac{x_4}{C_2 R} + \tfrac{u}{R} \\ \dot{x}_3 &= -\tfrac{x_2}{L2} + \tfrac{u}{L_2} \\ \dot{x}_4 &= \tfrac{x_1}{C_1} - \tfrac{x_2}{C_1 R} - \tfrac{x_4}{C_1 R} + \tfrac{u}{R} \end{aligned} \tag{3.16}$$

and

$$y = x_2 + x_4 - u \tag{3.17}$$

From equations (3.16) and (3.17), the state-space matrices are derived:

$$A = \begin{bmatrix} 0 & 0 & 0 & -\frac{1}{L_1} \\ 0 & -\frac{1}{C_2 R} & \frac{1}{C_2} & -\frac{1}{C_2 R} \\ 0 & -\frac{1}{L_2} & 0 & 0 \\ \frac{1}{C_1} & -\frac{1}{C_1 R} & 0 & -\frac{1}{C_1 R} \end{bmatrix}; B = \begin{bmatrix} \frac{1}{L_1} \\ \frac{1}{C_2 R} \\ \frac{1}{L_2} \\ \frac{1}{C_1 R} \end{bmatrix}; C = \begin{bmatrix} 0 & 1 & 0 & 1 \end{bmatrix}; D = -1 \tag{3.18}$$

Let us now consider $C = 1$, $L = 1$ and $R = 1$ (all expressed in dimensionless units), so that we have

$$A = \begin{bmatrix} 0 & 0 & 0 & -1 \\ 0 & -1 & 1 & -1 \\ 0 & -1 & 0 & 0 \\ 1 & -1 & 0 & -1 \end{bmatrix}; B = \begin{bmatrix} 1 \\ 1 \\ 1 \\ 1 \end{bmatrix}; C = \begin{bmatrix} 0 & 1 & 0 & 1 \end{bmatrix}; D = -1 \tag{3.19}$$

and let us calculate the minimal realization and the Kalman decomposition through MATLAB (it is in fact obvious that equations (3.19) are not a minimal realization of the system with transfer function (3.14)).
First of all, let us define the system:

```
>> A=[0 0 0 -1; 0 -1 1 -1; 0 -1 0 0; 1 -1 0 -1]
>> B=[1 1 1 1]'
>> C=[0 1 0 1]
>> D=-1
>> system=ss(A,B,C,D)
```

Let us now calculate the reachability, the observability and the Hankel matrix:

```
>> Mc=ctrb(A,B)
>> Mo=obsv(A,C)
>> H= Mo*Mc
```

One obtains:

$$M_c = \begin{bmatrix} 1 & -1 & 1 & -1 \\ 1 & -1 & 1 & -1 \\ 1 & -1 & 1 & -1 \\ 1 & -1 & 1 & -1 \end{bmatrix}; M_o = \begin{bmatrix} 0 & 1 & 0 & 1 \\ 1 & -2 & 1 & -2 \\ -2 & 3 & -2 & 3 \\ 3 & -4 & 3 & -4 \end{bmatrix}$$

and

$$H = \begin{bmatrix} 2 & -2 & 2 & -2 \\ -2 & 2 & -2 & 2 \\ 2 & -2 & 2 & -2 \\ -2 & 2 & -2 & 2 \end{bmatrix}$$

The ranks of these matrices are then computed:

```
>> rank(Mc)
>> rank(Mo)
>> rank(H)
```

M_c has rank one, M_o two and H one. It can be concluded that the minimal form of the system is first order and that the unobservable part is second order and the uncontrollable part is third order. The transfer function with the command `tf(system)` or `zpk(system)` is also calculated, obtaining $G(s) = \frac{1-s}{1+s}$, as in equation (3.14).
Let us now calculate the minimal realization with the MATLAB command:

```
>> [msystem,U] = minreal(system)
```

This gives the system in minimal form (first order):

$$A_B = \ -1 \ ; B_B = \ 2 \ ; C_B = \ 1 \ ; D_B = -1 \tag{3.20}$$

Matrix U is the inverse of the state transformation matrix from the original representation to the Kalman decomposition. To obtain this, we have thus to consider (since U is orthogonal):

```
>> T=U'
```

and then to compute:

```
>> Atilde=T'*A*T
>> Btilde=T'*B
>> Ctilde=C*T
```

One obtains

$$\tilde{A} = \begin{bmatrix} -1 & -2 & 0 & 0 \\ 0 & -1 & 0 & 0 \\ 0 & 0 & 0 & 1 \\ 0 & 0 & -1 & 0 \end{bmatrix}; \tilde{B} = \begin{bmatrix} 2 \\ 0 \\ 0 \\ 0 \end{bmatrix}; \tilde{C} = \begin{bmatrix} 1 & 1 & 0 & 0 \end{bmatrix}; \tilde{D} = -1 \tag{3.21}$$

It should be noted that the system is divided in three parts. They appear in $\tilde{A}, \tilde{B}, \tilde{C}, \tilde{D}$ in the following sequence: controllable and observable part (first order); uncontrollable and observable part (first order); uncontrollable and unobservable part (second order). In conclusion, in this example, a fourth-order non-minimal circuit realization of a first-order all-pass system that uses only passive components is obtained.

3.5 Remarks on Kalman Decomposition

The example in Figure 3.5 is very different to the example in Figure 3.1. In the first case, if we consider different values of the parameters (for example, $R_1 = 0.8$, $C_1 = 1.1$ and $L_2 = 1.2$), the rank of controllability matrix M_c and observability matrix M_o change. The same structure with different parameters (which may be due to perturbations) behaves very differently. In one case, the transfer function is second order and in the other fourth order. This is because the system consists of two symmetric parts when the values of resistors, capacitors and inductors are equal, but cease to be when those values change.

Instead, the circuit in Figure 3.1 is always an uncontrollable system, notwithstanding the parameter values. Perturbations in the values cannot make it controllable.

For this reason, we must also know how much a system is effectively controllable and observable, not only by evaluating controllability and observability matrices ranks, but also their singular values. If there are very small singular values compared to others, parametric uncertainties may change the structural properties of controllability and observability of the system.

In addition to whether a system is controllable and/or observable or not, we should also ask how much a system is controllable and observable.

In the next chapters, we will examine systems that while controllable and observable can also be decomposed (and so approximated) in a part that mostly influences the input/output properties of the system and in a less important part.

An interesting exercise may be to construct, for example, an unobservable and uncontrollable system of given order and with assigned properties of the controllable and observable subsystem. It is easy to show that, once found, an infinite number of such same order systems can be generated.

3.6 Exercises

1. Calculate the Kalman decomposition for the system with state-space matrices:

$$\mathrm{A} = \begin{bmatrix} -2 & 3 & 0 & 0 & 0 \\ 1 & 0 & 0 & 0 & 0 \\ 1 & -1 & 3 & 0 & 0 \\ -2 & 1 & -1 & -1 & 0 \\ -4 & 1 & 2 & -1 & -2 \end{bmatrix}; \mathrm{B} = \mathrm{C}^T = \begin{bmatrix} 1 \\ 1 \\ 1 \\ 1 \\ 1 \end{bmatrix}$$

2. Calculate the Kalman decomposition for the system with state-space matrices:

$$\mathrm{A} = \begin{bmatrix} -2 & 0 & 0 & 0 & 0 \\ 1 & 0 & 0 & 0 & 0 \\ 1 & -1 & 3 & 0 & 0 \\ -2 & 1 & -1 & -1 & 0 \\ -4 & 1 & 2 & -1 & -2 \end{bmatrix}; \mathrm{B} = \mathrm{C}^T = \begin{bmatrix} 1 \\ 1 \\ 1 \\ 1 \\ 1 \end{bmatrix}$$

3. Calculate the Kalman decomposition for the system with state-space matrices:

$$\mathrm{A} = \begin{bmatrix} -1 & 0 & 0 & 0 \\ 0 & -1 & 0 & 0 \\ 0 & 0 & 2 & 0 \\ 1 & 1 & -3 & 2 \end{bmatrix}; \mathrm{B} = \begin{bmatrix} 0 \\ 0 \\ 1 \\ 0 \end{bmatrix}; \mathrm{C} = \begin{bmatrix} 0 & 1 & 1 & 0 \end{bmatrix}$$

4. Calculate the Kalman decomposition for the system with state-space matrices:

$$\mathrm{A} = \begin{bmatrix} -1 & 0 & 0 & 0 \\ 0 & -2 & 0 & 0 \\ 0 & 0 & 3 & 0 \\ 0 & 0 & 0 & 2 \end{bmatrix}; \mathrm{B} = \begin{bmatrix} 1 \\ 0 \\ 1 \\ 0 \end{bmatrix}; \mathrm{C} = \begin{bmatrix} 1 & 1 & 1 & 1 \end{bmatrix}$$

5. Calculate the Kalman decomposition for the system with state-space matrices:

$$\mathrm{A} = \begin{bmatrix} -1 & 0 & 0 & 0 \\ 0 & -2 & 0 & 0 \\ 0 & 0 & 3 & 0 \\ 0 & 0 & 0 & 2 \end{bmatrix}; \mathrm{B} = \begin{bmatrix} 1 \\ 1 \\ 1 \\ 1 \end{bmatrix}; \mathrm{C} = \begin{bmatrix} 0 & 1 & 0 & 1 \end{bmatrix}$$

6. Calculate the Kalman decomposition for the system with state-space matrices:

$$A = \begin{bmatrix} 2 & 1 & 0 & 0 \\ 0 & -1 & 0 & 0 \\ 0 & 0 & 1 & 1 \\ 1 & 0 & -3 & 4 \end{bmatrix}; B = \begin{bmatrix} 1 \\ 0 \\ 1 \\ 0 \end{bmatrix}; C = \begin{bmatrix} 1 & 1 & 1 & 1 \end{bmatrix}$$

4

Singular Value Decomposition

CONTENTS

Some matrix numerical tools are developed in this part. The singular value decomposition (SVD) represents one of the more robust algorithms used in problems and procedures based on matrix numerical tools. The core of the MATLAB platform derives from robust algorithms devoted to solve systems of linear equations and find eigenvalues of matrices. The SVD allows us, in the view of the robust control theory, to introduce some fundamental topics like the spectral norm of a matrix and its condition number. The chapter is considered of particular importance as the book is self-contained and each argument is strictly related to each other. The effort made in the book is to highlight the links among the various subjects. In this respect, the quite transversal topic of the SVD is particularly important to remark.

4.1 Singular Values of a Matrix

Any matrix can be decomposed into three matrices with special properties. This unique decomposition is called singular value decomposition.

Consider matrix $\mathrm{A} \in \mathbb{R}^{m\times n}$ and for clarity suppose $m \geq n$, then there are always three matrices U, Σ and V^T such that matrix A can be written as

$$\mathrm{A} = \mathrm{U}\Sigma\mathrm{V}^T$$

where U is a unitary matrix of dimensions $m \times m$ (i.e., $\mathrm{U}^T\mathrm{U} = \mathrm{I}$), V is also a unitary matrix, but with dimensions $n \times n$, and Σ matrix $m \times n$ is defined as:

$$\Sigma = \begin{bmatrix} \bar{\Sigma} \\ \mathbf{0} \end{bmatrix}$$

In this last expression matrix $\mathbf{0}$ is a matrix of $(m - n) \times n$ null elements,

DOI: 10.1201/9781003196921-4

while matrix $\bar{\Sigma}$ is a $n \times n$ diagonal matrix. The elements of the diagonal of $\bar{\Sigma}$ in descending order are the *singular values* of matrix A:

$$\bar{\Sigma} = \begin{bmatrix} \sigma_1 & 0 & \dots & 0 \\ 0 & \sigma_2 & \dots & 0 \\ \vdots & \vdots & & \vdots \\ 0 & 0 & \dots & \sigma_n \end{bmatrix}$$

Calculating the singular values of matrix A is simple. They are the square roots of the eigenvalues of matrix A^TA. Note that, since the eigenvalues of A^TA are always non-negative real values, it is always possible to calculate their square root.

The eigenvalues of A^TA are always real and non-negative because A^TA is a symmetric positive semi-definite matrix. The symmetry of this matrix is immediately verifiable, in fact

$$(A^TA)^T = A^TA.$$

Moreover, if we consider the associated quadratic form $V(\mathbf{x}) = \mathbf{x}^TA^TA\mathbf{x}$ and apply Theorem 5 of Chapter 2, we deduce that A^TA is positive definite if A has no null eigenvalues, otherwise it is positive semi-definite.

To prove that the eigenvalues of A^TA are the squares of the singular values of matrix A, consider the singular value decomposition of its transpose:

$$A^T = V\Sigma^TU^T$$

Now consider A^TA, which is a symmetric matrix and therefore diagonalizable:

$$A^TA = V\Sigma^TU^TU\Sigma V^T$$

Since U is a unitary matrix, then:

$$A^TA = V\bar{\Sigma}^2V^T \tag{4.1}$$

Expression (4.1) represents the diagonalization of matrix A^TA. From this we can draw two important conclusions:

- $\sigma_i^2 = \lambda_i$ (where λ_i are the eigenvalues of A^TA);
- V is the matrix of the orthonormal eigenvectors of A^TA.

In the same way if we consider AA^T (which is a symmetric $m \times m$ matrix, with a maximum rank of n, since $m \geq n$), we obtain m eigenvalues, of which at most n are non-zero. These n eigenvalues are the squares of the singular values of matrix A:

$$AA^T = U \begin{bmatrix} \bar{\Sigma}^2 & \mathbf{0} \\ \mathbf{0} & \mathbf{0} \end{bmatrix} U^T \tag{4.2}$$

from which we deduce that U is the matrix of the orthonormal eigenvectors of AA^T. The columns of U and V are called left-singular and right-singular vectors.

Singular value decomposition can also be applied when matrix A is complex. In this case, instead of the transpose matrix, the conjugate transpose has to be considered. Even here, the singular values are always real and non-negative.

An example of a complex matrix is the transfer function matrix $G(s) = C(sI - A)^{-1}B$. The restriction of $G(s)$ to $s = j\omega$, $G(j\omega)$, is a complex matrix as ω varies. Later on, the importance of singular values of matrix $G(j\omega)$ will be discussed.

MATLAB® Exercise 4.1

Consider $A \in \mathbb{C}^{3\times 3}$:

$$A = \begin{bmatrix} j & 1 & 1+j \\ 1-j & 2 & j \\ 5+j & j & 5 \end{bmatrix}$$

The singular values of matrix A can be computed in MATLAB by first calculating its conjugate transpose with the command:

```
>> A'
```

One obtains:

$$A^* = \begin{bmatrix} -j & 1 & 1-j \\ 1+j & 2 & -j \\ 5-j & -j & 5 \end{bmatrix}$$

Then, the matrix A^*A is calculated:

```
>> A'*A
```

One obtains an Hermitian matrix:

$$A^*A = \begin{bmatrix} 29 & 3+6j & 25-5j \\ 3-6j & 6 & 1-2j \\ 25+5j & 1+2j & 28 \end{bmatrix}$$

Finally, the square root of the eigenvalues of A^*A are calculated:

```
>> sqrt(eig(A'*A))
```

One gets: $\sigma_1 = 7.4044$, $\sigma_2 = 2.7191$ and $\sigma_3 = 0.8843$.

The same result can be directly obtained by the command:

```
>> svd(A)
```

The `svd` command will be discussed in more details in the MATLAB exercise 4.2.

4.2 Spectral Norm and Condition Number of a Matrix

In this section we define what is meant by the spectral norm of a matrix. The norm of a matrix (similar to vector norms) is defined as a non-negative number with these properties:

- $\|A\| \geq 0$ for any matrix;
- $\|A\| = 0$ if and only if $A = 0$;
- $\|A + B\| \leq \|A\| + \|B\|$ (triangular inequality).

All the matrix norms which also have (in addition to the listed properties) the property that the norm of the product of two matrices is less than or equal to the product of the norms of the matrices, i.e.:

$$\|A \cdot B\| \leq \|A\| \cdot \|B\|$$

are defined as consistent norms.

Definition 8 (Spectral norm) *The spectral norm of a matrix is the largest singular value of the matrix:*

$$\|A\|_S = \sigma_1$$

The spectral norm is a consistent norm.

In addition, the spectral norm is an induced norm (from the Euclidean vectorial norm). This property allows us to clarify its meaning. Consider in fact matrix A as a linear operator mapping vector $\mathbf{x} \in \mathbb{R}^n$ into $\mathbf{y} \in \mathbb{R}^n$: $\mathbf{y} = A\mathbf{x}$. Consider all the vectors $\mathbf{x}$ with unitary norm, i.e., those for which $\|\mathbf{x}\| = 1$ (Euclidean norm, i.e., $\|\mathbf{x}\| = (\sum_{i=1}^n x_i^2)^{\frac{1}{2}}$), and calculate the norm of the vectors obtained by the mapping associated with matrix A. The spectral norm corresponds to the maximum of the resulting vector norm:

$$\|A\|_S = \max_{\|\mathbf{x}\|=1} \|A\mathbf{x}\|.$$

In the case of matrices $A \in \mathbb{R}^{2\times 2}$ the spectral norm can be interpreted geometrically. Figure 4.1 shows how a circumference with unitary radius defined by $\|\mathbf{x}\| = 1$ is mapped, using $\mathbf{y} = A\mathbf{x}$, into an ellipse in the plane $y_1 - y_2$. σ_1 represents the major semi-axis of this ellipse.

Example 4.1

If A is a 2 × 2 diagonal matrix, e.g., $A = \begin{bmatrix} \lambda_1 & 0 \\ 0 & \lambda_2 \end{bmatrix}$, then $y_1 = \lambda_1 x_1$ and $y_2 = \lambda_2 x_2$. So:

$$\frac{y_1^2}{\lambda_1^2} = x_1^2$$

and

$$\frac{y_2^2}{\lambda_2^2} = x_2^2$$

Then summing and remembering that $x_1^2 + x_2^2 = 1$ the equations of an ellipse are obtained:

$$\frac{y_1^2}{\lambda_1^2} + \frac{y_2^2}{\lambda_2^2} = 1$$

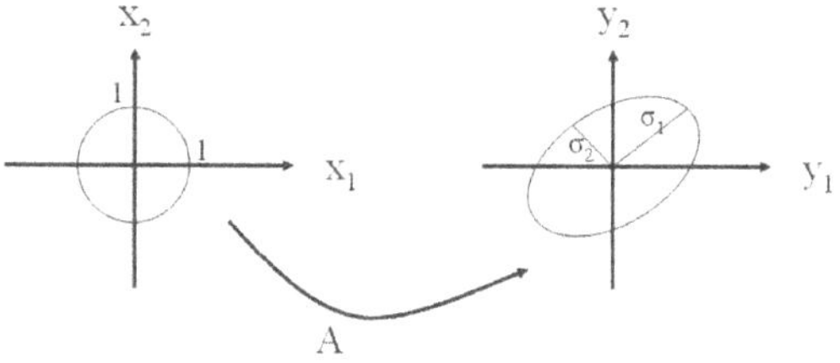

FIGURE 4.1
The unit circumference $\|\mathbf{x}\| = 1$ is mapped into an ellipse in the plane $y_1 - y_2$ by the linear operator A.

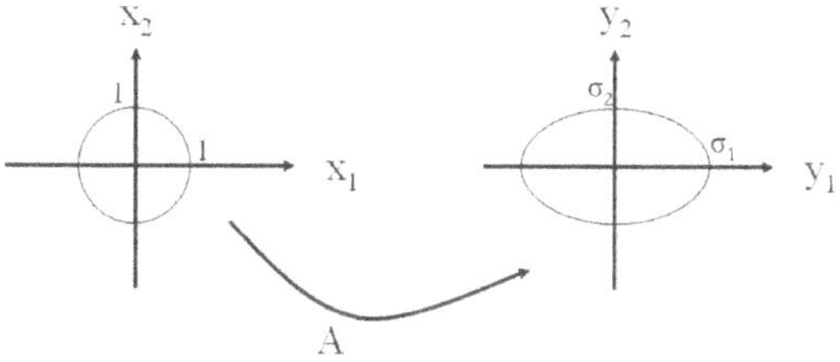

FIGURE 4.2
The unit circumference $\|\mathbf{x}\| = 1$ is mapped into an ellipse in the plane $y_1 - y_2$ by the linear operator A. Example with $A = \text{diag}(\lambda_1, \lambda_2)$.

with semi-major axis $\sigma_1 = |\lambda_1|$ and semi-minor axis $\sigma_2 = |\lambda_2|$ as shown in Figure 4.2. Note that the ratio between the maximum and minimum singular value $\frac{\sigma_1}{\sigma_2}$ accounts for the eccentricity of the ellipse. The larger σ_1, the thinner the ellipse. If $\sigma_2 = 0$ (i.e., if matrix A is rank deficient), the ellipse tends to a straight line.

Generally, the invertibility of matrix A depends on the smallest singular value. Since the determinant of a unitary matrix is 1, then

$$\det \text{A} = \det \text{U} \det \Sigma \det \text{V}^T = \det \Sigma = \sigma_1 \cdot \sigma_2 \cdot \ldots \cdot \sigma_n$$

so if the smallest singular value is $\sigma_n \neq 0$, then A is invertible, otherwise it is not.

Moreover, if the matrix rank is $rank(\text{A}) = k$, then k is the number of non-zero singular values.

The ratio between the maximum and minimum singular value of a matrix is therefore a measure of matrix invertibility and is defined as the condition number of a matrix.

Definition 9 (Condition number) *The condition number of an invertible* *matrix* $A \in \mathbb{R}^{n \times n}$ *is defined as the ratio between the maximum singular value and the minimum singular value of* A*:*

$$h = \frac{\sigma_1}{\sigma_n}.$$

If h is large, the matrix is ill-conditioned and it is difficult to calculate its inverse. The condition number denotes the invertibility of a matrix, unlike the determinant of a matrix which cannot be considered a measure of matrix invertibility.

Example 4.2

Consider matrices $A = \begin{bmatrix} 1 & 1 \\ 1 & 1 \end{bmatrix}$ and $\tilde{A} = \begin{bmatrix} 1.1 & 1 \\ 0.9 & 1.05 \end{bmatrix}$. A is not invertible since its determinant equals zero, whereas $\tilde{A}$ is invertible ($\det \tilde{A} = 0.255$). However, matrix A can be seen as a perturbation of matrix $\tilde{A}$, i.e.,

$$\tilde{A} + \Delta_A = A$$

with $\Delta_A = \begin{bmatrix} -0.1 & 0 \\ 0.1 & -0.05 \end{bmatrix}$. So, perturbing an invertible matrix ($\tilde{A}$) you obtain a matrix A which is no longer invertible.

Consider instead matrix $\bar{A} = \begin{bmatrix} 10^{-3} & 10^{-3} \\ 5 \cdot 10^{-4} & 10^{-5} \end{bmatrix}$, whose determinant is very small ($\det \bar{A} = -9 \cdot 10^{-8}$), yet the matrix is more robust than $\tilde{A}$ to perturbations that may make it non-invertible. In fact, $h_{\tilde{A}} \simeq 16$ when $h_{\bar{A}} \simeq 4$.

The condition number of a matrix is also related to the issue of the uncertainty of the solution of a system of linear equations. Consider the system of linear equations

$$A\mathbf{x} = \mathbf{c}$$

with $A \in \mathbb{R}^{n \times n}$, $\mathbf{c} \in \mathbb{R}^n$ known and $\mathbf{x} \in \mathbb{R}^n$ unknown. The solution depends on the invertibility of matrix A. Clearly, any uncertainty about the constant $\mathbf{c}$ will affect the solution. It can be proved that the condition number of matrix A is the link between uncertainty of the known terms and the uncertainty of the solution, i.e.,

$$\frac{\|\delta\mathbf{x}\|}{\|\mathbf{x}\|} \leq \frac{\sigma_1}{\sigma_n} \frac{\|\delta\mathbf{c}\|}{\|\mathbf{c}\|}$$

where $\|\delta\mathbf{x}\|$ is the (Euclidean) norm of the error with solution $\mathbf{x}$, caused by uncertainty $\delta\mathbf{c}$ of the known term.

If the condition number of the matrix is large, a small uncertainty in the constant term will cause a large perturbation in the solution.

The condition number of orthonormal matrices has one property easy to verify.

Theorem 9 *The condition number of a unitary matrix* U *is one.*

In fact, since $U^T U = I$, matrix U has singular values $\sigma_1 = \sigma_2 = \ldots = \sigma_n = 1$ so the condition number is one.

The singular values of a matrix provide upper and lower bounds for the eigenvalues of a matrix. In fact, if σ_1 and σ_n are the maximum and minimum singular values of a matrix A then:

$$\sigma_n \leq \min_i |\lambda_i| \leq \max_i |\lambda_i| \leq \sigma_1 \tag{4.3}$$

To conclude this brief overview of the properties linked to singular values of a matrix, note that, once the singular value decomposition has been calculated, the inverse of a matrix can be immediately computed.

Consider $A = U\Sigma V^T$ and suppose $\sigma_1 \geq \sigma_2 \geq \ldots \geq \sigma_n \neq 0$, the inverse matrix is given by:

$$A^{-1} = V\Sigma^{-1}U^T = V \begin{bmatrix} \frac{1}{\sigma_1} & 0 & \ldots & 0 \\ 0 & \frac{1}{\sigma_2} & \ldots & 0 \\ \vdots & \vdots & & \vdots \\ 0 & 0 & \ldots & \frac{1}{\sigma_n} \end{bmatrix} U^T \tag{4.4}$$

Inversion only requires calculating the inverse of n real numbers.

Finally, singular value decomposition can also be applied to calculating pseudo-inverse matrices. If A is not invertible, at least one singular value is zero: in equation (4.4) only the inverses of non-zero singular values are considered.

MATLAB® Exercise 4.2

This exercise explains how to use MATLAB to calculate the singular value decomposition of a matrix.

The MATLAB command for the singular value decomposition is

```
[U,S,V] = svd(A)
```

MATLAB uses an algorithm called the singular value decomposition algorithm which is similar to line elimination for calculating matrix rank. The algorithm is robust and works well also with large matrices. Define matrix A as follows

```
>> A=[0.001 0.001; 0.0001 0.00001]
```

singular values are calculated with command

```
>> svd(A)
```

You get $\sigma_1 = 0.0014$ and $\sigma_2 = 0.0001$.

Left and right singular vectors A are calculated with command

```
>> [U,S,V]=svd(A)
```

To complete the exercise, verify the following properties:

1. Orthonormality of matrices U and V, using commands:
   ```
   >> U'*U
   >> V'*V
   ```
2. Matrix A is given by $A = U\Sigma V^T$. Use command:
   ```
   >> U*S*V'
   ```

3. Calculate the inverse of A. Use command

```
>> V*inv(S)*U'
```

and compare the obtained result using command

```
>> inv(A)
```

Note. The numerical algorithm for obtaining singular value decomposition was proposed by Golub in 1975.

4.3 Exercises

1. Calculate the singular value decomposition of

$$A = \begin{bmatrix} 0 & 1 & 0 & 0.3 & -3 \\ -1 & -7 & 3 & -7 & -2 \\ 1 & 0.5 & 2 & 1 & 1 \\ 2 & 1 & 0 & 0 & 1 \end{bmatrix}$$

2. Calculate the singular value decomposition of $A = \begin{bmatrix} 2-j & j & 1 \\ -j & 3j & 1 \\ 7 & 1 & 6j \end{bmatrix}$.

3. Calculate the condition number of matrix $A = \begin{bmatrix} -1 & 0.5 & 3 \\ 0.1 & 7 & 1 \\ 3 & -4 & -5 \end{bmatrix}$.

4. Calculate the inverse of $A = \begin{bmatrix} 2 & 0 & 2 & 2 \\ 2 & 5 & -7 & -11 \\ 0 & -2 & 6 & -7 \\ 0 & 2 & -6 & -7 \end{bmatrix}$.

5. Calculate the eigenvalues of

$$A = \begin{bmatrix} 2.5000 & -1.0000 & -3.6416 & -1.6416 \\ -21.9282 & 2.0359 & 13.7439 & 6.2080 \\ 13.4641 & -1.7679 & -8.1927 & -2.9248 \\ -13.4641 & 1.7679 & 11.3343 & 6.0664 \end{bmatrix}$$

and verify that they satisfy (4.3).

5

Open-loop Balanced Realization

CONTENTS

In Chapter 3 we analyzed the Kalman decomposition, which allows us to determine the controllable and observable part of a system. In this chapter we deal with determining how controllable and observable a system is. In simplifying the poles and zeros of a system, the Kalman decomposition takes into account the scenario where there is an exact simplification between pole and zero, whereas the technique discussed in this chapter examines the case in which poles and zeros are very close to each other. The role of both the controllability and observability gramians is studied. The discussion leads to remark the concept of system invariants, in particular the singular values of a linear dynamical system are presented. The subject is considered one of the most important in the book and therefore both the theoretical aspects and the algorithms to derive the open-loop balanced representation are discussed. The case of discrete-time systems is also reported. In this part, the principal component analysis procedure is also discussed. The chapter includes several worked examples.

5.1 Controllability and Observability Gramians

Consider a generic Lyapunov equation

$$A^T X + XA = -Q$$

DOI: 10.1201/9781003196921-5

if A is the state matrix of an asymptotically stable system, then for Theorem 7 of Chapter 2 there is an integral solution

$$X = \int_0^{\infty} e^{A^T t} Q e^{At} dt.$$

Moreover if Q is positive definite, so is X.

Now let us consider two particular Lyapunov equations:

$$AX + XA^T = -BB^T \tag{5.1}$$

$$A^T X + XA = -C^T C \tag{5.2}$$

associated to the linear time-invariant system described by the state matrices (A, B, C). If we assume that the system is asymptotically stable, then each equation has a solution.

In particular, the solution to Lyapunov equation (5.1) is

$$W_c^2 = \int_0^{\infty} e^{At} BB^T e^{A^T t} dt$$

while the solution to Lyapunov equation (5.2) is

$$W_o^2 = \int_0^{\infty} e^{A^T t} C^T C e^{At} dt.$$

Note that, generally, for Theorem 5 the matrices BB^T and $C^T C$ are positive semi-definite matrices. Consider for example a single output system with $C = \begin{bmatrix} 1 & 0 \end{bmatrix}$, matrix

$$C^T C = \begin{bmatrix} 1 & 0 \\ 0 & 0 \end{bmatrix}$$

has one null and one positive eigenvalue, so it is positive semi-definite. Notwithstanding, it can be proved that if and only if the system is asymptotically stable and controllable, matrix W_c^2 is positive definite. Likewise, if the system is asymptotically stable and observable, matrix W_o^2 is positive definite. These two matrices are called controllability and observability gramians. Formally, the two gramians are defined as follows:

Definition 10 (Controllability gramian) *Given an asymptotically stable system, the controllability gramian*

$$W_c^2 = \int_0^{\infty} e^{At} BB^T e^{A^T t} dt$$

is the solution of the following Lyapunov equation:

$$AW_c^2 + W_c^2 A^T = -BB^T \tag{5.3}$$

Definition 11 (Observability gramian) *Given an asymptotically stable system, the observability gramian*

$$\mathrm{W}_o^2 = \int_0^\infty e^{\mathrm{A}^T t}\mathrm{C}^T\mathrm{C}e^{\mathrm{A}t}dt$$

is the solution of the following Lyapunov equation:

$$\mathrm{A}^T\mathrm{W}_o^2 + \mathrm{W}_o^2\mathrm{A} = -\mathrm{C}^T\mathrm{C} \tag{5.4}$$

The notation W_c^2 and W_o^2 (with the square) is adopted to highlight that the matrices are positive definite if the system is asymptotically stable, controllable and observable. In the following example we see the importance of asymptotic stability in obtaining the solution.

Example 5.1

Given a linear first-order system with $\mathrm{A} = 0$, $\mathrm{B} = 1$ and $\mathrm{C} = 1$, the Lyapunov equations (5.3) and (5.4) have no solution. For the system under consideration, in fact, they become $0 \cdot \mathrm{W}^2 = 1$ which has no solution. Indeed, this system is only marginally stable. Since there is an eigenvalue at $\lambda = 0$, the necessary conditions for solving a Lyapunov equation of type (5.3) and (5.4), i.e., the requirement that there are no eigenvalues on the imaginary axis, are not met.

Since the gramians are symmetric and positive definite, the singular values coincide with the eigenvalues of the matrix. This can be proven by considering a generic positive definite symmetric matrix W, and recalling that singular values are calculated from the eigenvalues of matrix $\mathrm{W}^T\mathrm{W}$. For symmetric matrices $\mathrm{W}^T\mathrm{W} = \mathrm{W}^2$. But since W^2 has λ_i^2 eigenvalues, where λ_i are the eigenvalues of W, it follows that $\sigma_i = \lambda_i$. For this reason, the singular values and eigenvalues of W_c^2 or W_o^2 are indistinguishable.

MATLAB® Exercise 5.1

This exercise illustrates the commands for calculating the controllability and observability gramians and discusses their properties.
As a first example, consider the linear time-invariant system described by the following state matrices:

$$\mathrm{A} = \begin{bmatrix} 0 & 1 & 0 \\ 0 & 0 & 1 \\ -5 & -4 & -3 \end{bmatrix}; \mathrm{B} = \begin{bmatrix} 0 \\ 0 \\ 1 \end{bmatrix}; \mathrm{C} = \begin{bmatrix} 0 & 1 & 0 \end{bmatrix}$$

Obviously, the system is in canonical control form, so completely controllable, and furthermore asymptotically stable. For this reason the controllability gramian has to be positive definite.
The state matrices are defined by these commands:

```
>> A=[0 1 0; 0 0 1; -5 -4 -3]
>> B=[0; 0; 1]
>> C=[0 1 0]
```

Let us study the observability of the system by calculating the rank of the system observability matrix, with the command:

```
>> rank(obsv(A,C))
```

Since the rank is maximum and the system completely observable, the observability gramian has to be positive definite.
Let us calculate the two gramians by solving the associated Lyapunov equations with the commands:

```
>> Wc2=lyap(A,B*B')
>> Wo2=lyap(A',C'*C)
```

Note that they are two symmetric matrices. Let us verify that the two gramians are solutions of the associated Lyapunov equations:

```
>> A*Wc2+Wc2*A'+B*B'
>> A'*Wo2+Wo2*A+C'*C
```

In both cases, as expected, we obtain a zero matrix.
Now let us verify that the two gramians are positive definite. To do this we can calculate the eigenvalues of the matrices:

```
>> eig(Wc2)
>> eig(Wo2)
```

and note that they are all positive. The same result can be obtained from the Sylvester test (a symmetric matrix is positive definite if all the leading principal minors are positive). For example, for the controllability gramian the test is applied by the commands:

```
>> det(Wc2(1,1))
>> det(Wc2(1:2,1:2))
>> det(Wc2)
```

Gramians can also be calculated with the command **gram**. In this case we have to define an LTI model, for example using the command:

```
>> system1=ss(A,B,C,0)
```

At this point we can calculate the gramians with the commands:

```
>> Wc2=gram(system1,'c')
>> Wo2=gram(system1,'o')
```

As a second example, consider the linear time-invariant system described by the state matrices:

$$\mathrm{A} = \begin{bmatrix} -1 & 0 & 0 \\ 1/2 & -1 & 0 \\ 1/2 & 0 & -1 \end{bmatrix}; \mathrm{B} = \begin{bmatrix} 1 & 0 \\ 0 & -1 \\ 0 & 1 \end{bmatrix}$$

In this case the (asymptotically stable) system has two inputs, u_1 and u_2.
Note that the fact that the system is completely controllable is immediately verifiable with the command:

```
>> rank(ctrb(A,B))
```

But if we consider only the input u_1 (i.e., $u_2 = 0$) then the system is no more completely controllable. The instruction

```
>> rank(ctrb(A,B(:,1)))
```

gives that the rank is 2. In other words, both inputs are strictly necessary to reach any state of $\mathbb{R}^3$.
Initially, let us suppose that both inputs are manipulable and calculate the controllability gramian with the command:

```
>> Wc2=lyap(A,B*B')
>> eig(Wc2)
```

We find a positive definite gramian.
Instead, when only u_1 can be used to act on the system ($u_2 = 0$), with the commands

```
>> Wc2=lyap(A,B(:,1)*B(:,1)')
>> eig(Wc2)
```

we find that the gramian has one zero eigenvalue. So it is positive semi-definite.

5.2 Principal Component Analysis

In this paragraph, we will recall the most important properties of principal component analysis.

Consider matrix $W \in \mathbb{R}^{n \times n}$ defined as:

$$W = \int_0^\infty F(t)F(t)^T dt$$

with $F : \mathbb{R} \to \mathbb{R}^{n \times m}$ (time function matrix). The implicit assumption in defining W is that the integral exists.

Consider the singular value decomposition of matrix W. This matrix is certainly positive semi-definite (by definition). So, let us consider a set of orthonormal vectors assigned to the non-negative eigenvalues to obtain the singular value decomposition:

$$W = V\Sigma V^T$$

where Σ is the diagonal matrix containing the singular values of W ($\Sigma = \mathtt{diag}\{\sigma_1, \sigma_2, \ldots, \sigma_n\}$) and $\mathbf{v_1}, \mathbf{v_2}, \ldots, \mathbf{v_n}$ constitute the set of orthonormal vectors of W. Since $\mathbf{v_1}, \mathbf{v_2}, \ldots, \mathbf{v_n}$ are orthonormal vectors, they can be used as the orthonormal base of $\mathbb{R}^n$ so that $F(t)$ represents the sum of n components:

$$F(t) = \mathbf{v_1}\mathbf{f_1}^T(t) + \mathbf{v_2}\mathbf{f_2}^T(t) + \ldots + \mathbf{v_n}\mathbf{f_n}^T(t)$$

where $\mathbf{f_1}(t), \mathbf{f_2}(t), \ldots, \mathbf{f_n}(t)$ are vectors of m time-dependent elements, and $\mathbf{v_1}, \mathbf{v_2}, \ldots, \mathbf{v_n}$ are vectors of n constant time-independent elements. $\mathbf{f_1}(t), \mathbf{f_2}(t), \ldots, \mathbf{f_n}(t)$ are called principal components of $F(t)$ and are given by:

$$\mathbf{f_i}^T(t) = \mathbf{v_i}^T F(t).$$

Principal components have certain properties:

$$\int_0^\infty \mathbf{f_i}^T(t)\mathbf{f_j}(t)dt = 0 \text{ for } i \neq j$$

$$\int_0^\infty \mathbf{f_i}^T(t)\mathbf{f_i}(t)dt = \int_0^\infty \|\mathbf{f_i}\|^2 dt = \sigma_i$$

$$\int_0^\infty \|F(t)\|_F^2 dt = \sum_{i=1}^n \sigma_i$$

where the Frobenius norm is defined by

$$\|A\|_F = \left(\sum_{i=1,j=1}^n a_{ij}^2 \right)^{\frac{1}{2}}$$

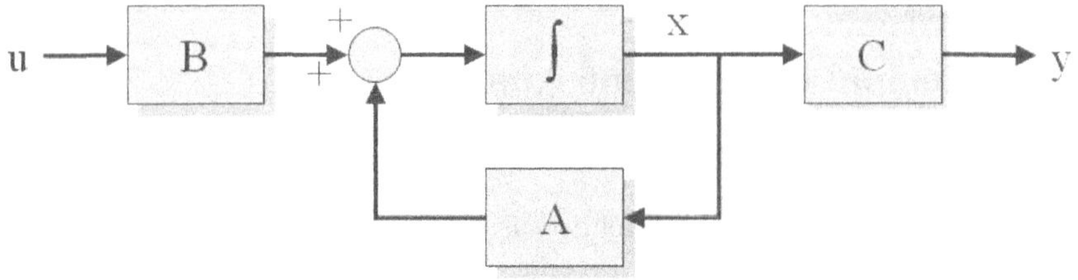

FIGURE 5.1
Linear time-invariant system for defining $\mathrm{F}(t)$ of the controllability gramian.

So, $\mathrm{F}(t)$ can be decomposed into n time function components whose energy can be calculated from the singular values of W.

5.3 Principal Component Analysis Applied to Linear Systems

Let us apply principal component analysis to linear time-invariant systems which is the same as fixing particular $\mathrm{F}(t)$ with a precise physical meaning in systems theory.

In particular, given the system in Figure 5.1, imagine applying an input consisting of m Dirac impulses and calculating the state evolution from zero intial conditions. This would be the same as calculating the forced output evolution with $\mathrm{C} = \mathrm{I}$. The state evolution is given by:

$$\mathbf{x}(t) = e^{\mathrm{A}t}\mathrm{B}.$$

By choosing this function as matrix $\mathrm{F}(t)$ in our principal component analysis, where $\mathrm{F}(t) = e^{\mathrm{A}t}\mathrm{B}$, we get the precise definition of the controllability gramian:

$$\mathrm{W}_c^2 = \int_0^{\infty} \mathrm{F}(t)\mathrm{F}^T(t)dt = \int_0^{\infty} e^{\mathrm{A}t}\mathrm{B}\mathrm{B}^T e^{\mathrm{A}^T t}dt.$$

In the integral which defines the controllability gramian, the matrix $\mathrm{F}(t)$ represents the state evolution in response to a Dirac impulse at the input. The controllability subspace X_c is the smallest subspace containing the image of $e^{\mathrm{A}t}\mathrm{B}$ with $T > 0$.

In the same way, the physical meaning of $\mathrm{F}(t)$ which appears in the observability gramian can be derived. Examine the system in Figure 5.2 with $\mathbf{u} = 0$, apply a vector of n Dirac impulses at the summing node and consider the system response with zero intial conditions. This is the same as considering

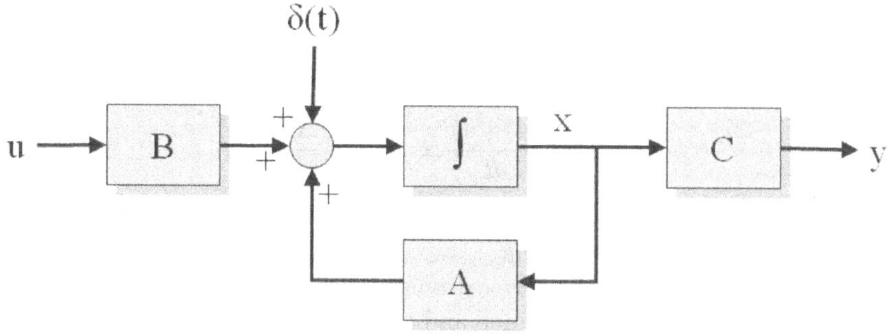

FIGURE 5.2
Linear time-invariant system for defining $F(t)$ of the observability gramian.

the impulse response of a system with $B = I$. The system response is given by:

$$\mathbf{y}(t) = Ce^{At}$$

Note that $F^T(t) = Ce^{At}$ is exactly the term which appears in the observability gramian. In fact the response $\mathbf{y}(t)$ plays a fundamental role in the analysis of the observability properties of the system.

The principal component analysis of the function $F(t) = e^{At}B$ in the controllability gramian W_c^2 and of the function $F^T(t) = Ce^{At}$ of the observability gramian W_o^2 allows us to evaluate, based on the size of the singular values σ_i relative to W_c^2 or to W_o^2, which components are associated to greater or lesser energy and, therefore, those terms which need more energy to be controlled and those that need more energy to be observed.

In particular, taking into account the controllability analysis and so the function $F(t) = e^{At}B$, the following result is particularly significant. Assuming that the system is controllable, if $\bar{\mathbf{x}}$ is the state obtained by applying input $\bar{\mathbf{u}}$, the required energy is given by $\int_0^\infty \|\bar{\mathbf{u}}\|^2 dt$. If instead we consider the state $\mathbf{x} = \bar{\mathbf{x}} + \Delta\mathbf{x}$, the necessary input would be $\mathbf{u} = \bar{\mathbf{u}} + \Delta\mathbf{u}$. The relation between the energy difference in the two cases and the associated energy at the input $\bar{\mathbf{u}}$ is directly proportional to the ratio between $\|\Delta\mathbf{x}\|$ and $\|\mathbf{x}\|$ and the proportionality constant is linked to the ratio between the maximum singular value (denoted by σ_{c1}) and the minimum singular value (σ_{cn}) of matrix W_c^2. So, the following expression is valid:

$$\frac{\int_0^\infty \|\bar{\mathbf{u}} - \mathbf{u}\|^2 dt}{\int_0^\infty \|\bar{\mathbf{u}}\|^2 dt} \propto \sqrt{\frac{\sigma_{1c}}{\sigma_{nc}}} \frac{\|\Delta\mathbf{x}\|}{\|\mathbf{x}\|} \tag{5.5}$$

where $\|\cdot\|$ is the Euclidean norm. If the ratio $\frac{\sigma_{1c}}{\sigma_{nc}}$ is large, a lot of energy is required to change the state even slightly. The value of the ratio $\frac{\sigma_{1c}}{\sigma_{nc}}$ gives an idea of the degree of controllability of the system.

Suppose now that the system is observable, an analogous result holds:

$$\frac{\int_0^\infty \|\mathbf{y}-\mathbf{y}^*\|^2 dt}{\int_0^\infty \|\mathbf{y}\|^2 dt} \propto \sqrt{\frac{\sigma_{no}}{\sigma_{1o}}} \frac{\|\Delta\mathbf{x}\|}{\|\mathbf{x}\|} \tag{5.6}$$

where $\|\cdot\|$ is the Euclidean norm and $\sigma_{1o}, \ldots, \sigma_{no}$ represent the singular values of matrix $\mathrm{W}_o^2 = \int_0^\infty e^{\mathrm{A}^T t}\mathrm{C}^T\mathrm{C}e^{\mathrm{A}t}dt$.

Note that in this case the proportionality constant is given by the ratio between the minimum singular value and the maximum singular value of matrix W_o^2. If the value of this ratio is small, i.e., if we have at least one singular value of W_o^2 which is smaller than the others, reconstructing the state from the output terms requires a lot of energy. In fact, ideally it would be good to be able to discriminate a small difference $\|\Delta\mathbf{x}\|$ in the initial condition, so it would be desirable that a small variation significantly affects the term in the left hand side of (5.6). If the ratio $\frac{\sigma_{no}}{\sigma_{1o}}$ is small, a small difference in the initial condition is attenuated. In this case, the variables are poorly observable.

5.4 State Transformations of Gramians

In this paragraph we consider an invertible state transformation $\tilde{\mathbf{x}} = \mathrm{T}^{-1}\mathbf{x}$ and see how the gramians change as the reference system changes. Recall that, given a system $(\mathrm{A},\mathrm{B},\mathrm{C})$ and applying state transformation $\tilde{\mathbf{x}} = \mathrm{T}^{-1}\mathbf{x}$ we obtain an equivalent system $(\tilde{\mathrm{A}},\tilde{\mathrm{B}},\tilde{\mathrm{C}})$ with

$$\begin{aligned}\tilde{\mathrm{A}} &= \mathrm{T}^{-1}\mathrm{AT}\\ \tilde{\mathrm{B}} &= \mathrm{T}^{-1}\mathrm{B}\\ \tilde{\mathrm{C}} &= \mathrm{CT}\end{aligned} \tag{5.7}$$

In the original reference system the gramians are solutions of the Lyapunov equations:

$$\begin{aligned}\mathrm{AW}_c^2 + \mathrm{W}_c^2\mathrm{A}^T &= -\mathrm{BB}^T\\ \mathrm{A}^T\mathrm{W}_o^2 + \mathrm{W}_o^2\mathrm{A} &= -\mathrm{C}^T\mathrm{C}\end{aligned} \tag{5.8}$$

In the new reference system $\tilde{\mathbf{x}}$ the gramians $\tilde{\mathrm{W}}_c^2$ and $\tilde{\mathrm{W}}_o^2$ are solutions of equations:

$$\begin{aligned}\tilde{\mathrm{A}}\tilde{\mathrm{W}}_c^2 + \tilde{\mathrm{W}}_c^2\tilde{\mathrm{A}}^T &= -\tilde{\mathrm{B}}\tilde{\mathrm{B}}^T\\ \tilde{\mathrm{A}}^T\tilde{\mathrm{W}}_o^2 + \tilde{\mathrm{W}}_o^2\tilde{\mathrm{A}} &= -\tilde{\mathrm{C}}^T\tilde{\mathrm{C}}\end{aligned} \tag{5.9}$$

To find the relation between $\tilde{\mathrm{W}}_c^2$ with W_c^2 and $\tilde{\mathrm{W}}_o^2$ with W_o^2, let us apply relations (5.7) to equations (5.9). First, let us consider the controllability gramian:

$$T^{-1}AT\tilde{W}_c^2 + \tilde{W}_c^2T^TA^T(T^T)^{-1} = -T^{-1}BB^T(T^T)^{-1}$$

Multiplying to the right both members by matrix T^T, we get:

$$T^{-1}AT\tilde{W}_c^2T^T + \tilde{W}_c^2T^TA^T = -T^{-1}BB^T$$

Multiplying left with matrix T, we get:

$$AT\tilde{W}_c^2T^T + T\tilde{W}_c^2T^TA^T = -BB^T \tag{5.10}$$

The expression obtained is equal to the first of the two Lyapunov equations (5.8) if

$$W_c^2 = T\tilde{W}_c^2T^T$$

from which we get:

$$\tilde{W}_c^2 = T^{-1}W_c^2(T^T)^{-1}$$

Note that $\tilde{W}_c^2$ and W_c^2, are both symmetric and positive semi-definite (or definite) matrices not related by any similarity relation. So, the two gramians do not have the same eigenvalues (nor the same singular values). Therefore, we have reached a very important conclusion: the singular values of the controllability gramian depend on the reference system.

An example in $\mathbb{R}^2$ may facilitate to gain some insight on the problem. In presence of a high condition number, we have a very distorted ellipsoid which reflects an interior unbalance of the system. In this case, strongly or weakly controllable variables exist. Note that this depends however on the reference system.

With regard to the transformation of the observability gramian, the same reasoning for the controllability gramian can be repeated. Consider the second Lyapunov equation (5.9), replacing relations (5.7), we obtain:

$$T^TA^T(T^T)^{-1}\tilde{W}_o^2 + \tilde{W}_o^2T^{-1}AT = -T^TC^TCT$$

Multiplying all the terms in this case on the right by T^{-1} and left for $(T^T)^{-1}$, we obtain:

$$A^T(T^T)^{-1}\tilde{W}_o^2T^{-1} + (T^T)^{-1}\tilde{W}_o^2T^{-1}A = -C^TC$$

So, equalling the resulting relation with the second Lyapunov equation (5.8) we obtain:

$$W_o^2 = (T^T)^{-1}\tilde{W}_o^2T^{-1}$$

and therefore

$$\tilde{W}_o^2 = T^TW_o^2T$$

Even with regard to the observability gramian, the relation that links $\tilde{W}_o^2$ with W_o^2, is not a relation of similarity. The observability gramian eigenvalues therefore depend on the reference system.

Summarizing, the relations linking the controllability and observability gramians in two reference systems related by a state transformation $\tilde{\mathbf{x}} = T^{-1}\mathbf{x}$ are:

$$\begin{aligned} \tilde{W}_c^2 &= T^{-1}W_c^2(T^T)^{-1} \\ \tilde{W}_o^2 &= T^T W_o^2 T \end{aligned} \tag{5.11}$$

Since the gramian eigenvalues depend on the reference system, also $\frac{\sigma_{1c}}{\sigma_{nc}}$ and $\frac{\sigma_{1o}}{\sigma_{no}}$ depend on the reference system. So, supposing that the system is asymptotically stable and minimal, the choice of the reference system for solving the controllability or observability issue is very significant as the degree of observability or controllability depends on the reference system. Referring to relations (5.5) and (5.6), there is a reference system in which all the variables are controllable in the same way. This occurs when the controllability gramian is equal to the identity matrix. In the same way there is a reference system in which all the variables are equally observable, thus requiring the same energy to reconstruct the initial condition. This occurs when the observability gramian equals the identity matrix. So, controllability and observability (as structural properties) in a system do not change as the reference system changes. What may change is degree of controllability and observability.

At this point we should ask ourselves if there is a reference system in which the two gramians are equal, that is, a reference system in which the degree of controllability and observability are equal. We will see that the answer to this question is positive, and that under proper hypotheses in this reference system it is possible to split the system into a strongly controllable and observable part and into a weakly controllable and observable part.

5.5 Singular Values of Linear Time-invariant Systems

We have seen that the controllability and observability gramian eigenvalues depend on the reference system, so they are not system invariants.

As regards the product of the two gramians $W_c^2 W_o^2$, by applying a transformation state, as the consequence of relations (5.11) we find that:

$$\tilde{W}_c^2 \tilde{W}_o^2 = T^{-1} W_c^2 W_o^2 T \tag{5.12}$$

Since $\tilde{W}_c^2 \tilde{W}_o^2$ and $W_c^2 W_o^2$ are linked by a relation of similarity, they have the same eigenvalues. The same is true if we consider the observability and controllability gramian product:

$$\tilde{W}_o^2 \tilde{W}_c^2 = T^T W_o^2 W_c^2 (T^T)^{-1} \quad (5.13)$$

The eigenvalues of the gramian product do not change as the reference system changes. Furthermore, since $(W_c^2 W_o^2)^T = W_o^2 W_c^2$, the two matrices (the product matrix of the controllability and observability gramian and the product matrix of the observability and controllability gramian) have the same eigenvalues. Ultimately, these two matrices are positive definite (or semi-definite), being the product of positive definite (or semi-definite) matrices. More precisely, the hypothesis of asymptotic stability of the system implies that the matrices are positive semi-definite; and if the system is also controllable and observable, the matrices are positive definite.

Notice that $W_c^2 W_o^2$ and $W_o^2 W_c^2$ are not symmetric, as the product of two symmetric matrices generally is not a symmetric matrix. Effectively, in the case of the gramian product, without considering particular states:

$$(W_c^2 W_o^2)^T = W_o^2 W_c^2 \neq W_c^2 W_o^2$$

Since the gramian product matrix does not have negative eigenvalues their roots may be found. As before, these roots are system invariants. They take the name of singular values of the system.

Definition 12 (Singular values of the system) *Given a linear asymptotically stable system, the singular values of the system, $\sigma_1 \geq \sigma_2 \geq \dots \sigma_n$, are the square roots of the eigenvalues of the product of the controllability and observability gramians. Singular values are system invariants.*

These quantities are also known as the Hankel singular values.

5.6 Computing the Open-loop Balanced Realization

Once all the required mathematical tools have been defined, we can deal with the problem of determining a system in which the two gramians are equal and so their controllability and observability are measured by the same parameters since the eigenvalues of W_c^2 coincide with those of W_o^2. Given an asymptotically stable system which is controllable and observable (as we will consider in the rest of this chapter, unless otherwise specified), it is therefore possible to distinguish a strongly controllable and observable part as well as a weakly controllable and observable part. Once this system is broken down, we will only consider the strongly controllable and observable part of the system as a lower order model for the system. So, let us apply principal component analysis to functions $e^{At}B$ and $(Ce^{At})^T$ making sure to omit the $\mathbf{f}_i^T(t)$ terms associated with (relatively) small singular values. Since singular values represent the energy associated with those components $\mathbf{f}_i^T(t)$ the aim is to be able to identify the low-energy components and neglect them. When we apply principal component analysis we have to be sure that the various contributions are weighted only by components $\mathbf{f}_i^T(t)$ and do not depend on $\mathbf{v}_i$. This leads to better specifying the characteristic conditions of this system: the gramians must be equal and diagonal as the state variables associated with the strongly controllable and observable part can be associated with the weakly controllable and observable part. Ultimately, since the eigenvalues of the gramian product equal the squares of the system singular values, what characterizes it is that the two gramians are diagonal matrices containing the system singular values. This particular realization is called open-loop balanced.

Definition 13 (Open-loop balanced realization) *Given a controllable and observable, linear and asymptotically stable system, the open-loop balanced realization is the realization in which the controllability and observability gramian are equal, diagonal and the diagonal contains the singular values of the system* $W_c^2 = W_o^2 = \text{diag}(\sigma_1, \sigma_2, \ldots, \sigma_n)$. $W_c^2 = W_o^2 = \text{diag}(\sigma_1, \sigma_2, \ldots, \sigma_n)$.

Now let us discuss how to construct such a realization, or, equivalently, which transformation matrix T from the initial reference system to the new balanced one is required.

Consider the Lyapunov solution to the controllability gramian equation

$$AW_c^2 + W_c^2A^T = -BB^T$$

Once the solution W_c^2 is found (we know it exists because the system is asymptotically stable), let us consider its singular value decomposition:

$$W_c^2 = V_c\Sigma_c^2V_c^T$$

where for convenience $\sigma_{1c}^2, \sigma_{2c}^2, \ldots, \sigma_{nc}^2$ indicate the singular values of W_c^2. Note also that $U_c = V_c$ given that the matrix is symmetric. Recall also that if M is a symmetric matrix, then $M^TM = MM^T = M^2$ and the eigenvectors of M^TM coincide with those of MM^T.

Consider the transformation state defined by matrix $T_1 = V_c\Sigma_c$ (Σ_c the diagonal matrix which contains the roots of the singular values of W_c^2).

Since $T_1^{-1} = \Sigma_c^{-1}V_c^T$, taking into account equation (5.11), we obtain:

$$\tilde{W}_c^2 = \Sigma_c^{-1}V_c^TW_c^2V_c\Sigma_c^{-1} = \Sigma_c^{-1}\Sigma_c^2\Sigma_c^{-1} = I$$

With $\tilde{W}_c^2 = I$ we obtain $\tilde{W}_o^2$ whose eigenvalues equal the square of the singular values of the system, since $\tilde{W}_c^2\tilde{W}_o^2 = \tilde{W}_o^2$ (this intermediate representation is called *input normal form*).

In this state representation, the system singular values can be calculated using the Lyapunov equation:

$$\tilde{A}^T\tilde{W}_o^2 + \tilde{W}_o^2\tilde{A} = -\tilde{C}^T\tilde{C}$$

At this point, considering the singular value decomposition of the observability gramian, we have:

$$\tilde{W}_o^2 = \tilde{V}_o \Sigma^2 \tilde{V}_o^T$$

where Σ is exactly the diagonal matrix which contains the Hankel singular values $\sigma_1, \sigma_2, \ldots, \sigma_n$.

Consider a second state transformation $T_2 = \tilde{V}_o \Sigma^{-\frac{1}{2}}$. Since $T_2^T = \Sigma^{-\frac{1}{2}} \tilde{V}_o^T$, in the new reference system (let $\bar{\mathbf{x}}$), we have:

$$\bar{W}_o^2 = \Sigma^{-\frac{1}{2}} \tilde{V}_o^T \tilde{V}_o \Sigma^2 \tilde{V}_o^T \tilde{V}_o \Sigma^{-\frac{1}{2}} = \Sigma$$

and

$$\bar{W}_c^2 = \Sigma$$

To balance the gramians, two different state transformations were applied. The overall state transformation is given by $\bar{\mathbf{x}} = T^{-1}\mathbf{x}$. Since $\bar{\mathbf{x}} = T_2^{-1}\tilde{\mathbf{x}} = T_2^{-1}T_1^{-1}\mathbf{x}$, then $T = T_1 T_2$.

So, applying the state transformation $T = T_1 T_2$, a reference system can be obtained in which:

$$\begin{array}{l} \bar{W}_c^2 = \bar{W}_o^2 = \Sigma = \mathrm{diag}(\sigma_1, \sigma_2, \ldots, \sigma_n) \\ \bar{W}_c^2 \bar{W}_o^2 = \Sigma^2 \end{array} \quad (5.14)$$

MATLAB® Exercise 5.2

Now let us find the open-loop balanced realization of an asymptotically stable controllable and observable system. Here is an example of the procedure for the following system in state-space form:

$$A = \begin{bmatrix} -1 & 0 & 0 \\ 1/2 & -1 & 0 \\ 1/2 & 0 & -1 \end{bmatrix}; B = \begin{bmatrix} 1 & 0 \\ 0 & -1 \\ 0 & 1 \end{bmatrix}; C = \begin{bmatrix} 0 & 0 & 1 \\ 1 & 1 & 0 \end{bmatrix}$$

Let us follow these steps:

1. Define the system:
```
>> A=[-1 0 0; 1/2 -1 0; 1/2 0 -1]
>> B=[1 0 0; 0 -1 1]'
>> C=[0 0 1; 1 1 0]
>> system=ss(A,B,C,0)
```
2. Verify the hypotheses so the open-loop balanced realization (asymptotic stability, controllability, observability) can be calculated:
```
>> eig(A)
>> rank(ctrb(system))
>> rank(obsv(system))
```
3. Calculate the state transformation matrix P_1
```
>> Wc2=lyap(A,B*B')
```
(alternatively `Wc2=gram(system,'c')`)
```
>> [Uc,Sc2,Vc]=svd(Wc2)
>> P1=Vc*sqrt(Sc2)
```

4. Calculate the state-space representation $(\tilde{A}, \tilde{B}, \tilde{C})$ (input normal form)

```
>> Atilde=inv(P1)*A*P1
>> Btilde=inv(P1)*B
>> Ctilde=C*P1
```

To test it we can calculate $\tilde{W}_c^2$ and see if it equals the identity matrix

```
>> Wtildec2=lyap(Atilde,Btilde*Btilde')
```

5. Calculate the transformation matrix T_2

```
>> Wtildeo2=lyap(Atilde',Ctilde'*Ctilde)
>> [Uo,So2,Vo]=svd(Wtildeo2)
>> T2=Vo*(So2)^(-1/4)
```

6. Calculate the open-loop balanced realization

```
>> Abal=inv(T2)*Atilde*T2
>> Bbal=inv(T2)*Btilde
>> Cbal=Ctilde*T2
>>systembal=ss(Abal,Bbal,Cbal,0)
```

To test it, note that the gramians are diagonal and equal to the singular values of the system:

```
>> Wbalc2=gram(systembal,'c')
>> Wbalo2=gram(systembal,'o')
```

The singular values of the system can be found using these instructions:

```
>> Wc2=lyap(A,B*B')
>> Wo2=lyap(A',C'*C)
>> sqrt(eig(Wc2*Wo2))
```

Note that, instead of using this procedure, shown mainly for educational purposes, the balanced form of the system can be found with the instruction:

```
>> [systemb,G,T,Ti]=balreal(system)
```

Note (typing `help balreal`) that in this case the transformation of command `balreal` is $\mathbf{x}_{bil} = T\mathbf{x}$, i.e., the inverse of the one found in the discussed procedure.

Example 5.2

Consider the linear time-invariant system:

$$A = \begin{bmatrix} -2 & 4\alpha \\ -\frac{4}{\alpha} & -1 \end{bmatrix}; B = \begin{bmatrix} 2\alpha \\ 1 \end{bmatrix}; C = \begin{bmatrix} \frac{2}{\alpha} & -1 \end{bmatrix}$$

with $\alpha \neq 0$. By calculating the transfer function of this system ($G(s) = \frac{3s+18}{s^2+3s+18}$) we can see that the system is controllable, observable and asymptotically stable for any value of α (the transfer function does not depend on α). So it makes sense to calculate the balanced form of this system.
Consider, for example, $\alpha = 10$ and calculate the balanced form according to the procedure in Exercise 5.2. For $\alpha = 10$ the system becomes:

$$A = \begin{bmatrix} -2 & 40 \\ -0.4 & -1 \end{bmatrix}; B = \begin{bmatrix} 20 \\ 1 \end{bmatrix}; C = \begin{bmatrix} 0.2 & -1 \end{bmatrix}$$

The controllability and observability gramians (calculated by the MATLAB command `gram`) are:

$$W_c^2 = \begin{bmatrix} 100 & 0 \\ 0 & 0.5 \end{bmatrix}; W_o^2 = \begin{bmatrix} 0.01 & 0 \\ 0 & 0.5 \end{bmatrix}$$

The singular values of the system are the roots of the eigenvalues of the product between the two gramians and can be calculated through the command `sqrt(eig(Wc2*Wo2))`. So, we obtain $\sigma_1 = 1$ and $\sigma_2 = 0.5$.

Note that in this case the two gramians are already diagonal matrices. Since they are not equal the system is unbalanced. Applying the balancing procedure, since the controllability gramian is already diagonal, we obtain $V_c = I$ and so:

$$T_1 = I\Sigma_c = \sqrt{W_c^2} = \begin{bmatrix} 10 & 0 \\ 0 & \sqrt{0.5} \end{bmatrix}$$

At this point consider system $(\tilde{A}, \tilde{B}, \tilde{C})$. In this reference system $\tilde{W}_c^2 = I$, while $\tilde{W}_o^2$ is equal to:

$$\tilde{W}_o^2 = \begin{bmatrix} 1 & 0 \\ 0 & 0.25 \end{bmatrix}$$

Matrix T_2 is given by:

$$T_2 = V_o\Sigma_o^{-\frac{1}{2}} = I\Sigma_o^{-\frac{1}{2}} = (\tilde{W}_o^2)^{-1/4} = \begin{bmatrix} 1 & 0 \\ 0 & \frac{1}{\sqrt{0.5}} \end{bmatrix}$$

which yields $T = T_1T_2 = \begin{bmatrix} 10 & 0 \\ 0 & 1 \end{bmatrix}$. The particular form of the transformation matrix highlights the fact that in this case it is sufficient to change scale to obtain the open-loop balanced form. Applying the transformation $\bar{\mathbf{x}} = T^{-1}\mathbf{x}$ we obtain the balanced form:

$$\bar{A} = \begin{bmatrix} -2 & 4 \\ -4 & -1 \end{bmatrix}; \bar{B} = \begin{bmatrix} 2 \\ 1 \end{bmatrix}; \bar{C} = \begin{bmatrix} 2 & -1 \end{bmatrix}$$

In the most trivial case, the balanced form is obtained by appropriately scaling the state variables (in this case, with $\bar{x}_1 = \frac{1}{10}x_1$ and $\bar{x}_2 = x_2$). In the general case, to determine the balanced form, a reference system change is usually needed. In this case, with diagonal gramians it is sufficient to scale the state variables.
Let us return to the more general case with $\alpha \neq 0$. In this case the gramians are:

$$W_c^2 = \begin{bmatrix} \alpha^2 & 0 \\ 0 & 0.5 \end{bmatrix}; W_o^2 = \begin{bmatrix} \frac{1}{\alpha^2} & 0 \\ 0 & 0.5 \end{bmatrix}$$

Consider the case in which α is very small. The degree of controllability of the variable which corresponds to the singular value of W_c^2 equalling $\sigma_2 = \alpha$ is very small. But the larger the degree of observability of this variable, the smaller its controllability.
This example confirms the importance of having balanced controllability and observability which also helps understand at the same time which variables are less controllable and observable.
In this system the parameter α represents the imbalance between observability and controllability. The balanced form is equal to:

$$\bar{A} = \begin{bmatrix} -2 & 4 \\ -4 & -1 \end{bmatrix}; \bar{B} = \begin{bmatrix} 2 \\ 1 \end{bmatrix}; \bar{C} = \begin{bmatrix} 2 & -1 \end{bmatrix}$$

So we have a form independent of α which obviously coincides with the case in which $\alpha = 1$, that is when the system variables are observable and controllable in the same way.

5.7 Balanced Realization for Discrete-time Linear Systems

The results presented so far for linear continuous-time systems can be easily extended to linear discrete-time systems:

$$\begin{cases} \mathbf{x}(k+1) = \tilde{\mathrm{A}}\mathbf{x}(k) + \tilde{\mathrm{B}}\mathbf{u}(k) \\ \mathbf{y}(k) = \tilde{\mathrm{C}}\mathbf{x}(k) \end{cases} \tag{5.15}$$

The Lyapunov equations of the gramians take a slightly different form:

$$\tilde{\mathrm{A}}\tilde{\mathrm{W}}_c^2\tilde{\mathrm{A}}^T - \tilde{\mathrm{W}}_c^2 = -\tilde{\mathrm{B}}\tilde{\mathrm{B}}^T \tag{5.16}$$

$$\tilde{\mathrm{A}}^T\tilde{\mathrm{W}}_o^2\tilde{\mathrm{A}} - \tilde{\mathrm{W}}_o^2 = -\tilde{\mathrm{C}}^T\tilde{\mathrm{C}} \tag{5.17}$$

We are still dealing with linear equations for which if a system is controllable, $\tilde{\mathrm{B}}\tilde{\mathrm{B}}^T$ is positive semi-definite and $\tilde{\mathrm{W}}_c^2$ is positive definite, then the system is asymptotically stable.

Let us assume that the system is asymptotically stable, the expression of $\tilde{\mathrm{W}}_c^2$ and $\tilde{\mathrm{W}}_o^2$, which represent the dual with respect to the integral form assumed in the continuous-time case, is the following:

$$\tilde{\mathrm{W}}_c^2 = \sum_{i=0}^{\infty} \tilde{\mathrm{A}}^i\tilde{\mathrm{B}}\tilde{\mathrm{B}}^T(\tilde{\mathrm{A}}^T)^i \tag{5.18}$$

$$\tilde{\mathrm{W}}_o^2 = \sum_{i=0}^{\infty} (\tilde{\mathrm{A}}^T)^i\tilde{\mathrm{C}}^T\tilde{\mathrm{C}}\tilde{\mathrm{A}}^i \tag{5.19}$$

The two series converge if the system is asymptotically stable. Defining the balanced form for discrete-time systems is otherwise quite similar to the continuous-time case.

There are some algorithms which can solve the Lyapunov equation for discrete-time systems, but usually we use the bilinear transformation $z = \frac{1+s}{1-s}$ to obtain from the discrete-time system a fictitious continuous-time system which is analogous to the original one in terms of stability. By applying the bilinear transformation to a discrete-time system in state-space form ($\tilde{\mathrm{A}}$, $\tilde{\mathrm{B}}$, $\tilde{\mathrm{C}}$, $\tilde{\mathrm{D}}$) we obtain a continuous-time equivalent system (A, B, C, D) with:

$$\begin{aligned} \mathrm{A} &= (\mathrm{I} + \tilde{\mathrm{A}})^{-1}(\tilde{\mathrm{A}} - \mathrm{I}) \\ \mathrm{B} &= \sqrt{2}(\mathrm{I} + \tilde{\mathrm{A}})^{-1}\tilde{\mathrm{B}} \\ \mathrm{C} &= \sqrt{2}\tilde{\mathrm{C}}(\mathrm{I} + \tilde{\mathrm{A}})^{-1} \\ \mathrm{D} &= \tilde{\mathrm{D}} - \tilde{\mathrm{C}}(\mathrm{I} + \tilde{\mathrm{A}})^{-1}\tilde{\mathrm{B}} \end{aligned} \tag{5.20}$$

The singular values are also invariant according to this transformation.

Once the balanced representation of the equivalent continuous-time system is calculated, we apply the inverse transformations of (5.20) as

$$\begin{aligned} \tilde{\mathrm{A}} &= (\mathrm{I} - \tfrac{1}{2}\mathrm{A})^{-1}(\tfrac{1}{2}\mathrm{A} + \mathrm{I}) \\ \tilde{\mathrm{B}} &= (\mathrm{I} - \tfrac{1}{2}\mathrm{A})^{-1}\mathrm{B} \\ \tilde{\mathrm{C}} &= \mathrm{C}(\mathrm{I} - \tfrac{1}{2}\mathrm{A})^{-1} \\ \tilde{\mathrm{D}} &= \mathrm{D} + \tfrac{1}{2}\mathrm{C}(\mathrm{I} - \tfrac{1}{2}\mathrm{A})^{-1}\mathrm{B} \end{aligned} \tag{5.21}$$

to obtain the discrete-time balanced realization.

MATLAB® Exercise 5.3

In this MATLAB® exercise the open-loop balanced realization of a discrete-time linear system is computed. Let the system be described by the following state-space matrices:

$$A = \begin{bmatrix} 0.5 & 0 & 0 \\ 0 & -0.7 & 0 \\ 0 & 0 & 0.3 \end{bmatrix}; B = \begin{bmatrix} 1 \\ 1 \\ 1 \end{bmatrix}; C = \begin{bmatrix} 1 & 1 & 1 \end{bmatrix}; D = 0 \tag{5.22}$$

System (5.22) is stable (having all eigenvalues inside the unitary circle), controllable and observable and therefore the open-loop balanced realization can be calculated.
Let us define the system with the MATLAB commands:

```
>> A=[0.5 0 0; 0 -0.7 0; 0 0 0.3]
>> B=[1; 1; 1]
>> C=[1 1 1]
>> D=0
>> system=ss(A,B,C,D,-1)
```

Notice that, since the sampling time is unspecified, in the MATLAB command $Ts = -1$ has been set. The open-loop balanced realization can be calculated with the `balreal` command as follows:

```
>> [systembal,S]=balreal(system)
```

One obtains:

$$\begin{array}{ll} \bar{A} = \begin{bmatrix} -0.04133 & 0.5468 & -0.02004 \\ 0.5468 & -0.2342 & -0.1298 \\ -0.02004 & -0.1298 & 0.3755 \end{bmatrix}; & \bar{B} = \begin{bmatrix} -1.727 \\ -0.122 \\ -0.04216 \end{bmatrix}; \\ \bar{C} = \begin{bmatrix} -1.727 & -0.122 & -0.04216 \end{bmatrix}; & \bar{D} = 0 \end{array} \tag{5.23}$$

5.8 Exercises

1. Given the continuous-time LTI system with state matrices

$$A = \begin{bmatrix} 0 & 1 \\ -3 & -2 \end{bmatrix}; B = \begin{bmatrix} 0 \\ 1 \end{bmatrix}; C = \begin{bmatrix} -1 & 1 \end{bmatrix}$$

calculate the gramians and study system controllability and observability.

2. Calculate analytically the singular values of the system with transfer function $G(s) = \frac{s+3}{(s+1)(s+7)}$.
3. Calculate analytically the singular values of the system with transfer function $G(s) = \frac{s^2+1}{s^2+s+1}$.
4. Calculate the singular values of the system with transfer function $G(s) = -10 + \frac{60s}{s^2+3s+2}$.
5. Calculate the open-loop balanced realization of the system with state-space matrices:

$$A = \begin{bmatrix} -0.5 & -1 & 0 & 0 \\ 1 & -0.5 & 0 & 0 \\ 0 & 0 & -3 & 0 \\ 0 & 0 & 0 & -4 \end{bmatrix}; B = \begin{bmatrix} 1 \\ -1 \\ -1 \\ 1 \end{bmatrix}; C = \begin{bmatrix} 0 & 1 & -1 & 1 \end{bmatrix}$$

6

Reduced Order Models and Symmetric Systems

CONTENTS

In this chapter we look at constructing a reduced order model from a dynamical system. This represents the practical application of the open-loop balanced representation studied in the previous chapter. In fact, there are various ways of obtaining a reduced order model and in the chapter we discuss the method based on the open-loop balanced realization. Here, we first calculate a system representation where the state variables are ordered according to a particular system characteristic (controllability or observability or by both properties as in the open-loop balanced realization). The next step is to eliminate the less important state variables from the original system so as to obtain a reduced order model with a lower number of state variables. This can be done by *direct truncation* or by *singular perturbation approximation.* Both techniques will be examined. Another general aspect of the methods for constructing reduced order models is the requirement to obtain a small error, the error being a certain norm between the original model and the reduced order one (the chosen norm helps classify the approximation methods). The two techniques for

DOI: 10.1201/9781003196921-6

constructing a reduced order model will be examined from the point of view of the errors they produce (so cut the quality of the model). This chapter deals with reduced order models with an open-loop balanced form, but other chapters will deal with reduced order models based on other techniques.

Finally, this chapter also includes a few results regarding symmetric systems. The links between dynamical systems and electrical networks are presented. The dichotomy between circuits and systems is emphasized in several aspects. The discussion regards both continuous and discrete time systems. The various subjects of this chapter are complemented by several MATLAB problems that are critically discussed.

6.1 Reduced Order Models Based on the Open-loop Balanced Realization

The first step in building a reduced order model is to calculate a state-space representation highlighting certain system characteristics. In Chapter 5 we saw that an open-loop balanced form is a state representation in which controllability and observability were measured by the same parameters. This highlights that parts of the system are strongly controllable and observable and other parts are weakly controllable and observable.

Let us then consider an asymptotically stable system constructed in a state-space form:

$$\begin{cases} \dot{\mathbf{x}} = \mathrm{A}\mathbf{x} + \mathrm{B}\mathbf{u} \\ \mathbf{y} = \mathrm{C}\mathbf{x} \end{cases} \tag{6.1}$$

such that the controllability and observability gramians, that is, the two solutions of Lyapunov equations

$$\begin{array}{c} \mathrm{AW}_c^2 + \mathrm{W}_c^2\mathrm{A}^T = -\mathrm{BB}^T \\ \mathrm{A}^T\mathrm{W}_o^2 + \mathrm{W}_o^2\mathrm{A} = -\mathrm{C}^T\mathrm{C} \end{array} \tag{6.2}$$

coincide and are diagonal $\mathrm{W}_c^2 = \mathrm{W}_o^2 = \mathrm{diag}(\sigma_1, \sigma_2, \dots, \sigma_n)$.

Let us suppose that

$$\sigma_1 \geq \sigma_2 \geq \cdots \geq \sigma_r \gg \sigma_{r+1} \geq \cdots \geq \sigma_n$$

In this case, the fact that $\sigma_r \gg \sigma_{r+1}$ reveals the presence of a group of strongly controllable and observable variables associated with large singular values compared to those remaining, and a group of weakly controllable and observable variables associated with small singular values. This subdivision of state variables suggests a partition of the system:

$$\begin{bmatrix} \dot{\mathbf{x}}_1 \\ \dot{\mathbf{x}}_2 \end{bmatrix} = \begin{bmatrix} A_{11} & A_{12} \\ A_{21} & A_{22} \end{bmatrix} \begin{bmatrix} \mathbf{x}_1 \\ \mathbf{x}_2 \end{bmatrix} + \begin{bmatrix} B_1 \\ B_2 \end{bmatrix} \mathbf{u}$$
$$\mathbf{y} = \begin{bmatrix} C_1 & C_2 \end{bmatrix} \begin{bmatrix} \mathbf{x}_1 \\ \mathbf{x}_2 \end{bmatrix} \tag{6.3}$$

where $\mathbf{x}_1 = \begin{bmatrix} x_1 \\ x_2 \\ \vdots \\ x_r \end{bmatrix}$ are the strongly controllable and observable variables and

$\mathbf{x}_2 = \begin{bmatrix} x_{r+1} \\ x_{r+2} \\ \vdots \\ x_n \end{bmatrix}$ are the weakly controllable and observable variables.

Given this partition, there are two methods to define the reduced order model: by direct truncation or by singular perturbation approximation.

6.1.1 Direct Truncation Method

In this method, the weakly controllable and observable part is neglected, using the strongly controllable and observable part as the reduced order model. This model has an order r, and is obtained from equations (6.3) taking $\mathbf{x}_2 = 0$:

$$\begin{cases} \dot{\mathbf{x}}_1 = A_{11}\mathbf{x}_1 + B_1\mathbf{u} \\ \mathbf{y} = C_1\mathbf{x}_1 \end{cases} \tag{6.4}$$

Note that the reduced order model is asymptotically stable (the original system is asymptotically stable). If we look at Lyapunov equation for the subblock $W_{c1}^2 = \begin{bmatrix} \sigma_1 & 0 & \cdots & 0 \\ 0 & \sigma_2 & \cdots & 0 \\ \vdots & \vdots & & \vdots \\ 0 & 0 & \cdots & \sigma_r \end{bmatrix}$, then we have

$$A_{11}W_{c1}^2 + W_{c1}^2A_{11}^T = -B_1B_1^T \tag{6.5}$$

From this, we deduce that matrix A_{11} satisfies a Lyapunov equation with $B_1B_1^T$ positive semi-definite positive and a solution W_{c1}^2 positive definite. So, according to the property of gramians discussed in Section 5.1, the reduced order model is asymptotically stable. For the same reason, even the neglected subsystem is asymptotically stable.

In terms of block diagram (see Figure 6.1), it may be noted that this method neglects the coupling (given by matrices A_{21} and A_{12}) between the weakly controllable and observable parts and the strongly controllable and observable parts, considering the reduced order model in terms only of the strongly controllable and observable part which does not change the system

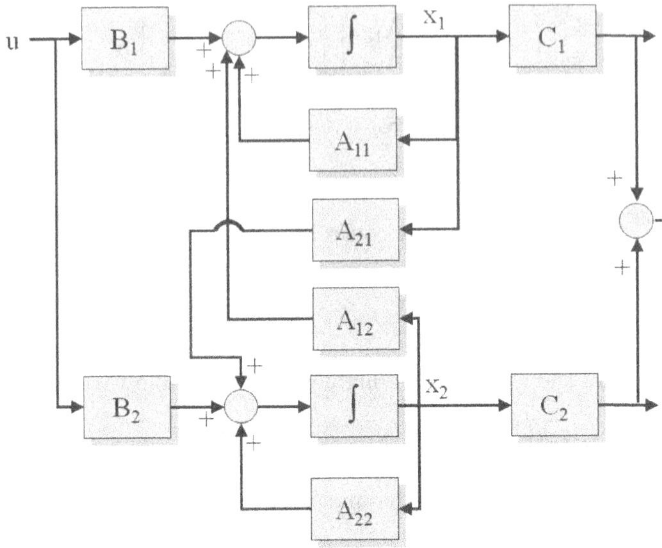

FIGURE 6.1
Block diagram for a partitioned system as in equation (6.3).

stability. So, one property of the direct truncation method is that the reduced order model is asymptotically stable.

We saw that, to evaluate the quality of a reduced order model, the error between the original and the reduced order model must be calculated. Let $\mathrm{G}(s) = \mathrm{C}(s\mathrm{I}-\mathrm{A})^{-1}\mathrm{B}$ be the transfer matrix of the original system and $\tilde{\mathrm{G}}(s) = \mathrm{C}_1(s\mathrm{I}-\mathrm{A}_{11})^{-1}\mathrm{B}_1$ be the transfer matrix of the approximated model and let us consider the frequency response of the two models: $\mathrm{G}(j\omega)$ and $\tilde{\mathrm{G}}(j\omega)$. The difference in the frequency response can be quantified by the matrix given by the difference between $\mathrm{G}(j\omega)$ and $\tilde{\mathrm{G}}(j\omega)$. Since this matrix is a function of ω, calculating the spectral norm will produce a result which depends on ω. Let us then consider the maximum singular value (i.e., the maximum of the spectral norm) with respect to ω: $\max_{\omega} \|\mathrm{G}(j\omega) - \tilde{\mathrm{G}}(j\omega)\|_S$. The following result holds:

$$\max_{\omega} \|\mathrm{G}(j\omega) - \tilde{\mathrm{G}}(j\omega)\|_S \leq 2 \sum_{i=r+1}^{n} \sigma_i \tag{6.6}$$

The norm defined by the first member of equation (6.6) is also known as the H_∞ norm of a transfer matrix.

Definition 14 (H_∞ norm of a system) *The H_∞ norm of a system is defined as the maximum value of the spectral norm of matrix $\mathrm{G}(j\omega)$ with respect to ω:*

$$\|G(s)\|_{\infty} = \max_{\omega} \|\mathrm{G}(j\omega)\|_S$$

The error produced by the direct truncation method is less than twice the sum of singular values associated with the variables not considered in the reduced order model. The quality of the model is greater the smaller the singular values $\sigma_{r+1} + \cdots + \sigma_n$ of the neglected part.

6.1.2 Singular Perturbation Method

In the direct TR method the weakly controllable and observable state variables are totally discarded, their contribution assumed of no relevance. In the singular perturbation method it is supposed that the weakly controllable and observable part is faster than the strongly controllable and observable part (furthermore, we known it is asymptotically stable). What is neglected is the subsystem dynamic, supposing that the variables quickly reach the steady-state. So, imagine that the weakly controllable and observable subsystem evolves so rapidly as to make the state variables $\mathbf{x}_2$ at the steady-state. This is the same as supposing $\dot{\mathbf{x}}_2 = 0$.

Considering $\dot{\mathbf{x}}_2 = 0$ in the second equation (6.3), then:

$$\mathbf{x}_2 = \mathrm{A}_{22}^{-1}(-\mathrm{A}_{21}\mathbf{x}_1 - \mathrm{B}_2\mathbf{u}) \tag{6.7}$$

and by substituting in the first equation (6.3) we get

$$\dot{\mathbf{x}}_1 = \mathrm{A}_{11}\mathbf{x}_1 + \mathrm{A}_{12}\mathrm{A}_{22}^{-1}(-\mathrm{A}_{21}\mathbf{x}_1 - \mathrm{B}_2\mathbf{u}) + \mathrm{B}_1\mathbf{u} \tag{6.8}$$

and so

$$\dot{\mathbf{x}}_1 = (\mathrm{A}_{11} - \mathrm{A}_{12}\mathrm{A}_{22}^{-1}\mathrm{A}_{21})\mathbf{x}_1 + (-\mathrm{A}_{12}\mathrm{A}_{22}^{-1}\mathrm{B}_2 + \mathrm{B}_1)\mathbf{u} \tag{6.9}$$

Proceeding in the same way for the output $\mathbf{y}$ we get:

$$\mathbf{y} = (\mathrm{C}_1 - \mathrm{C}_2\mathrm{A}_{22}^{-1}\mathrm{A}_{21})\mathbf{x}_1 - \mathrm{C}_2\mathrm{A}_{22}^{-1}\mathrm{B}_2\mathbf{u} \tag{6.10}$$

The reduced order model obtained is thus not strictly proper:

$$\begin{cases} \dot{\mathbf{x}}_1 = \bar{\mathrm{A}}_{11}\mathbf{x}_1 + \bar{\mathrm{B}}_1\mathbf{u} \\ \mathbf{y} = \bar{\mathrm{C}}_1\mathbf{x}_1 + \bar{\mathrm{D}}_1\mathbf{u} \end{cases} \tag{6.11}$$

with

$$\begin{array}{l} \bar{\mathrm{A}}_{11} = \mathrm{A}_{11} - \mathrm{A}_{12}\mathrm{A}_{22}^{-1}\mathrm{A}_{21} \\ \bar{\mathrm{B}}_1 = -\mathrm{A}_{12}\mathrm{A}_{22}^{-1}\mathrm{B}_2 + \mathrm{B}_1 \\ \bar{\mathrm{C}}_1 = \mathrm{C}_1 - \mathrm{C}_2\mathrm{A}_{22}^{-1}\mathrm{A}_{21} \\ \bar{\mathrm{D}}_1 = -\mathrm{C}_2\mathrm{A}_{22}^{-1}\mathrm{B}_2 \end{array} \tag{6.12}$$

$(\bar{A}_{11}, \bar{B}_1, \bar{C}_1)$ is an open-loop balanced system as can be verified by calculating the gramians from equations $\bar{A}_{11}W_{c1}^2 + W_{c1}^2\bar{A}_{11}^T = -\bar{B}_1\bar{B}_1^T$ and $\bar{A}_{11}^T W_{o1}^2 + W_{o1}^2\bar{A}_{11} = -\bar{C}_1^T\bar{C}_1$.

So, the system has the same singular values as the strongly controllable and observable part except that it takes into account the asymptotic contribution of the $\mathbf{x}_2$ state variables, which the direct truncation method does not.

Note that the singular perturbation method is a general method and valid for linear and nonlinear systems. Let us consider a generic nonlinear system:

$$\begin{aligned} \dot{\mathbf{x}}_1 &= f_1(\mathbf{x}_1, \mathbf{x}_2) \\ \varepsilon\dot{\mathbf{x}}_2 &= f_2(\mathbf{x}_1, \mathbf{x}_2) \end{aligned} \tag{6.13}$$

If ε is very small, we can suppose that $\varepsilon\dot{\mathbf{x}}_2 \simeq 0$ (i.e., variables evolve much more rapidly than $\mathbf{x}_1$). So $f_2(\mathbf{x}_1, \mathbf{x}_2) = 0$ from which we get $\mathbf{x}_2 = g(\mathbf{x}_1)$ to obtain $\mathbf{x}_1 = f_1(\mathbf{x}_1, g(\mathbf{x}_1))$.

The singular perturbation method can also be applied to discrete-time systems, but the system with state matrix $\bar{A}_{11}$ is not open-loop balanced.

The error encountered with the singular perturbation method is given by:

$$\max_{\omega} \|G(j\omega) - \tilde{G}(j\omega)\|_S \leq 2 \sum_{i=r+1}^{n} \sigma_i \tag{6.14}$$

where $G(s)$ is the transfer matrix of the original system, whereas $\tilde{G}(s)$ is the transfer matrix of the reduced order system.

The error takes the same expression in the two methods (direct truncation and singular perturbation). There is, however, a significant difference. It can be demonstrated that the direct truncation method does not preserve the static gain whereas the singular perturbation method does. In the former, $G(j\omega) = \tilde{G}(j\omega)$ for $\omega \to \infty$, whereas in the latter $G(0) = \tilde{G}(0)$. The two methods differ because one better approximates the low frequency behavior (the singular perturbation method) and the other the high frequency behavior (the direct truncation method).

6.2 Reduced Order Model Exercises

In this section we discuss a series of exercises on reduced order models for linear time-invariant systems that are stable and compare the results of the different approximation methods.

MATLAB® Exercise 6.1

In this MATLAB® exercise the procedure for obtaining a reduced order model is illustrated through a fourth order system defined by its transfer function:

$$G(s) = \frac{0.5s^4 + 9s^3 + 47.5s^2 + 95s + 62}{(s+1)(s+2)(s+3)(s+4)}$$

The first step is always to define the system:

```
>> n=4
>> s=tf('s')
>> G=(0.5*s^4+9*s^3+47.5*s^2+95*s+62)/((s+1)*(s+2)*(s+3)*(s+4))
```

Then, the balanced form of the system is calculated:

```
>> [system_bal,S]=balreal(G)
```

To verify it, the gramians can be calculated to see if they are diagonal and equal:

```
>> gram(system_bal,'o')
>> gram(system_bal,'c')
```

Now, let us calculate the reduced order model (6.4):

```
>> r=2
>> reducedordersystem=ss(system_bal.a(1:r,1:r),...
   system_bal.b(1:r),system_bal.c(1:r),system_bal.d)
```

and its transfer function:

```
>> tf(reducedordersystem)
```

The approximation error is given by

```
>> DTerror=G-tf(reducedordersystem)
>> directtruncerror=normhinf(G-tf(reducedordersystem))
```

It can be verified that effectively expression (6.6) is valid, by calculating the sum of the discarded singular values:

```
>> error=2*sum(S(r+1:n))
```

Table 6.1 shows the errors by using a reduced order model as r varies.

TABLE 6.1
Errors as order r varies for the reduced order model in Exercise 6.1.

r	$\max_\omega \|\mathrm{G}(j\omega) - \tilde{\mathrm{G}}(j\omega)\|_S$
1	0.1283
2	0.0037
3	$4.2637 \cdot 10^{-5}$

MATLAB® Exercise 6.2

With reference to the system in Exercise 6.1, two reduced order models will be constructed using the two methods described and the results will be compared.
Having defined the system as in Exercise 6.1, a reduced order model will be constructed where $r = 2$ given that $\sigma_2 = 0.0623 >> \sigma_3 = 0.0018$.
The direct truncation reduced order model is constructed using the following commands:

```
>> reducedordersystem=ss(system_bal.a(1:2,1:2),...
    system_bal.b(1:2),system_bal.c(1:2),system_bal.d)
```

Now, the transfer function of the reduced order model can be calculated as well as the approximation error:

```
>> tf(reducedordersystem)
>> DTerror=G-tf(reducedordersystem)
>> directtruncationerror=normhinf(G-tf(reducedordersystem))
```

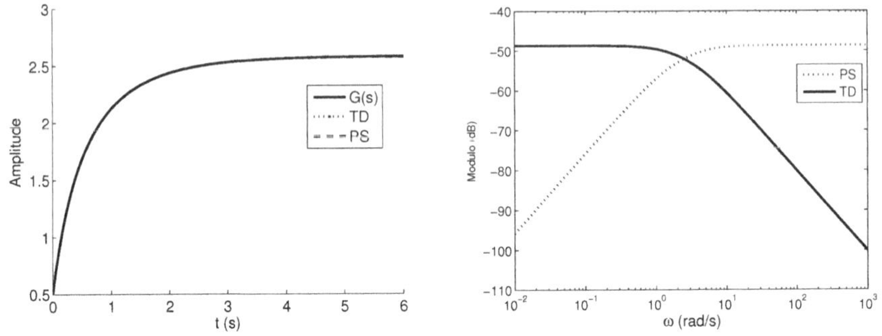

(a) (b)

FIGURE 6.2
(a) Unit step response of the original system $G(s)$ and of the reduced order models. (b) Magnitude Bode diagram for the error between the original model and the reduced order model.

Instead, the singular perturbation reduced order model is constructed using the following commands:

```
>> A11=system_bal.a(1:2,1:2);
>> A12=system_bal.a(1:2,3:4);
>> A21=system_bal.a(3:4,1:2);
>> A22=system_bal.a(3:4,3:4);
>> B1=system_bal.b(1:2);
>> B2=system_bal.b(3:4);
>> C1=system_bal.c(1:2);
>> C2=system_bal.c(3:4);
>> D=system_bal.d;
>> A11r=A11-A12*inv(A22)*A21;
>> B1r=-A12*inv(A22)*B2+B1;
>> C1r=C1-C2*inv(A22)*A21;
>> D1r=-C2*inv(A22)*B2+D;
>> SPreducedordersystem=ss(A11r,B1r,C1r,D1r);
```

Analogously, the transfer function of the singular perturbation approximate model can be calculated as well as the error with the following method:

```
>> tf(SPreducedordersystem)
>> SPerror=G-tf(SPreducedordersystem)
>> singularperterror=normhinf(G-tf(SPreducedordersystem))
```

Finally, with command

```
>> ltiview
```

the two reduced order models can be compared. For example, Figure 6.2(a) shows the unit-step response of the original system and the two reduced order models (the three responses are almost indistinguishable).

From the magnitude Bode diagram of the error between the original model and the reduced order model, shown in Figure 6.2(b), we observe that the error obtained in each case is very small, but the approximation of the direct truncation method is better at high frequencies, whereas the singular perturbation method is better at low frequencies.

MATLAB® Exercise 6.3

In this further MATLAB® exercise, the use of the command `modred` is described and the properties of the truncated reduced model in the case of continuous-time and discrete-time systems are briefly discussed.

Consider the continuous-time system with state-space matrices:

$$\mathrm{A} = \begin{bmatrix} -1 & 0 & 0 \\ 0 & -2 & 0 \\ 0 & 0 & -3 \end{bmatrix}; \mathrm{B} = \begin{bmatrix} 1 \\ 1 \\ 1 \end{bmatrix}; \mathrm{C} = \begin{bmatrix} 1 & 1 & 1 \end{bmatrix} \tag{6.15}$$

Define the system in MATLAB:

```
>> A=[-1  0 0 ; 0 -2 0; 0 0 -3]
>> B=[1; 1; 1]
>> C=[1 1 1]
>> D=0
>> system=ss(A,B,C,D)
```

and compute the open-loop balanced realization:

```
>> [systembal,S]=balreal(system)
```

Now, let us use the command `modred`. Given a system in balanced form and given a vector indicating which state variables have to be eliminated, the command `modred` gives the reduced order model. To use direct truncation, the option 'Truncate' is used. To use singular perturbation approximation, the option 'MatchDC' (forcing equal DC gains) is used.

Consider a second-order reduced model:

```
>> elim=(S<1e-2)
>> systemred=modred(systembal,elim,'Truncate')
```

One gets:

$$\bar{\mathrm{A}} = \begin{bmatrix} -1.617 & 0.7506 \\ 0.7506 & -2.042 \end{bmatrix}; \bar{\mathrm{B}} = \begin{bmatrix} -1.682 \\ 0.4087 \end{bmatrix}; \bar{\mathrm{C}} = \begin{bmatrix} -1.682 & 0.4087 \end{bmatrix}; \bar{\mathrm{D}} = 0 \tag{6.16}$$

Now, let us compute the two gramians

```
>> Wo2=gram(systemred,'o')
>> Wc2=gram(systemred,'c')
```

We get:

$$\bar{\mathrm{W}}_c^2 = \begin{bmatrix} 0.8751 & 0.0000 \\ 0.0000 & 0.0409 \end{bmatrix}; \bar{\mathrm{W}}_o^2 = \begin{bmatrix} 0.8751 & 0.0000 \\ 0.0000 & 0.0409 \end{bmatrix} \tag{6.17}$$

The singular values of the reduced order system are exactly the same as the original system, as can be observed by comparing the diagonal elements of the gramians with the first two singular values listed in $\mathrm{S} = \begin{bmatrix} 0.8751 \\ 0.0409 \\ 0.0006 \end{bmatrix}$.

We now consider a discrete-time example. Let us come back to system (5.22) and apply the command `modred`:

```
>> A=[0.5 0 0; 0 -0.7 0; 0 0 0.3]
>> B=[1; 1; 1]
>> C=[1 1 1]
>> D=0
>> system=ss(A,B,C,D,-1)
>> [systembal,S]=balreal(system)
>> elim=(S<1e-1)
>> systemred=modred(systembal,elim,'Truncate')
>> Wc2=gram(systemred,'c')
>> Wo2=gram(systemred,'o')
```

We get:

$$\bar{W}_c^2 = \begin{bmatrix} 3.3064 & 0.0000 \\ 0.0000 & 1.0616 \end{bmatrix}; \bar{W}_o^2 = \begin{bmatrix} 3.3064 & 0.0000 \\ 0.0000 & 1.0616 \end{bmatrix} \tag{6.18}$$

In this case, the singular values of the reduced order system are not exactly the same as the original system. In fact, the second singular value of the reduced order system is $\sigma_2 = 1.0616$, while the second singular value of the original system (listed in $S = \begin{bmatrix} 3.3065 \\ 1.0621 \\ 0.0244 \end{bmatrix}$) is $\sigma_2 = 1.0621$.

6.3 Symmetric Systems

There is a particular class of linear time-invariant systems characterized by having a symmetric transfer function matrix $G(s)$. They are called symmetric systems. Obviously, the definition makes sense if the transfer matrix is square, that is, the number of inputs equals the number of outputs.

Definition 15 (Symmetric system) *A linear time-invariant system with the number of inputs equalling the number of outputs ($m = p$) is symmetric if its transfer matrix $G(s)$ is such that $G(s) = G^T(s)$.*

SISO systems are an example of symmetric systems for which $m = p = 1$ and the transfer matrix is such that $G(s) = G^T(s)$ as it is a scalar function.

6.3.1 Reduced Order Models for SISO Systems

The issue of reduced order models of symmetric systems can be faced by reasoning in terms of the energy of impulse response. Given an asymptotically stable symmetric SISO system

$$\begin{cases} \dot{\mathbf{x}} = \mathrm{A}\mathbf{x} + \mathrm{B}\mathbf{u} \\ \mathbf{y} = \mathrm{C}\mathbf{x} \end{cases} \tag{6.19}$$

and considering the impulse response $y(t) = \mathrm{C}e^{\mathrm{A}t}\mathrm{B}$, the energy associated is defined by $E = \int_0^\infty y(t)^T y(t)dt$. This is a finite integral since the system is asymptotically stable and so the impulse response tends to zero.

Furthermore, this energy is linked to the observability gramian:

$$E = \int_0^\infty \mathrm{B}^T e^{\mathrm{A}^T t}\mathrm{C}^T \mathrm{C}e^{\mathrm{A}t}\mathrm{B}dt = \mathrm{B}^T \int_0^\infty e^{\mathrm{A}^T t}\mathrm{C}^T \mathrm{C}e^{\mathrm{A}t}dt\mathrm{B} = \mathrm{B}^T W_o^2 \mathrm{B}$$

Since the impulse response does not vary as the reference system varies, the energy does not depend on the reference system used. In particular, in the case of an open-loop balanced form $(\bar{\mathrm{A}}, \bar{\mathrm{B}}, \bar{\mathrm{C}})$ the energy is given by:

$$E = \bar{\mathrm{B}}^T \bar{W}_o^2 \bar{\mathrm{B}} = \sum_{i=1}^{n} \sigma_i \bar{b}_i^2 \tag{6.20}$$

Expression (6.20) suggests that to properly reduce the system order, it is better to consider conditions on the various terms of the impulse response energy $E = \sigma_1 \bar{b}_1^2 + \sigma_2 \bar{b}_2^2 + \cdots + \sigma_n \bar{b}_n^2$ rather than the condition $\sigma_r \gg \sigma_{r+1}$, neglecting terms with relations of the type $\sigma_r \bar{b}_r^2 \gg \sigma_{r+1} \bar{b}_{r+1}^2$, or in other words variables which contribute with less energy.

MATLAB® Exercise 6.4

Let us consider two SISO systems: $G_1(s) = \frac{\frac{s}{0.9}+1}{(s+1)(\frac{s}{10}+1)}$ and $G_2(s) = \frac{\frac{s}{9}+1}{(s+1)(\frac{s}{10}+1)}$. Before looking at the open-loop balanced reduced order model, let us make some preliminary considerations. The two systems are asymptotically stable and have a pole at $s = -1$ and another at $s = -10$. For both systems, the impulse response is of the type:

$$G(s) = \frac{A}{s+1} + \frac{B}{s+10} \Rightarrow y(t) = Ae^{-t} + Be^{-10t} \tag{6.21}$$

Obtaining a reduced order model requires neglecting one of the two system modes. The choice of which mode to neglect is based not only on the fastest eigenvalue, but also on the associated residue, the A and B values. Besides, notice that system $G_1(s)$ has a zero in $z = -0.9$ the effect of which is to cancel out nearly all the dynamics at pole $s = -1$, whereas the zero of system $G_2(s)$ is very close to the pole $s = -10$ making residue B small. Effectively, $A = -0.1235$ and $B = 11.2346$ for system $G_1(s)$, whereas for $G_2(s)$ $A = 0.9877$ and $B = 0.1235$.

Now, let us consider the open-loop balanced system using MATLAB® commands:

```
>> s=tf('s')
>> G1=(s/0.9+1)/(s+1)/(s/10+1)
>> G2=(s/9+1)/(s+1)/(s/10+1)
>>  [system1b,S1]=balreal(G1)
>>  [system2b,S2]=balreal(G2)
```

and consider the reduced order models obtained by direct truncation

```
>> reducedsystem1=ss(system1b.a(1,1),system1b.b(1),
        system1b.c(1),0)
>> tf(reducedsystem1)
>> reducedsystem2=ss(system2b.a(1,1),system2b.b(1),
        system2b.c(1),0)
>> tf(reducedsystem2)
```

The transfer functions of the approximated models are: $\tilde{G}_1(s) = \frac{11.17}{s+10.29}$ and $\tilde{G}_2(s) = \frac{1.029}{s+1.038}$. Note that in the first case the reduced order model pole is very close to $s = -10$, whereas in the second case the pole is very close to $s = -1$ in agreement with the considerations above. Note also that, notwithstanding the two original systems had the same static (unitary) gain, the reduced order models (obtained by direct truncation) no longer have the same gain.

As regards the system singular values, they are $\sigma_1 = 0.5428$, $\sigma_2 = 0.0428$ in the first case and $\sigma_1 = 0.4959$, $\sigma_2 = 0.0041$ in the second, may it be concluded that the approximation is better in the second case? Let us look at the impulse response energy. The various $\sigma_i \bar{b}_i^2$ terms can be obtained from MATLAB® commands:

```
>> S1(:).*(sistema1b.B(:).^2)
>> S2(:).*(sistema2b.B(:).^2)
```

We get $\sigma_1 \bar{b}_1^2 = 6.0636$, $\sigma_2 \bar{b}_2^2 = 0.0026$ for $G_1(s)$ and $\sigma_1 \bar{b}_1^2 = 0.5103$, $\sigma_2 \bar{b}_2^2 = 0.0003$ for $G_2(s)$. It may be concluded that, in terms of the impulse response energy, the two approximations are equivalent.

6.3.2 Properties of Symmetric Systems

Given that $G(s) = C(sI - A)^{-1}B$ and $G^T(s) = B^T((sI - A)^T)^{-1}C^T$, it can be verified that, for a system to be symmetric, it suffices that $B = C^T$ and $A^T = A$.

More generally, we will see that if a system is symmetric then matrices B and C and matrices A^T and A are linked by certain relations which involve an invertible and symmetric matrix, T. To obtain these relations let us consider $G^T(s)$:

$$G^T(s) = B^T((sI - A)^T)^{-1}C^T$$

If a matrix is invertible and symmetric ($I = TT^{-1}$), then:

$$G^T(s) = B^T(sTT^{-1} - A^T)^{-1}C^T = B^T T(sI - T^{-1}A^T T)^{-1}T^{-1}C^T$$

Equalling this expression with $G(s) = C(sI - A)^{-1}B$ we find that:

$$\begin{array}{l} C = B^T T \\ B = T^{-1}C^T \\ A = T^{-1}A^T T \end{array} \tag{6.22}$$

The first two relations (6.22) are equivalent if T is symmetric. The transpose of the first relation is:

$$C^T = T^T B \Rightarrow B = (T^T)^{-1}C^T \Rightarrow B = T^{-1}C^T$$

For symmetric controllable and observable systems there exists a matrix T which links matrices B and C and which can easily be obtained from the observability and controllability matrices. Assuming that the system is symmetric and in minimal form, in fact one gets:

$$M_o^T = \begin{bmatrix} C \\ CA \\ \vdots \\ CA^{n-1} \end{bmatrix}^T = \begin{bmatrix} C^T & A^T C^T & \cdots & (A^T)^{n-1}C^T \end{bmatrix} =$$

$$= \begin{bmatrix} TB & TAB & \cdots & T(A^T)^{n-1}B \end{bmatrix} = TM_c$$

where we have used the fact that $B = T^{-1}C^T$ and so $C = TB$.

Now, if the system is SISO, then

$$T = M_o^T M_c^{-1}$$

If the system is MIMO, instead, we have:

$$T = M_o^T M_c^T (M_c M_c^T)^{-1}$$

To show that T is symmetric, we can notice that from the previous considerations we have obtained that $M_o^T = TM_c$. Now, if we consider again M_o^T and now plug $C = B^T T$, we get:

$$M_o^T = \begin{bmatrix} C \\ CA \\ \vdots \\ CA^{n-1} \end{bmatrix}^T = \begin{bmatrix} C^T & A^T C^T & \cdots & (A^T)^{n-1} C^T \end{bmatrix} =$$

$$= \begin{bmatrix} T^T B & T^T AB & \cdots & T^T (A^T)^{n-1} B \end{bmatrix} = T^T M_c$$

Hence, $M_o^T = T^T M_c$. Comparing this result with the relationship previously found, i.e., $M_o^T = TM_c$, we derive that $T = T^T$.

MATLAB® Exercise 6.5

Consider the continuous-time LTI system:

$$A = \begin{bmatrix} -4 & -1.5 & -1.5 \\ -5 & -5.5 & -0.5 \\ -1 & 1.5 & -3.5 \end{bmatrix}; B = \begin{bmatrix} 0.5 & 0.55 \\ 1.5 & -1.35 \\ -1.5 & 0.45 \end{bmatrix}; \quad C = \begin{bmatrix} 3 & 0 & -1 \\ 5.4 & -1.8 & -0.8 \end{bmatrix}; D = \begin{bmatrix} 0 & 0 \\ 0 & 0 \end{bmatrix} \tag{6.23}$$

```
>> A=[-4 -1.5 -1.5; -5 -5.5 -0.5; -1 1.5 -3.5]
>> B=[0.5 0.55;  1.5 -1.35; -1.5 0.45]
>> C=[3 0 -1; 5.4 -1.8 -0.8]
>> D=zeros(2)
>> system=ss(A,B,C,D)
```

By calculating the transfer matrix of the system with the command

```
>> tf(system)
```

one obtains

$$G(s) = \begin{bmatrix} \frac{3s^2+26s+47}{s^3+13s^2+47s+35} & \frac{1.2s^2+17.2s+64}{s^3+13s^2+47s+35} \\ \frac{1.2s^2+17.2s+64}{s^3+13s^2+47s+35} & \frac{5.04s^2+56.24s+147.2}{s^3+13s^2+47s+35} \end{bmatrix} \tag{6.24}$$

So, since $G(s) = G^T(s)$, the system is symmetric.
Let us now calculate the matrix T, using $T = M_o^T M_c^T (M_c M_c^T)^{-1}$. We first calculate M_c and M_o

```
>> Mc=ctrb(A,B)
>> Mo=obsv(A,C)
```

and then

```
>> T=Mo'*Mc'*inv(Mc*Mc')
```

One obtains a symmetric matrix:

$$\mathrm{T} = \begin{bmatrix} 9.0000 & 0 & 1.0000 \\ 0.0000 & 2.0000 & 2.0000 \\ 1.0000 & 2.0000 & 3.0000 \end{bmatrix} \tag{6.25}$$

We can now verify that equations (6.22) hold:

```
>> B'*T
>> C
>> inv(T)*A'*T
>> A
```

Alternatively, we can extract from the controllability and observability matrices two invertible 3×3 blocks taking the first three columns in M_c, or the first three rows in M_o (which are linear independent):

```
>> Mcr=Mc(1:3,1:3)
>> Mor=Mo(1:3,1:3)
```

and then calculate T as

```
>> T=Mor'*inv(Mcr)
```

The same matrix T is found.

6.3.3 The Cross-gramian Matrix

For symmetric systems, the product of the matrices B and C, which is an $n \times n$ matrix, can be defined. For symmetric systems, another Lyapunov equation, called the cross-gramian equation, can be introduced:

$$\mathrm{AW}_{co} + \mathrm{W}_{co}\mathrm{A} = -\mathrm{BC} \tag{6.26}$$

In the most general case, BC is not a symmetric matrix, so the solution to the Lyapunov equation (6.26) may not be symmetric. Furthermore, nothing is known about whether it is defined positive or not.

If the system is asymptotically stable, the solution W_{co} to the Lyapunov equation (6.26) can be expressed in integral form:

$$\mathrm{W}_{co} = \int_0^\infty e^{\mathrm{A}t}\mathrm{BC}e^{\mathrm{A}t}dt$$

6.3.4 Relations Between W_c^2, W_o^2 and W_{co}

The cross-gramian is linked to the controllability and observability gramians by the matrix T. To obtain these relations, let us consider the Lyapunov equations for the gramians:

$$\begin{aligned} \mathrm{AW}_c^2 + \mathrm{W}_c^2\mathrm{A}^T &= -\mathrm{BB}^T \\ \mathrm{A}^T\mathrm{W}_o^2 + \mathrm{W}_o^2\mathrm{A} &= -\mathrm{C}^T\mathrm{C} \end{aligned}$$

and in the second equation let us substitute the relationships (6.22):

$$\mathrm{TAT}^{-1}\mathrm{W}_o^2 + \mathrm{W}_o^2\mathrm{A} = -\mathrm{TBC}$$

Multiplying left by matrix T^{-1} we obtain

$$AT^{-1}W_o^2 + T^{-1}W_o^2A = -BC$$

Comparing the result with the cross-gramian equation (6.26) we obtain:

$$W_{co} = T^{-1}W_o^2$$

Analogously, starting with the Lyapunov equation for the controllability gramian, we find:

$$AW_c^2 + W_c^2TAT^{-1} = -BCT^{-1}$$

$$AW_c^2T + W_c^2TA = -BC$$

$$\Rightarrow W_{co} = W_c^2T$$

At this point notice that, since $W_{co} = T^{-1}W_o^2 = W_c^2T$, we have:

$$W_c^2W_o^2 = W_{co}^2$$

This relation produces an important result for the cross-gramian eigenvalues.

Theorem 10 *The eigenvalues of* W_{co} *in modulus are equal to the system singular values, i.e.,*

$$|\lambda_i(W_{co})| = \sigma_i$$

In fact, the relationship $W_c^2W_o^2 = W_{co}^2$ yields that the eigenvalues of the square of W_{co} equal the square of the system singular values and therefore the eigenvalues of W_{co}, which may be positive or negative, are in modulus equal to the system singular values, that is $\lambda_i(W_{co}) = \pm\sigma_i$.

Often this relation is expressed in a matrix form. If $\Sigma = \begin{bmatrix} \sigma_1 & & & \\ & \sigma_2 & & \\ & & \ddots & \\ & & & \sigma_n \end{bmatrix}$ is the diagonal matrix formed by the system singular values, and Λ is the matrix formed by the eigenvalues (in decreasing order of their modulus) $\Lambda = \begin{bmatrix} \lambda_1 & & & \\ & \lambda_2 & & \\ & & \ddots & \\ & & & \lambda_n \end{bmatrix}$, then we can write:

$$\Lambda = S\Sigma$$

where S is an appropriate matrix (also diagonal), named the signature matrix, whose diagonal components are either +1 o −1.

Since the eigenvalues of W_{co} are in modulus equal to the system singular values, also the eigenvalues of W_{co} are invariants of the systems. The invariance of the W_{co} eigenvalues can be also proved by seeing how the cross-gramian varies as the reference system varies.

Consider the cross-gramian equation in the new reference system defined by the state transformation $\tilde{\mathbf{x}} = \bar{T}^{-1}\mathbf{x}$

$$\tilde{A}\tilde{W}_{co} + \tilde{W}_{co}\tilde{A} = -\tilde{B}\tilde{C}$$

and then apply relation (5.7):

$$\bar{T}^{-1}A\bar{T}\tilde{W}_{co} + \tilde{W}_{co}\bar{T}^{-1}A\bar{T} = -\bar{T}^{-1}BC\bar{T}$$

$$\Rightarrow A\bar{T}\tilde{W}_{co}\bar{T}^{-1} + \bar{T}\tilde{W}_{co}\bar{T}^{-1}A = -BC$$

from which we obtain:

$$W_{co} = \bar{T}\tilde{W}_{co}\bar{T}^{-1} \tag{6.27}$$

As opposed to what happens with controllability and observability gramians, in this case we find a relation of similitude. So, the eigenvalues of W_{co} do not depend on the reference system.

We have seen that for a symmetric system in whatever state representation the matrices B and C are linked by relations which depend on the matrix T.

In the reference system where the system is open-loop balanced this is the signature matrix. Let us now prove this important result. Preliminarily note that $S^{-1} = S$.

Theorem 11 *In a symmetric open-loop balanced SISO system,* $T = S$ *in (6.22), i.e.:*

$$\begin{aligned} C &= B^T S \\ B &= SC^T \\ A &= SA^T S \end{aligned}$$

Proof *The Lyapunov equations for controllability and observability gramians in an open-loop balanced reference system ($W_c^2 = W_o^2 = \Sigma$) become:*

$$\begin{aligned} A\Sigma + \Sigma A^T &= -BB^T \\ A^T\Sigma + \Sigma A &= -C^T C \end{aligned}$$

Subtracting the second equation from the first we obtain:

$$(A - A^T)\Sigma + \Sigma(A^T - A) = -(BB^T - C^T C)$$

The terms on the diagonal in the first member are all zero, whereas for those on the second, the i-th diagonal term is $d_{ii} = -b_i^2 + c_i^2$. *From this we find that* $b_i^2 = c_i^2$ *and so* $b_i = \pm c_i$, *a relation which in matrix form can be expressed through the signature matrix* $B = SC^T$. *It therefore follows, given the uniqueness of* T, *that* $T = S$, *i.e., the theorem thesis.*

The case is particularly interesting when the signature matrix equals the identity matrix, $S = I$, or rather when all W_{co} eigenvalues are positive. In this case, $B = C^T$ and $A = A^T$. At the beginning of this chapter, we saw that this was a sufficient condition (but not necessary) for a system to be symmetric.

Generally, the signature matrix S has a certain number of components equal to +1 (indicated by n_+) and a certain number equal to −1 (n_-). Obviously the sum of the two equals the system order. It can be demonstrated that the difference between these two numbers is related to an important index, which will be now introduced.

Definition 16 (Cauchy index) *The Cauchy index of a real rational function* $f(\lambda)$ *defined in the interval* $[\alpha, \beta]$ *is:*

$$I_\alpha^\beta = N_1 - N_2 \tag{6.28}$$

where N_1 *is the number of jumps of the function from* $-\infty$ *to* $+\infty$ *in the interval* $[\alpha, \beta]$ *and* N_2 *is the number of jumps of the function from* $+\infty$ *to* $-\infty$ *in the same interval.*

The definition of the Cauchy index can be extended to SISO systems with transfer function $F(s)$ when $s = \sigma$, with $\sigma \in \mathbb{R}$. When the interval $(-\infty, \infty)$ is considered, the Cauchy index I_C is given by the number of positive eigenvalues of W_{co} minus the number of negative eigenvalues of W_{co} of a minimal realization of $F(s)$. Also the Cauchy index is a system invariant as follows from its definition.

Example 6.1

Consider the system with transfer function $F(s) = \frac{1}{s^2+s+1}$. Consider now $s = \sigma$ with $\sigma \in \mathbb{R}$ and $F(\sigma) = \frac{1}{\sigma^2+\sigma+1}$. The plot of $F(\sigma)$ vs. σ is shown in Figure 6.3, from which it is immediate to derive that $I_c = 0$.
Moreover, a minimal realization of $F(s)$ is:

$$A = \begin{bmatrix} 0 & 1 \\ -1 & -1 \end{bmatrix}; B = \begin{bmatrix} 0 \\ 1 \end{bmatrix}; C = \begin{bmatrix} 1 & 0 \end{bmatrix}$$

From this realization we calculate the cross-gramian W_{co}:

$$W_{co} = \begin{bmatrix} 0.5 & 0.5 \\ 0.5 & 0 \end{bmatrix}$$

Since its eigenvalues are $\lambda_1 = 0.8090$ and $\lambda_2 = -0.3090$, one gets $I_C = 0$.

The Cauchy index of a symmetric system can be also calculated from the signature matrix. It is given by the number of positive elements n_+ minus the negative elements n_- in the diagonal of the signature matrix, that is, $I_C = n_+ - n_-$.

The Cauchy index is also linked to the difference between positive and

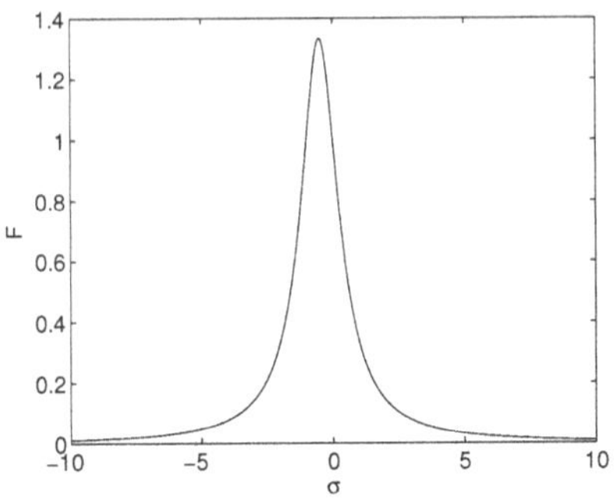

FIGURE 6.3
Plot of $F(\sigma) = \frac{1}{\sigma^2+\sigma+1}$ vs. σ.

negative residues. In fact, if you consider a stable symmetric SISO system in Jordan form with distinct and real eigenvalues, that is with $\mathrm{A} = \mathrm{diag}(\lambda_1, \lambda_2, \ldots, \lambda_n)$, $\mathrm{B}^T = \begin{bmatrix} 1 & 1 & \cdots & 1 \end{bmatrix}$ and $\mathrm{C} = \begin{bmatrix} R_1 & R_2 & \cdots & R_n \end{bmatrix}$, then

$$G(s) = \frac{R_1}{s-\lambda_1} + \frac{R_2}{s-\lambda_2} + \cdots + \frac{R_n}{s-\lambda_n}$$

and the Cauchy index is actually equal to the difference between the number of positive and negative residues.

When the Cauchy index equals the system order, that is, when $\mathrm{S} = \mathrm{I}$, all the residues are positive. Systems with this property are called *relaxation systems*, for the following reason. The impulse response $y(t) = R_1 e^{-\lambda_1 t} + R_2 e^{-\lambda_2 t} + \cdots + R_n e^{-\lambda_n t}$ of a relaxation system is given by the sum of all positive terms which tend monotonically to zero, that is they relax toward zero. These systems have an exact electrical analogue in circuits which are series of elementary impedances, each formed by a resistor and a capacitor in parallel as in Figure 6.4. The generic block formed by a resistor R_i and a capacitor C_i in parallel has an impedance $Z_i(s) = \frac{R_i}{sC_iR_i+1}$, whereas the impedance of the entire circuit is given by the sum of the impedances of the single blocks: $Z(s) = Z_1(s) + Z_2(s) + \cdots + Z_n(s)$.

From Theorem 11, it follows that relaxation systems have a balanced form of type $\mathrm{A} = \mathrm{A}^T$ and $\mathrm{B} = \mathrm{C}^T$.

Another property of the cross-gramian W_{co} is that it is a diagonal matrix in the reference system where the system is balanced. If $\mathrm{S} = \mathrm{I}$ (so $\mathrm{A} = \mathrm{A}^T$ and $\mathrm{B} = \mathrm{C}^T$), this follows from the fact that equation $\mathrm{AW}_{co} + \mathrm{W}_{co}\mathrm{A} = -\mathrm{BC}$ equals $\mathrm{AW}_c^2 + \mathrm{W}_c^2\mathrm{A}^T = -\mathrm{BB}^T$ and $\mathrm{A}^T\mathrm{W}_o^2 + \mathrm{W}_o^2\mathrm{A} = -\mathrm{C}^T\mathrm{C}$, whereas in the more general case of $\mathrm{S} \neq \mathrm{I}$ this can be demonstrated by considering that in balanced form $\mathrm{W}_{co}^2 = \mathrm{W}_c^2\mathrm{W}_o^2 = \Sigma^2$.

From such considerations, a simplified procedure to obtain the balanced form for a symmetric system can be derived. Obviously, the procedure in

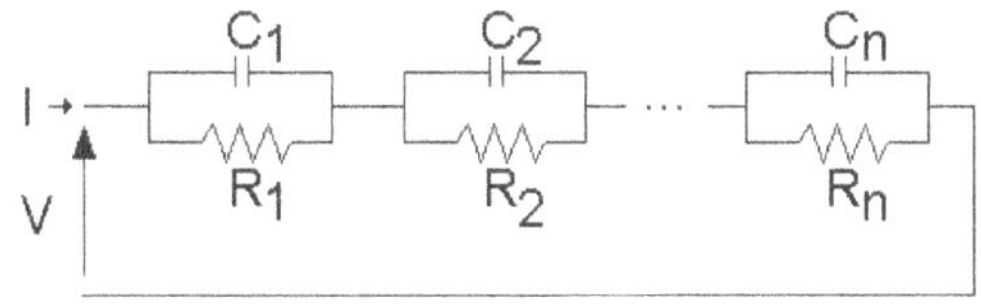

FIGURE 6.4
A relaxation circuit.

Chapter 5 can be applied, but for symmetric systems you can start by diagonalizing the cross-gramian W_{co}, and ordering the eigenvalues in decreasing order with respect to their absolute value. Thus, a reference system is obtained in which W_{co} is diagonal and in which the controllability and observability gramians are diagonal but not equal. At this point, proceed exactly as for Example 5.2, to obtain the balanced form by rescaling the state variables.

MATLAB® Exercise 6.6

Consider system $G(s) = \frac{s+2}{s^2+3s+5}$.
First, let us define the system:
```
>> s=tf('s')
>> G=(s+2)/(s^2+3*s+5)
```
and consider the canonical controllability form of the system
```
>> A=[0 1; -5 -3];
>> B=[0; 1];
>> C=[2 1];
>> system=ss(A,B,C,0)
```
Let us calculate the cross-gramian from Lyapunov equation (6.26)
```
>> Wco=lyap(A,A,B*C)
```
and calculate the singular values of the system
```
>> eig(Wco)
```
the singular values of the system are $\sigma_1 = 0.2633$ and $\sigma_2 = 0.0633$.
It can be proven that the same result is obtained by calculating the singular values from the product of controllability and observability gramians:
```
>> Wc2=gram(system,'c')
>> Wo2=gram(system,'o')
>> sqrt(eig(Wc2*Wo2))
```
To calculate the balanced form, let us diagonalize the cross-gramian:
```
>> [T,L]=eig(Wco)
```
Matrix T is such that `inv(T)*Wco*T` is diagonal. Let us consider the system obtained applying the state transformation $\tilde{\mathbf{x}} = \mathrm{T}^{-1}\mathbf{x}$:
```
>> Atilde=inv(T)*A*T
>> Btilde=inv(T)*B
>> Ctilde=C*T
```
In this reference system, the cross-gramian is diagonal:
```
>> Wcotilde=lyap(Atilde,Atilde,Btilde*Ctilde)
```
The gramians are also diagonal:
```
>> systemtilde=ss(Atilde,Btilde,Ctilde,0)
>> Wc2tilde=gram(systemtilde,'c')
>> Wo2tilde=gram(systemtilde,'o')
```

Note that this property derives from the fact that the matrix T is diagonal:

```
>> Tsim=obsv(systemtilde)'*inv(ctrb(systemtilde))
```

So, now we can proceed as in Example 5.2. However, it should be noted that since MATLAB® diagonalization procedure does not sequence the eigenvalues, the order of the two state variables must first be inverted using the transformation $\bar{x}_1 = \tilde{x}_2$ and $\bar{x}_2 = \tilde{x}_1$, or via the matrix:

```
>> Tbis=[0 1; 1 0];
>> Atilde=inv(Tbis)*Atilde*Tbis
>> Btilde=inv(Tbis)*Btilde
>> Ctilde=Ctilde*Tbis
```

6.3.5 Open-loop Parameterization

The main issue with the open-loop parameterization is determining which systems admit a certain set of assigned singular values. In the particular case of symmetric SISO systems the issue is finding a system with an assigned signature matrix $S = \mathrm{diag}(s_1, s_2, \dots, s_n)$, a certain set of assigned singular values $\Sigma = \mathrm{diag}(\sigma_1, \sigma_2, \dots, \sigma_n)$ and assigned b_i coefficients of matrix $B^T = \begin{bmatrix} b_1 & b_2 & \cdots & b_n \end{bmatrix}$. Below, we will always assume that assigning singular values also means assigning a certain signature matrix.

Once the coefficients b_i with $i = 1, \dots, n$ and the signature matrix S are known, the matrix C is obtained from $C = B^T S$. As regards the matrix A, the coefficients a_{ij} can be obtained from the following formula:

$$a_{ij} = -\left\{\frac{\sigma_i s_i s_j - \sigma_j}{\sigma_i^2 - \sigma_j^2}\right\} * BB^T \tag{6.29}$$

where the symbol $*$ is the Hadamard product, i.e., the component by component product of two square matrices, $F * G = \{f_{ij} g_{ij}\}$.

In particular, the diagonal terms of matrix A are given by the formula:

$$a_{ii} = -\frac{1}{2\sigma_i} b_i^2$$

Example 6.2

Construct a system with signature matrix $S = \begin{bmatrix} 1 & 0 & 0 \\ 0 & 1 & 0 \\ 0 & 0 & -1 \end{bmatrix}$, singular values σ_1, σ_2 and σ_3 and matrix $B = \begin{bmatrix} b_1 \\ b_2 \\ b_3 \end{bmatrix}$.

By applying the parameterization formula (6.29), we obtain: $a_{11} = -\frac{1}{2\sigma_1} b_1^2$, $a_{12} = -\frac{b_1 b_2}{\sigma_1 + \sigma_2}$, $a_{13} = -\frac{b_1 b_3}{\sigma_1 - \sigma_3}$, $a_{21} = -\frac{b_1 b_2}{\sigma_1 + \sigma_2}$, $a_{22} = -\frac{1}{2\sigma_2} b_2^2$, $a_{23} = -\frac{b_2 b_3}{\sigma_2 - \sigma_3}$, $a_{31} = -\frac{b_1 b_3}{\sigma_3 - \sigma_1}$, $a_{32} = -\frac{b_2 b_3}{\sigma_3 - \sigma_2}$, $a_{33} = -\frac{1}{2\sigma_3} b_3^2$.

Formula (6.29) can be obtained (to simplify, consider the SISO case) from the equations of the gramians expressed in the open-loop balanced realization: $W_o^2 = W_c^2 = \Sigma = \mathrm{diag}(\sigma_1, \sigma_2, \dots, \sigma_n)$:

$$\begin{aligned} A^T\Sigma + \Sigma A &= -C^T C \\ A\Sigma + \Sigma A^T &= -BB^T \end{aligned} \tag{6.30}$$

Summing the two matrix equations (6.30), we obtain

$$(A^T + A)\Sigma + \Sigma(A + A^T) = -C^T C - BB^T$$

and so

$$(A^T + A)\Sigma + \Sigma(A + A^T) = -SBB^T S - BB^T \tag{6.31}$$

The generic diagonal term of this expression is given by:

$$4a_{ii}\sigma_i = -2b_i^2$$

from which we obtain the expression which is valid for the diagonal terms of the parameterization (6.29).

Subtracting the second of (6.30) from the first:

$$(A^T - A)\Sigma + \Sigma(A - A^T) = -C^T C + BB^T$$

and so

$$(A^T - A)\Sigma + \Sigma(A - A^T) = -SBB^T S + BB^T \tag{6.32}$$

The diagonal terms of this expression are zero. The general terms a_{ij} of the parameterization (6.29) can be obtained from (6.31) and (6.32).

From equation (6.32), let us obtain the generic ij term:

$$(a_{ji} - a_{ij})\sigma_j + \sigma_i(a_{ij} - a_{ji}) = -s_i b_i b_j s_j + b_i b_j \tag{6.33}$$

whereas, from (6.31), we obtain:

$$(a_{ji} + a_{ij})\sigma_j + \sigma_i(a_{ij} + a_{ji}) = -s_i b_i b_j s_j - b_i b_j \tag{6.34}$$

Equations (6.33) and (6.34) represent a system with two equations and two unknowns, a_{ij} and a_{ji}, which can be found by substitution obtaining a_{ji} from the relation which is the sum of the two prior relations (that is considering the generic term ji of the first equation (6.30)):

$$a_{ji} = \frac{-s_i s_j b_i b_j - \sigma_i a_{ij}}{\sigma_j} \tag{6.35}$$

Substituting in (6.33), we get:

$$-s_i s_j b_i b_j - \sigma_i a_{ij} - \sigma_j a_{ij} + \frac{\sigma_i}{\sigma_j}(s_i s_j b_i b_j + \sigma_i a_{ij}) + \sigma_i a_{ij} = -s_i s_j b_i b_j + b_i b_j \tag{6.36}$$

from which we obtain $a_{ij} = -\{\frac{\sigma_i s_i s_j - \sigma_j}{\sigma_i^2 - \sigma_j^2}\} * BB^T$.

Note that specifying a SISO system of order n through its transfer function $G(s)$ requires $2n$ parameters. Even for the open-loop parameterization the number of assigned parameters is $2n$: n are singular values and n are the coefficients of the matrix B.

If matrix B is not specified, there is an infinite number of systems with matrix S and assigned singular values. To fix the system unequivocally, other n parameters must be specified, but they cannot just be any parameters. For example, it would not be possible to know a priori whether it is possible to find a system with n singular values and n assigned eigenvalues: such a problem could have a solution, several solutions or none at all. In fact, in the equations which could provide a solution (the unknowns are b_1, b_2, ..., b_n) we find nonlinear terms. By contrast, the solution to a similar problem has the advantage of being a system with known stability properties (including the stability margins) and controllability and observability levels.

Instead, notice that finding a system with S = I and complex eigenvalues has no solution, as $A = A^T$ is symmetric and has real eigenvalues.

6.3.6 Relation Between the Cauchy Index and the Hankel Matrix

Cauchy index I_c correlates with the properties of the Hankel matrix H. Recall that Hankel matrix is defined as $H = M_o M_c$ and is symmetric by construction.

Suppose, we have a SISO system with distinct eigenvalues and with this transfer function:

$$G(s) = \frac{\alpha_1}{s+\lambda_1} + \frac{\alpha_2}{s+\lambda_2} + \cdots + \frac{\alpha_n}{s+\lambda_n}$$

In this case, the Cauchy index can be calculated by counting the number of positive and negative residues: $I_c = N_{pos.res.} - N_{neg.res.}$. By highlighting the residue signs, each residue can be written as $\alpha_i = s_i|\alpha_i|$. Notice how this system can be associated with a Jordan form of type:

$$A = \begin{bmatrix} -\lambda_1 & & & \\ & -\lambda_2 & & \\ & & \ddots & \\ & & & -\lambda_n \end{bmatrix}; B = \begin{bmatrix} \sqrt{|\alpha_1|} \\ \sqrt{|\alpha_2|} \\ \vdots \\ \sqrt{|\alpha_n|} \end{bmatrix}; C^T = \begin{bmatrix} s_1\sqrt{|\alpha_1|} \\ s_2\sqrt{|\alpha_2|} \\ \vdots \\ s_n\sqrt{|\alpha_n|} \end{bmatrix}$$

Since Hankel matrix H is:

$$H = \begin{bmatrix} C \\ CA \\ \vdots \\ CA^{n-1} \end{bmatrix} \begin{bmatrix} B & AB & \cdots A^{n-1}B \end{bmatrix} =$$

$$= \begin{bmatrix} \mathrm{B}^T\mathrm{S} \\ \mathrm{B}^T\mathrm{SA} \\ \vdots \\ \mathrm{B}^T\mathrm{SA}^{n-1} \end{bmatrix} \begin{bmatrix} \mathrm{B} & \mathrm{AB} & \cdots \mathrm{A}^{n-1}\mathrm{B} \end{bmatrix}$$

Given that A is diagonal then:

$$\mathrm{H} = \begin{bmatrix} \mathrm{B}^T\mathrm{S} \\ \mathrm{B}^T\mathrm{A}^T\mathrm{S} \\ \vdots \\ \mathrm{B}^T(\mathrm{A}^\mathrm{T})^{n-1}\mathrm{S} \end{bmatrix} \begin{bmatrix} \mathrm{B} & \mathrm{AB} & \cdots \mathrm{A}^{n-1}\mathrm{B} \end{bmatrix} = \mathrm{M}_c^T\mathrm{SM}_c$$

The Hankel matrix H is symmetric. Whether it is positive definite or not depends exclusively on the signature matrix S. The signs of the eigenvalues (which are real) of the Hankel matrix reflect the number of positive and negative elements in the signature matrix. Furthermore, the Cauchy index can be calculated by counting the positive and negative eigenvalues in the Hankel matrix:

$$I_c = N_{\lambda_+(\mathrm{H})} - N_{\lambda_-(\mathrm{H})}$$

The Hankel matrix then helps understand the residue signs and the structure of the open-loop balanced form.

As we have already seen, for symmetric systems $\mathrm{W}_{co}^2 = \mathrm{W}_c^2\mathrm{W}_o^2$ can be defined.

In this case, the hypothesis of asymptotic stability, usually needed for non-symmetric systems to perform the open-loop balancing, can be removed. Here, however, it is no longer guaranteed that the eigenvalues of W_{co}^2 are real and positive. This condition in asymptotically stable minimal systems is assured by the fact that W_c^2 and W_o^2 matrices are positive definite. For symmetric systems without the hypothesis of asymptotic stability, W_{co}^2 can even have negative or complex and conjugated eigenvalues. When, having calculated W_{co}^2, real and positive eigenvalues are found, then the singular values of the system can be defined as $\sigma_i = \sqrt{\lambda_i(\mathrm{W}_{co}^2)}$.

It can be demonstrated that if the Hankel matrix is positive definite (i.e., if $I_c = n$ or $\mathrm{S} = \mathrm{I}$), then its eigenvalues $\lambda_i(\mathrm{W}_{co}^2)$ are real and positive. Therefore, for non-asymptotically stable symmetric systems, with a Hankel matrix positive definite, an open-loop balanced form can be defined. Clearly, the case where matrix A has purely imaginary eigenvalues should a priori be excluded since there is no solution to the Lyapunov equations and so a balanced form cannot be written.

6.3.7 Singular Values for a FIR Filter

Let us consider now discrete-time SISO systems. The transfer function $G(z) = \frac{Y(z)}{U(z)}$, as it is known, is the ratio of two polynomials in z. Two cases exist. If

$G(z)$ can be written as $G(z) = h_1 z^{-1} + h_2 z^{-2} + \cdots + h_n z^{-n}$, that is as the sum of a finite number of terms of type $\frac{h_i}{z^i}$, then the system is said to be FIR (finite impulse response). Otherwise, if $G(z) = \frac{a_1 z^{-1} + a_2 z^{-2} + \cdots + a_{n-1} z^{-n+1}}{1 + b_1 z^{-1} + b_2 z^{-2} + \cdots + b_n z^{-n}}$, then the system is IIR (infinite impulse response).

FIR filters are distinguished by having all their poles at the origin so their transfer function is $G(z) = \frac{p(z)}{z^n}$ where $p(z)$ is an $n-1$ order polynomial. Therefore, FIR systems are asymptotically stable.

Furthermore, their name derives from the fact that they have an impulse response which cancels itself out after a finite number (n) of samples. In fact, the output of a FIR system can be calculated from its transfer function:

$$Y(z) = G(z)U(z) = (h_1 z^{-1} + h_2 z^{-2} + \cdots + h_n z^{-n})U(z)$$

which, by applying the inverse z-transform, gives

$$y(k) = h_1 u(k-1) + h_2 u(k-2) + \cdots + h_n u(k-n) \tag{6.37}$$

As one can see from the formula (6.37), if an impulse input is applied ($U(z) = 1$, i.e., $u(0) = 1$ and $u(k) = 0 \ \forall k \neq 0$), then the output will vanish at sample $k = n+1$, that is, after n steps. The formula (6.37) clarifies another important property of this class of systems: FIR systems have finite memory. In fact, the output depends entirely on input regression so it is unnecessary to know the system state to calculate the output at time k. This does not happen in IIR systems where the output can be expressed by a model which includes regressions of the output itself:

$$y(k) = -b_1 y(k-1) - b_2 y(k-2) - \cdots - b_n y(k-n) + a_1 u(k-1) + \cdots + a_n u(k-n)$$

Therefore, to calculate the value of the output at time k in IIR systems, n prior samples of the output itself need to be memorized.

FIR filters do not bring about phase distortion. In continuous-time systems, as long as there is no phase distortion, the propagation time must be the same for all frequencies, so the phase must linearly decrease with ω (so, the phase Bode diagram has to be a straight line with negative slope). With all their poles at the origin, FIR systems show in the discrete-time domain an analogous behavior.

The fundamental parameters of a FIR filter are $h_1, h_2, \ldots, h_n$, which can be found in the Hankel matrix which has a finite number of non-zero coefficients:

$$\mathrm{H} = \begin{bmatrix} h_1 & h_2 & \cdots & h_{n-1} & h_n \\ h_2 & h_3 & \cdots & h_n & 0 \\ h_3 & h_4 & \cdots & 0 & 0 \\ \vdots & & & & \vdots \\ h_n & 0 & \cdots & 0 & 0 \end{bmatrix}$$

Example 6.3

The FIR filter $G(z) = z^{-1} + 5z^{-2}$ has Hankel matrix

$$H = \begin{bmatrix} 1 & 5 \\ 5 & 0 \end{bmatrix}$$

Example 6.4

The FIR filter $G(z) = z^{-1} + 5z^{-2} + 3z^{-3}$ has Hankel matrix

$$H = \begin{bmatrix} 1 & 5 & 3 \\ 5 & 3 & 0 \\ 3 & 0 & 0 \end{bmatrix}$$

To calculate the balanced form for a discrete-time system the bilinear transformation $z = \frac{1+s}{1-s}$ is applied to obtain an equivalent continuous-time system. The theory valid for continuous-time systems is then applied to the equivalent system.

For FIR filters the procedure can be simplified. The following theorem can give the singular values for a FIR filter without recourse to the equivalent system.

Theorem 12 *The singular values of a FIR filter are given by the eigenvalues in absolute value of the Hankel matrix* H.

Since the Hankel matrix H for a FIR filter is symmetric, its eigenvalues are always real and their absolute value can be always computed.

Example 6.5

System $G(z) = h_2 z^{-2}$ is a second order FIR filter whose Hankel matrix is

$$H = \begin{bmatrix} 0 & h_2 \\ h_2 & 0 \end{bmatrix}$$

The characteristic polynomial of matrix H is $p(\lambda) = \lambda^2 - h_2^2$, from which we obtain $\sigma_1 = \sigma_2 = |h_2|$.

Example 6.6

Now let us calculate the singular values of a FIR filter with transfer function $G(z) = h_3 z^{-3}$. The Hankel matrix of this system is:

$$H = \begin{bmatrix} 0 & 0 & h_3 \\ 0 & h_3 & 0 \\ h_3 & 0 & 0 \end{bmatrix}$$

and singular values of the system are $\sigma_1 = \sigma_2 = \sigma_3 = |h_3|$.

After having shown several examples on singular values of a FIR filter, we will now demonstrate Theorem 12, which requires using the property that, given two matrices A and B, the eigenvalues of their product AB are equal to those of matrix BA: $\lambda_i(AB) = \lambda_i(BA)$. In fact:

$$\det(\lambda I - AB) = \det(A(A^{-1}\lambda I - B)) = \det(A)\det(A^{-1}\lambda I - B) =$$

$$= \det(\mathrm{A}^{-1}\lambda\mathrm{I} - \mathrm{B})\det(\mathrm{A}) = \det((\mathrm{A}^{-1}\lambda\mathrm{I} - \mathrm{B})\mathrm{A}) = \det(\lambda\mathrm{I} - \mathrm{BA})$$

On this premise, let us consider the gramians of a FIR filter. The generic expression of gramians in discrete-time systems, reported in equations (5.18) and (5.19), for FIR filters becomes a finite sum:

$$\mathrm{W}_c^2 = \sum_{i=0}^{n-1} \mathrm{A}^i\mathrm{BB}^T(\mathrm{A}^T)^i \tag{6.38}$$

$$\mathrm{W}_o^2 = \sum_{i=0}^{n-1} (\mathrm{A}^T)^i\mathrm{C}^T\mathrm{CA}^i \tag{6.39}$$

and therefore,

$$\mathrm{W}_c^2 = \mathrm{BB}^T + \mathrm{ABB}^T\mathrm{A}^T + \cdots + \mathrm{A}^{n-1}\mathrm{BB}^T(\mathrm{A}^T)^{n-1} = \mathrm{M}_c\mathrm{M}_c^T \tag{6.40}$$

$$\mathrm{W}_o^2 = \mathrm{C}^T\mathrm{C} + \mathrm{A}^T\mathrm{C}^T\mathrm{CA} + \cdots + (\mathrm{A}^T)^{n-1}\mathrm{C}^T\mathrm{CA}^{n-1} = \mathrm{M}_o^T\mathrm{M}_o \tag{6.41}$$

which yields that

$$\lambda_i(\mathrm{W}_c^2\mathrm{W}_o^2) = \lambda_i(\mathrm{M}_c\mathrm{M}_c^T\mathrm{M}_o^T\mathrm{M}_o)$$

But, for the eigenvalue property of the product matrix:

$$\lambda_i(\mathrm{W}_c^2\mathrm{W}_o^2) = \lambda_i(\mathrm{M}_c^T\mathrm{M}_o^T\mathrm{M}_o\mathrm{M}_c)$$

and therefore, given that $\mathrm{H} = \mathrm{M}_o\mathrm{M}_c$, then

$$\lambda_i(\mathrm{W}_c^2\mathrm{W}_o^2) = \lambda_i(\mathrm{H}^T\mathrm{H})$$

from which it is clear that the singular values of a FIR filter are given by the eigenvalues in absolute value of the Hankel matrix.

6.3.8 Singular Values of All-pass Systems

Let us consider a continuous-time system with transfer function $G(s) = k\frac{(5-s)^3}{(5+s)^3}$. The frequency response $G(j\omega)$ of this system has a very particular behavior. In fact, $\forall\omega\ |G(j\omega)| = 1$, and the magnitude Bode diagram of this system is flat.

All systems of this type are all-pass systems because the only difference between the input and output signals is the phase, whereas the input amplitude is neither attenuated nor amplified. Consider now inner systems, which are all-pass systems that are stable. These systems are stable (all their poles are in the closed left half plane), not-strictly proper ($G(s) = D + \overline{G}(s)$), and not minimum phase (each pole has a zero symmetric with respect to the imaginary axis and so with positive real part).

The singular values for these systems can be easily calculated.

Theorem 13 *The singular values for a stable all-pass system* $G(s) = k\frac{(a-s)^n}{(a+s)^n}$ *with* $k > 0$ *and* $a > 0$ *are* $\sigma_1 = \sigma_2 = \cdots = \sigma_n = k$.

This result derives from the fact that a bilinear transformation as well as transformations of type $\overline{s} = \frac{s}{a}$ does not change the singular values of a system. Bearing in mind these two considerations, and applying to $G(z) = kz^{-n}$ (whose singular values are $\sigma_1 = \cdots = \sigma_n = k$) at first the bilinear transformation and then a transformation of type $\overline{s} = \frac{s}{a}$, the result expressed by the theorem is obtained.

The importance of this result is that reduced order modeling cannot be applied to an all-pass system since all its singular values are equal.

MATLAB® Exercise 6.7

Consider the system with transfer function:

$$G(s) = \frac{25s^4 - 352.5s^3 + 1830s^2 - 4120s + 3360}{s^4 + 14.1s^3 + 73.2s^2 + 164.8s + 134.4} \tag{6.42}$$

Let us define the system in MATLAB:

```
>> s=tf('s')
>> G=(25*s^4-352.5*s^3+1830*s^2-4120*s+3360)/...
          ...(s^4+14.1*s^3+73.2*s^2+164.8*s+134.4)
```

The transfer function can be factorized as follows

```
>> zpk(G)
```

to obtain:

$$G(s) = 25\frac{(4-s)^3(2.1-s)}{(4+s)^3(2.1+s)} \tag{6.43}$$

which clearly shows that the system is all-pass (and stable) with static gain equal to $k = 25$.

Calculating the singular values of the system with the command:

```
>> [systembal,S]=balreal(G)
```

one obtains: $\sigma_1 = \sigma_2 = \sigma_3 = \sigma_4 = 25$.

6.4 Exercises

1. Given the system with transfer function

$$G(s) = 1 - \frac{56s^6 + 2676s^4 + 8736s^2 + 1600}{(s+10)^2(s+2)^3(s+1)^2}$$

determine an open-loop balanced realization and a suitable reduced order model.

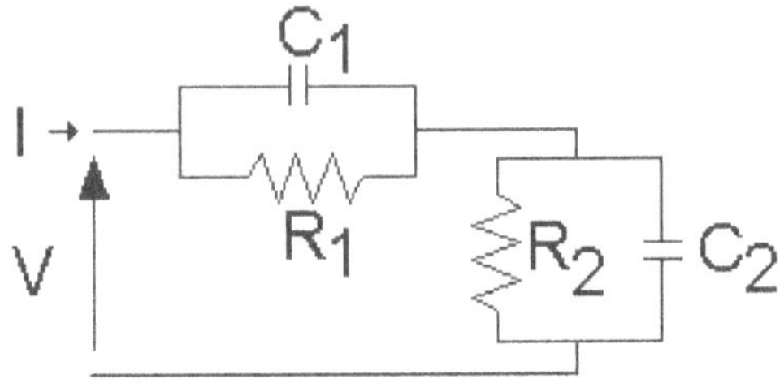

FIGURE 6.5
Circuit for Exercise 8.

2. Calculate a reduced order model of the system $G(s) = 200\frac{(s+10)(s^2+s+1)}{(s+5)^3(s+4)(s+2)^2}$.
3. Given the system $G(s) = \frac{s+1}{s^4+4.3s^3+7.92s^2+7.24s+2.64}$ calculate the reduced order models with direct truncation and singular perturbation approximation and compare the models obtained.
4. Given the continuous-time system with state-space matrices:

$$A = \begin{bmatrix} -2 & 1 & 1 & 0 \\ -3 & -5 & 6 & 1 \\ 0 & -1 & -5 & 0 \\ -4 & -5 & -7 & -1 \end{bmatrix}; \quad B = \begin{bmatrix} 0 \\ 1 \\ 0 \\ 1 \end{bmatrix}; \quad C = \begin{bmatrix} 1 & 2 & 1 & 2 \end{bmatrix}$$

   calculate the reduced order models with direct truncation and singular perturbation approximation and the error and compare the rise and settling time of the reduced order models and original system.

5. Given system $G(s) = \frac{s}{s^4+3s^3+4.1s^2+4.1s+0.2}$, choose the order of the reduced model to guarantee an error between nominal model and approximation not larger than 0.02.
6. Determine, if possible, the system with eigenvalues $\lambda_1 = -1$, $\lambda_2 = -2$ and with singular values $\sigma_1 = 5$ and $\sigma_2 = 2$.
7. Calculate analytically the Cauchy index of the system $G(s) = \frac{s+1}{s(s^2+s+1)}$.
8. Given the system in Figure 6.5 with $R_1 = R_2 = 1$, $C_1 = 1$ and $C_2 = \frac{1}{2}$ determine the transfer function and the Cauchy index.
9. Write down an example of a relaxation system and verify the value of the Cauchy index.

10. Calculate the singular values of system $G(z) = \frac{3+2z+5z^2+6z^3}{z^4}$.
11. Consider an open-loop balanced representation of a discrete-time system. Derive a model of order $r < n$ and verify that its singular values do not coincide with the first r singular values of the original system.

7

Variational Calculus and Linear Quadratic Optimal Control

CONTENTS

This chapter discusses the main concepts of variational theory and then optimal control, a technique for synthesizing a linear quadratic regulator (LQR) which is able to determine the closed-loop eigenvalues on the basis of an optimized criteria. Several problems regarding the variational calculus are first presented with emphasis on the formulation of general optimization problems. The second part of the chapter is devoted to optimal control, first presented in the context of variational theory and then studied in relation to the practical aspects related to the definition of suitable performance indices. The optimal gains are defined and a classical procedure to obtain them is presented. The nonlinear matrix algebraic Riccati equation is discussed. Several methods to find its positive definite solution are discussed. The dual problem of the optimal observer is also dealt with in the chapter. The LQR problem in the frequency domain is also discussed. In order to approach the dual problems of the optimal controller and the optimal observer, Hamiltonian matrices are introduced.

DOI: 10.1201/9781003196921-7

7.1 Variational Calculus: An Introduction

The problem faced in optimal control is to find a control law that guarantees optimal performance. To quantify the performance of a control law, one often resorts to consider an integral performance index of this type:

$$PI = \int_{t_i}^{t_f} L(x, u, t)dt \tag{7.1}$$

where t_i and t_f define the time interval, namely $[t_i, t_f]$, over which control is evaluated. Solving an optimal control problem, therefore, means finding the function u that minimizes PI. This can be viewed as a particular instance of a more general problem which is the subject of variational calculus.

In its general formulation, variational calculus is concerned with the solution of the problem of determining the maximum or the minimum of a given functional, that is a mathematical entity establishing a mapping between functions of a certain class and real numbers. In particular, let us consider the problem of finding the function $x(t)$ which minimizes the following functional

$$V(x) = \int_a^b L(x, \dot{x}, t)dt \tag{7.2}$$

where the initial and final time for the integration have been indicated as a and b, respectively, and the function $x(t)$ takes the following values at the extremes: $x(a) = A$ and $x(b) = B$.

A very important result established by variational calculus is that, if a continuous differentiable function $x(t)$ minimizes a functional $V(x)$ of the form (7.2), then it satisfies the so-called *Euler equation*:

$$\frac{\partial L(x, \dot{x}, t)}{\partial x} - \frac{d}{dt}\frac{\partial L(x, \dot{x}, t)}{\partial \dot{x}} = 0 \tag{7.3}$$

Notice that the Euler equation is a necessary condition for the solution, and, in general, not a sufficient one. However, when, for a specific engineering problem, physical considerations indicate the existence of a unique solution, then this can be obtained by the variational problem.

Consider now the case that $\mathbf{x}$ is a vector of n components, then the problem of variational calculus can be generalized to minimizing the functional

$$V(\mathbf{x}) = \int_a^b L(\mathbf{x}, \dot{\mathbf{x}}, t)dt \tag{7.4}$$

with $\mathbf{x}(a) = \mathbf{x}_a$ and $\mathbf{x}(b) = \mathbf{x}_b$. Also in this more general case, if the function minimizes the functional, then it satisfies the Euler equation, which now reads:

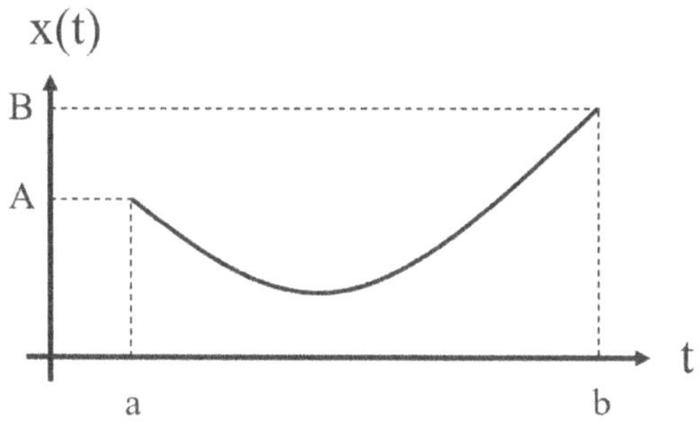

FIGURE 7.1
A simple example of the application of variational calculus to find the curve, with minimum length, joining the point (a, A) to (b, B).

$$\frac{\partial L(\mathbf{x}, \dot{\mathbf{x}}, t)}{\partial x_i} - \frac{d}{dt}\frac{\partial L(\mathbf{x}, \dot{\mathbf{x}}, t)}{\partial \dot{x}_i} = 0 \tag{7.5}$$

with $i = 1, \ldots, n$.

Example 7.1

Let us consider the following simple example. What is the function $x(t)$ that minimizes the path between the points A at time a and B at time b in Figure 7.1?
The answer is clearly a straight line going from (a, A) to (b, B), but now we prove this using variational calculus.
The length of the curve can be calculated as follows:

$$V(x) = \int_l dl \tag{7.6}$$

where, taking into account the coordinates considered, dl is given by:

$$dl = \sqrt{(dt)^2 + (dx)^2} = dt\sqrt{1 + \dot{x}^2} \tag{7.7}$$

with $\dot{x} = \frac{dx}{dt}$.
Hence, one can consider the problem of minimizing the following functional:

$$V(x) = \int_a^b \sqrt{1 + \dot{x}^2}dt \tag{7.8}$$

that corresponds to the general form (7.2) if

$$L(x, \dot{x}, t) = \sqrt{1 + \dot{x}^2} \tag{7.9}$$

Since $\frac{\partial L(x,\dot{x},t)}{\partial x} = 0$, the Euler equation (7.10) becomes

$$\frac{d}{dt}\frac{\partial L(x, \dot{x}, t)}{\partial \dot{x}} = 0 \tag{7.10}$$

and thus

$$\frac{\partial L(x, \dot{x}, t)}{\partial \dot{x}} = const. \tag{7.11}$$

It follows that:

$$\frac{\partial\sqrt{1+\dot{x}^2}}{\partial \dot{x}} = \frac{\dot{x}}{\sqrt{1+\dot{x}^2}} = const. \tag{7.12}$$

The solution of the differential equation (7.12) is given by:

$$x(t) = k_1 t + k_2 \tag{7.13}$$

where the values of k_1 and k_2 are determined by taking into account the boundary conditions $x(a) = A$ and $x(b) = B$.

Example 7.2

Consider now the problem of finding the function $\mathbf{x}(t)$ that minimizes the following functional:

$$V(x) = \int_0^{\frac{\pi}{2}} \left(\dot{x}_1^2 + \dot{x}_2^2 + 2x_1x_2\right) dt \tag{7.14}$$

with $x_1(0) = x_2(0) = 0$, $x_1(\frac{\pi}{2}) = 1$, and $x_1(\frac{\pi}{2}) = -1$.
To solve this problem, the vectorial form of the Euler equation (7.5) should be used. Since $\frac{\partial L}{\partial x_1} = 2x_2$, $\frac{\partial L}{\partial \dot{x}_1} = 2\dot{x}_1$, $\frac{\partial L}{\partial x_2} = 2x_1$, and $\frac{\partial L}{\partial \dot{x}_2} = 2\dot{x}_2$, the vectorial Euler equation becomes:

$$\begin{aligned} 2x_2 - 2\ddot{x}_1 = 0 \\ 2x_1 - 2\ddot{x}_2 = 0 \end{aligned} \tag{7.15}$$

that can be rewritten as:

$$\begin{aligned} \ddot{x}_1 = x_2 \\ \ddot{x}_2 = x_1 \end{aligned} \tag{7.16}$$

Solving (7.16) is equivalent to solve the following differential equation:

$$\frac{d^{(4)}x_1}{dt^4} = x_1 \tag{7.17}$$

whose solution is given by:

$$x_1(t) = c_1e^t + c_2e^{-t} + c_3 \sin t + c_4 \cos t \tag{7.18}$$

The constants c_1, c_2, c_3 and c_4 are determined by considering the given boundary conditions.

7.2 The Lagrange Method

The Lagrange method provides a systematic way to determine the differential equations governing a system. It makes use of the Lagrangian state function that is defined starting from coordinates obtained generalizing the variables appearing in classical mechanics problems, i.e., position and velocity. For this reason, these variables are called *generalized coordinates*. The Lagrange method demonstrates to be very effective in many engineering problems as the derivation of the system representation is directly linked to study of the

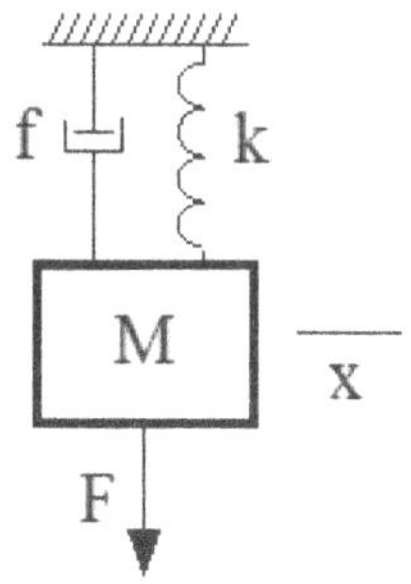

FIGURE 7.2
A mass-spring-damper system.

energy terms involved in its dynamics. In fact, the generalized position vector $\mathbf{x}$ is related to the potential energy, and its time derivative $\dot{\mathbf{x}}$ is linked to the kinematic energy. In mechanical systems these variables effectively represent the position and the velocity of a mass, in an electrical circuit they may represent charge and current associated to a capacitor or flux linkage and voltage associated to an inductor.

To briefly illustrate the method, we begin with the definition of the Lagrangian state function which combines the kynetic energy of the system, $I(\mathbf{x}, \dot{\mathbf{x}})$, and the potential energy, $V(\mathbf{x})$. In more detail, the Lagrangian state function is the difference between kynetic and potential energy:

$$\mathcal{L}(\mathbf{x}, \dot{\mathbf{x}}) = I(\mathbf{x}, \dot{\mathbf{x}}) - V(\mathbf{x}) \tag{7.19}$$

The Lagrangian state function can be used to derive the equations of motion of a dynamical system, which are obtained as follows:

$$\frac{d}{dt}\frac{\partial \mathcal{L}(\mathbf{x}, \dot{\mathbf{x}})}{\partial \dot{x}_i} - \frac{\partial \mathcal{L}(\mathbf{x}, \dot{\mathbf{x}})}{\partial x_i} + \frac{\partial D(\dot{\mathbf{x}})}{\partial \dot{x}_i} = F_i \tag{7.20}$$

where $D(\dot{\mathbf{x}})$ is the dissipation term and F_i is the generalized force.

For instance, in the case of a mechanical system, equation (7.20) correspond to the Newton's equations as shown in the following example.

Example 7.3

Consider the mass-spring-damper system shown in Figure 7.2. For this system the Lagrangian function is

$$\mathcal{L}(x, \dot{x}) = \frac{1}{2}M\dot{x}^2 - \frac{1}{2}kx^2 \tag{7.21}$$

and the dissipation energy is:

$$D(\dot{x}) = \frac{1}{2}f\dot{x}^2 \tag{7.22}$$

Using together these two expressions in equation (7.20) one obtains the Newton's equation for the system:

$$M\ddot{x} + f\dot{x} + kx = F \tag{7.23}$$

If the system is unforced and conservative, that is, the system does not have dissipative elements, then the Lagrange's equations (7.20) become:

$$\frac{d}{dt}\frac{\partial \mathcal{L}(\mathbf{x}, \dot{\mathbf{x}})}{\partial \dot{x}_i} - \frac{\partial \mathcal{L}(\mathbf{x}, \dot{\mathbf{x}})}{\partial x_i} = 0 \tag{7.24}$$

Note that, if we consider $L(\mathbf{x}, \dot{\mathbf{x}}, t) = \mathcal{L}(\mathbf{x}, \dot{\mathbf{x}})$ in the Euler equations (7.5), then we obtain exactly equation (7.24). This important property derives from the Hamilton's principle, which states that the path of motion of a system from $\mathbf{x}(t_i)$ to $\mathbf{x}(t_f)$ is such to minimize the functional:

$$V(\mathbf{x}) = \int_{t_i}^{t_f} \mathcal{L}(\mathbf{x}, \dot{\mathbf{x}})dt \tag{7.25}$$

7.3 Towards Optimal Control

Given a dynamical system described by

$$\dot{\mathbf{x}} = f(\mathbf{x}, \mathbf{u}, t) \tag{7.26}$$

and the following performance index

$$PI = \int_{t_i}^{t_f} L(\mathbf{x}, \mathbf{u}, t)dt \tag{7.27}$$

the optimal control problem consists of finding the optimal control $\mathbf{u}(t)$ or the optimal control law $\mathbf{u}(t) = k(\mathbf{x}(t), t)$ such that the performance index is minimum, which, in turn, indicates that the system performance in the interval $[t_i, t_f]$ is optimal.

Notice that implementing the optimal control $\mathbf{u}(t)$ or the optimal control law $\mathbf{u}(t) = k(\mathbf{x}(t), t)$ requires two different control schemes. Having at disposal the mathematical expression for the optimal control law $\mathbf{u}(t) = k(\mathbf{x}(t), t)$ makes possible the use of a feedback configuration as the dependence of the control law on the state is explicit.

Example 7.4

Consider the following functional:

$$PI = \int_0^t (x^2 + u^2)dt \tag{7.28}$$

with

$$\dot{x} = -x + u \tag{7.29}$$

Minimizing this functional with the constraint (7.29) can be viewed as a control problem where we want to find $u(t) = k(x(t), t)$ that optimizes the performance index (7.28).

In order to be able to apply variational calculus to the problem of optimal control, two issues need to be addressed. The first issue is related to the fact that the performance index for optimal control is of the type $PI = \int_{t_i}^{t_f} L(\mathbf{x}, \mathbf{u}, t)dt$, while, so far, we have discussed functionals of the type $V(\mathbf{x}) = \int_{t_i}^{t_f} L(\mathbf{x}, \dot{\mathbf{x}}, t)dt$. To deal with this issue, a new augmented vector can be considered, i.e., $\mathbf{x}' = [\mathbf{x}^T, \mathbf{u}^T]^T$, and then the mathematical approach discussed in the previous sections can be applied taking into account this new vector. However, for the sake of notation, to highlight the presence of the input vector we will still refer to these two vectors in a separate way.

The second issue concerns how to account for the dynamics of the system. The idea here is to rewrite equation (7.26) as:

$$g(\mathbf{x}, \dot{\mathbf{x}}, \mathbf{u}, t) = f(\mathbf{x}, \mathbf{u}, t) - \dot{\mathbf{x}} = 0 \tag{7.30}$$

For this reason, we now discuss how to solve a problem of variational calculus in presence of constraints. The main result is expressed in the following theorem.

Theorem 14 *Given a functional of the form*

$$V(\mathbf{x}) = \int_{t_i}^{t_f} L(\mathbf{x}, \dot{\mathbf{x}}, t)dt \tag{7.31}$$

with the constraints

$$g_i(\mathbf{x}, \dot{\mathbf{x}}, t) = 0, i = 1, \ldots, m \leq n \tag{7.32}$$

and state vector $\mathbf{x}$, *then, if* $\mathbf{x}(t)$ *minimizes* $V(\mathbf{x})$, *there exists a set of Lagrange multipliers* $\lambda(t) = [\lambda_1(t), \lambda_2(t), \ldots, \lambda_m(t)]^T$ *such that the state vector* $\mathbf{x}(t)$ *minimizes the scalar function:*

$$\begin{aligned} V^*(\mathbf{x}) &= \int_{t_0}^{t_f} \left[L(\mathbf{x}, \dot{\mathbf{x}}, t) + \sum_{i=1}^{m} \lambda_i(t) g_i(\mathbf{x}, \dot{\mathbf{x}}, t) \right] dt = \\ &= \int_{t_0}^{t_f} L^*(\mathbf{x}, \dot{\mathbf{x}}, t) dt \end{aligned} \tag{7.33}$$

where $L^*(\mathbf{x}, \dot{\mathbf{x}}, t) = L(\mathbf{x}, \dot{\mathbf{x}}, t) + \sum_{i=1}^{m} \lambda_i(t) g_i(\mathbf{x}, \dot{\mathbf{x}}, t)$ *satisfies the Euler equations:*

$$\frac{\partial L^*(\mathbf{x}, \dot{\mathbf{x}}, t)}{\partial x_i} - \frac{d}{dt} \frac{\partial L^*(\mathbf{x}, \dot{\mathbf{x}}, t)}{\partial \dot{x}_i} = 0 \tag{7.34}$$

for $i = 1, \ldots, n$.

Let us then consider the performance index $PI = \int_{t_i}^{t_f} L(\mathbf{x}, \mathbf{u}, t) dt$ with the constraints given by the state equations $\dot{\mathbf{x}} = f(\mathbf{x}, \mathbf{u}, t)$ and boundary conditions $\mathbf{x}_i = \mathbf{x}(t_i)$ and $\mathbf{x}_f = \mathbf{x}(t_f)$.

We want to apply the approach of Theorem 14 by considering the following functional:

$$V^*(\mathbf{x}, \mathbf{u}) = \int_{t_i}^{t_f} L^*(\mathbf{x}, \dot{\mathbf{x}}, \mathbf{u}, \dot{\mathbf{u}}) dt \tag{7.35}$$

where

$$L^*(\mathbf{x}, \dot{\mathbf{x}}, \mathbf{u}, \dot{\mathbf{u}}) = L(\mathbf{x}, \mathbf{u}, t) + \sum_{j=1}^{n} \lambda_j (f_j(\mathbf{x}, \mathbf{u}, t) - \dot{x}_j) \tag{7.36}$$

Replacing (7.36) into the Euler equations:

$$\frac{\partial L^*(\mathbf{x}, \dot{\mathbf{x}}, \mathbf{u}, \dot{\mathbf{u}})}{\partial x_i} - \frac{d}{dt} \frac{\partial L^*(\mathbf{x}, \dot{\mathbf{x}}, \mathbf{u}, \dot{\mathbf{u}})}{\partial \dot{x}_i} = 0 \tag{7.37}$$

we get:

$$\begin{aligned} &\frac{\partial}{\partial x_i} \left\{ L(\mathbf{x}, \mathbf{u}, t) + \sum_{j=1}^{n} \lambda_j \left[f_j(\mathbf{x}, \mathbf{u}, t) - \dot{x}_j \right] \right\} \\ &- \frac{d}{dt} \frac{\partial}{\partial \dot{x}_i} \left\{ L(\mathbf{x}, \mathbf{u}, t) + \sum_{j=1}^{n} \lambda_j \left[f_j(\mathbf{x}, \mathbf{u}, t) - \dot{x}_j \right] \right\} = 0 \end{aligned} \tag{7.38}$$

with $i = 1, \ldots, n$.

Taking into account that some of the derivatives appearing in (7.38) are zero, we obtain:

$$\frac{\partial}{\partial x_i}\left[L(\mathbf{x},\mathbf{u},t)+\sum_{j=1}^{n}\lambda_j f_j(\mathbf{x},\mathbf{u},t)\right]+\frac{d}{dt}\frac{\partial}{\partial \dot{x}_i}\sum_{j=1}^{n}\lambda_j\dot{x}_j=0 \tag{7.39}$$

Now, considering that $\frac{\partial}{\partial \dot{x}_i}\sum_{j=1}^{n}\lambda_j\dot{x}_j=\lambda_i$, we get:

$$\dot{\lambda}_i=-\frac{\partial}{\partial x_i}\left[L(\mathbf{x},\mathbf{u},t)+\sum_{j=1}^{n}\lambda_j f_j(\mathbf{x},\mathbf{u},t)\right] \tag{7.40}$$

for $i=1,\ldots,n$.

Similarly, if we start from the Euler equation with respect to the variables $\mathbf{u}$:

$$\begin{aligned}\frac{\partial}{\partial u_k}&\left\{L(\mathbf{x},\mathbf{u},t)+\sum_{j=1}^{n}\lambda_j\left[f_j(\mathbf{x},\mathbf{u},t)-\dot{x}_j\right]\right\}\\ -\frac{d}{dt}\frac{\partial}{\partial \dot{u}_k}&\left\{L(\mathbf{x},\mathbf{u},t)+\sum_{j=1}^{n}\lambda_j\left[f_j(\mathbf{x},\mathbf{u},t)-\dot{x}_j\right]\right\}=0\end{aligned} \tag{7.41}$$

for $k=1,\ldots,m$, with similar calculations, we get:

$$\frac{\partial}{\partial u_k}\left[L(\mathbf{x},\mathbf{u},t)+\sum_{j=1}^{n}\lambda_j f_j(\mathbf{x},\mathbf{u},t)\right]=0 \tag{7.42}$$

for $k=1,\ldots,m$.

Now, define the so-called state function of Pontryagin:

$$H(\mathbf{x},\mathbf{u},\lambda,t)=L(\mathbf{x},\mathbf{u},t)+\sum_{j=1}^{n}\lambda_j f_j(\mathbf{x},\mathbf{u},t) \tag{7.43}$$

or, equivalently, in matrix notation:

$$H(\mathbf{x},\mathbf{u},\lambda,t)=L(\mathbf{x},\mathbf{u},t)+\lambda^T f(\mathbf{x},\mathbf{u},t) \tag{7.44}$$

where $\lambda=[\lambda_1,\ldots,\lambda_n]^T$.

The state function of Pontryagin allows the Euler equations (7.40) and (7.42) to be rewritten as follows:

$$\dot{\lambda}_i=-\frac{\partial H(\mathbf{x},\mathbf{u},\lambda,t)}{\partial x_i} \tag{7.45}$$

for $i=1,\ldots,n$ and

$$\frac{\partial H(\mathbf{x},\mathbf{u},\lambda,t)}{\partial u_k}=0 \tag{7.46}$$

for $k = 1, \ldots, m$.

Alternatively, they can also be rewritten in a particularly convenient compact form:

$$\dot{\lambda} = -\frac{\partial H(\mathbf{x}, \mathbf{u}, \lambda, t)}{\partial \mathbf{x}} \tag{7.47}$$

and

$$\frac{\partial H(\mathbf{x}, \mathbf{u}, \lambda, t)}{\partial \mathbf{u}} = 0 \tag{7.48}$$

It can be checked by direct calculation that, using the state function of Pontryagin, the plant equations can be written as:

$$\dot{\mathbf{x}} = \frac{\partial H(\mathbf{x}, \mathbf{u}, \lambda, t)}{\partial \lambda} \tag{7.49}$$

In summary, to solve the optimal control problem the following steps need to be performed:

1. Given L, the performance index and the system dynamical equations $\dot{\mathbf{x}} = f(\mathbf{x}, \mathbf{u}, t)$, with the boundary conditions $\mathbf{x}_i$ and $\mathbf{x}_f$, the function H can be computed.
2. Solving the algebraic equation (7.48), one derives $\mathbf{u}^{optimal} = \mathbf{u}^{optimal}(\mathbf{x}, \lambda, t)$ and, in correspondence, $H^{optimal} = H(\mathbf{x}, \mathbf{u}^{optimal}, \lambda, t)$.
3. At this point, equations (7.49) and (7.47) can be used to derive $\mathbf{x}(t)$ and $\lambda(t)$, which are then substituted into $\mathbf{u}^{optimal} = \mathbf{u}^{optimal}(\mathbf{x}, \lambda, t)$.

Example 7.5

Consider the first-order system

$$\dot{x} = -2x + u \tag{7.50}$$

and the functional

$$PI = \int_0^1 (x^2 + 5u^2)dt \tag{7.51}$$

In this case, the function H becomes:

$$H = \lambda(-2x + u) + x^2 + 5u^2 \tag{7.52}$$

Equation (7.48) yields:

$$\frac{\partial H}{\partial u} = \lambda + 10u = 0 \tag{7.53}$$

and hence

$$u^{optimal} = -\frac{\lambda}{10} \tag{7.54}$$

It follows that:

$$H^{optimal} = -\frac{\lambda^2}{10} - 2\lambda x + x^2 + \frac{1}{20}\lambda^2 \tag{7.55}$$

For the system dynamics, we have:

$$\dot{x} = \frac{\partial H^{optimal}}{\partial \lambda} = -2x - \frac{\lambda}{10} \tag{7.56}$$

which can be also obtained replacing (7.54) into (7.50). In addition, we have that:

$$\dot{\lambda} = -\frac{\partial H^{optimal}}{\partial x} = 2\lambda - 2x \tag{7.57}$$

In summary, we have:

$$\begin{bmatrix} \dot{x} \\ \dot{\lambda} \end{bmatrix} = \begin{bmatrix} -2 & -\frac{1}{10} \\ -2 & 2 \end{bmatrix} \begin{bmatrix} x \\ \lambda \end{bmatrix} \tag{7.58}$$

that must be solved taking into account that $x(0) = 0$ and $x(1) = 1$.

In the next section we discuss the optimal control problem for linear systems.

7.4 LQR Optimal Control

Let us consider a continuous-time system

$$\begin{aligned} \dot{\mathbf{x}} &= \mathrm{A}\mathbf{x} + \mathrm{B}\mathbf{u} \\ \mathbf{y} &= \mathrm{C}\mathbf{x} \end{aligned} \tag{7.59}$$

and let us suppose that it is completely controllable and observable (alternatively consider only the part of the system which is controllable and observable). Using the control law $\mathbf{u} = -\mathrm{K}\mathbf{x}$, the system dynamic is governed by:

$$\dot{\mathbf{x}} = (\mathrm{A} - \mathrm{BK})\mathbf{x} \tag{7.60}$$

Usually, $\mathrm{A}_c = (\mathrm{A} - \mathrm{BK})$ indicates the closed-loop matrix of the system. As we know, the issue of linear state regulators is in selecting the proper gains K so as to arbitrarily fix the eigenvalues of A_c. Under the hypothesis that the system is observable and controllable, there is always a solution.

Clearly, there are infinite ways to choose these eigenvalues. One is on the basis of criteria which include various specifications which a closed-loop system must satisfy. Once the criterion is defined, the choice of the eigenvalues and so the gains K must be done to optimize the adopted criteria.

In the case of $\mathbb{L}_2$ or $\mathbb{H}_2$ optimal control the index (to minimize) is defined by the functional:

$$J = \int_0^\infty (\mathbf{x}^T \mathrm{Q}\mathbf{x} + \mathbf{u}^T \mathrm{R}\mathbf{u})dt \tag{7.61}$$

where the matrices $Q \in \mathbb{R}^{n \times n}$ and $R \in \mathbb{R}^{m \times m}$ are fixed weight matrices which define, as we will shortly see, the specifications for optimal control.

The optimal control issue lies in finding the gains K_{opt} so as to minimize functional (7.61).

As regards weight matrices, Q and R are generally positive definite. Q can eventually be selected as positive semi-definite, but R must always be positive definite.

This problem is referred to as the linear quadratic regulator (LQR) optimal control.

The physical meaning of optimal control can be clarified through an example where the matrix Q is chosen in a particular way. Consider $\mathbf{z}(t) = C_1\mathbf{x}(t)$ and the functional

$$J = \int_0^\infty (\mathbf{z}^T\mathbf{z} + \mathbf{u}^T R\mathbf{u})dt$$

obtained from functional (7.61) making $Q = C_1^T C_1$. Variables $\mathbf{z}(t)$ are not system output but other variables which we want to keep small. In other words, the control objective is to enable the variables $\mathbf{z}(t)$ to be very close to system equilibrium point ($\mathbf{x} = 0$ and so $\mathbf{z} = 0$). Supposing the system is initially excited (this is reflected by an initial condition $\mathbf{x}_0$ which is an undesired deviation from the equilibrium position $\mathbf{x} = 0$), the objective of the optimal control is finding the input so that the system reaches the equilibrium in the shortest possible time. As soon as the system is supposed controllable, this objective can always be obtained. Reaching the equilibrium in the shortest possible time generally requires a considerable control signal which is unacceptable from certain points of view. Firstly, such a control signal could saturate the real system. Secondly, it could stimulate an unmodeled high frequency dynamic in the original system. Therefore, two different costs need balancing $\mathbf{z}^T(t)\mathbf{z}(t) \geq 0$ and $\mathbf{u}^T(t)R\mathbf{u}(t) > 0 \ \forall t$.

The significance of matrices Q and R can be therefore clarified by considering that the functional J represents the energy to minimize (for this reason it is often called the quadratic index). The functional takes into account the two energy terms: the first, $J_1 = \int_0^\infty \mathbf{x}^T Q\mathbf{x}dt$, deals with the energy associated with the closed-loop state variables weighted by matrix Q, whereas the other deals with the energy associated with input $\mathbf{u}$ and weighted by matrix R.

Obviously the choice of the closed-loop eigenvalues has to guarantee that the system is always asymptotically stable so that the zero-input response of the system tends to zero ($\mathbf{x}(t) \to 0$). From this consideration, minimizing the functional $J_1 = \int_0^\infty \mathbf{x}^T Q\mathbf{x}dt$ means ensuring that the state variables tend to zero as soon as possible. The smaller the energy associated with the state variables the more rapidly they tend to zero.

Minimizing the energy means ensuring a fast transitory. This happens at the cost of input energy as we will see in the next example.

Example 7.6

Consider $G(s) = \frac{1}{s+1}$. The unit step response tends asymptotically to a value of one. Suppose output $y(t) = 1$ in the shortest time possible. By applying a step input with an amplitude of 10, the steady-state output value is exactly ten, and given the initial null conditions, this means that $y(t) = 1$ is verified for a given t, less than for the previous case (unitary step input). In the limit case, choosing as input a Dirac impulse $u(t) = \delta(t)$, one obtains $y(t) = e^{-t}$ (impulse response) which is one for $t = 0$. So, at the cost of growing input energy, it is possible to reduce the time required for the output to reach a determined value.

Generally, the input energy required to obtain certain specifications for a closed-loop system should always be evaluated. For this reason, the quadratic index (7.61) also accounts for the input energy. Matrix R therefore assigns a relative weight to the two energy terms dealt with by the quadratic index. Matrix R weights the input therefore establishing if according to the control objective, it is more important to minimize the first or second contribution. Matrix Q establishes the weight of the state variables, taking also into account for instance that it is not given that all the state variables have the same scale factor in the measures.

Now, let us see how the gains K_{opt} are determined. They are calculated from the following matrix equation:

$$PA + A^T P - PBR^{-1}B^T P + Q = 0 \tag{7.62}$$

with $P \in \mathbb{R}^{n \times n}$.

If $\overline{P}$ is the solution, the optimum gains are given by:

$$K_{opt} = R^{-1}B^T\overline{P} \tag{7.63}$$

Note that the inverse of R always exists, as it is a positive definite matrix.

The matrix equation (7.62), by contrast to the Lyapunov equations, is a nonlinear equation in P. In fact, a quadratic term ($PBR^{-1}B^T P$) appears in the equation.

This equation is called the algebraic Riccati equation from the name of Jacopo Francesco, the Count of Riccati (1676–1754), who was the first to study equations of this type. The equation is known as algebraic because matrix P does not depend on time. There is a differential Riccati equation which is a function of time and it comes into play when, rather than defining the integral of the J index between zero and infinity, a finite horizon $[t_1, t_2]$ is considered in which to reach the control objective.

For systems with only one input (R is a scalar quantity) the quadratic term has a weight which is inversely proportional to increasing R.

Generally, the Riccati equation does not have a single solution. For example, if $n = 1$, it produces a second order equation with two solutions. Among the possible solutions for Riccati equations however, there is only one positive definite. It is this matrix $\overline{P}$ which will provide the optimal gains K_{opt}.

The optimal gains have another fundamental property: they can guarantee closed-loop stability of the system, in other words $A_c = A - BK_{opt}$ has all eigenvalues with negative real part.

To demonstrate that the gains K_{opt} guarantee closed-loop system stability, let us consider the Riccati equation (7.62) with $P = \overline{P}$ and add and subtract $\overline{P}BR^{-1}B^T\overline{P}$:

$$\overline{P}A + A^T\overline{P} - \overline{P}BR^{-1}B^T\overline{P} + \overline{P}BR^{-1}B^T\overline{P} - \overline{P}BR^{-1}B^T\overline{P} + Q = 0$$

Rearranging, we obtain:

$$\overline{P}(A - BR^{-1}B^T\overline{P}) + (A^T - \overline{P}BR^{-1}B^T)\overline{P} + \overline{P}BR^{-1}B^T\overline{P} + Q = 0$$

Considering $K_{opt} = R^{-1}B^T\overline{P}$, then

$$\overline{P}(A - BK_{opt}) + (A^T - K_{opt}^T B^T)\overline{P} = -K_{opt}^T RK_{opt} - Q$$

and so

$$\overline{P}A_c + A_c^T\overline{P} = -K_{opt}^T RK_{opt} - Q$$

Since the right-hand term is a positive definite matrix because Q is positive semi-definite (or definite) and $K_{opt}^T RK_{opt}$ is positive definite, then the closed-loop system satisfies the Lyapunov equation for stability and, according to Lyapunov second theorem, it is asymptotically stable.

At variance with the Lyapunov equation that can be solved in closed form, calculating P requires an iterative method. The algorithm is based on the Kleinman method, that relies on the property that the gains K are stabilizing (so the corresponding closed-loop state matrix satisfies the Lyapunov equation) and on the relation linking K and P (i.e., $K = R^{-1}B^TP$).

In the Kleinman method, at the first iteration K_1 is fixed so that $A - BK_1$ is stable (if the closed-loop system with state matrix A is already stable, then $K_1 = 0$). For generic iterations K_i is fixed by $K_i = R^{-1}B^TP_{i-1}$.

Once K_i is fixed, P_i is obtained by solving the linear Lyapunov equation:

$$P_i(A - BK_i) + (A^T - K_i^T B^T)P_i = -K_i^T RK_i - Q \tag{7.64}$$

Next, the gains K_{i+1} are obtained by $K_{i+1} = R^{-1}B^TP_i$ iterating the procedure. The algorithm converges when $K_{i+1} \simeq K_i$.

Kleinman showed that starting from a matrix K which guarantees closed-loop stability (matrix A_c) and iterating the procedure, the result is always a stable matrix, so the various matrices P_i, obtained in this way, are all positive definite. The procedure converges to the positive definite solution of the Riccati equation: $P_i \simeq P_{i+1} = \overline{P}$. Moreover, Kleinman showed that the convergence of this method is monotonic, that is the error norm at each step decreases monotonically. If instead, at the first iteration K_1 is not such that the

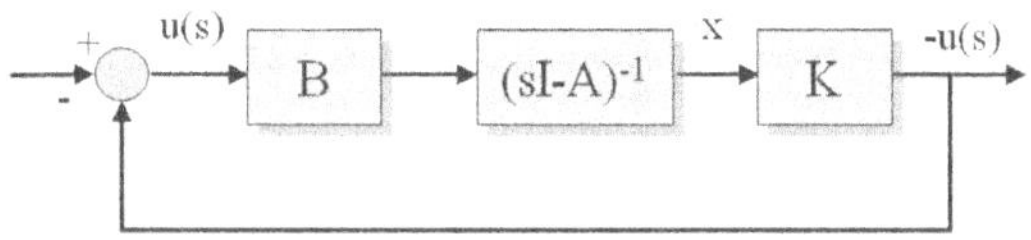

FIGURE 7.3
Block scheme of the linear quadratic regulator.

closed-loop system with state matrix A_c is asymptotically stable, the method does not converge or it does to a non-definite matrix.

The advantage of the method is solving Lyapunov equations iteratively, that is, solving linear equations with closed solutions rather than nonlinear ones.

It can be shown that the index J when $\mathbf{u} = -K_{opt}\mathbf{x}$ is $J = \frac{1}{2}\mathbf{x}^T(0)\overline{P}\mathbf{x}(0)$, which is therefore the smallest value the index can have $\forall\ \mathbf{x}(0)$.

Optimal control has however some drawbacks. First, it is based on feedback of all the state variables $\mathbf{x}$. This requires that the whole state is available for feedback. Optimal control therefore requires a sensor for every state variable, or it requires an observer to re-construct the variables which are not directly measurable. Optimal control is difficult to apply to flexible systems which require a high-order model (if not infinite).

Another drawback of optimal control is the gap with classical control specifications such as disturbance rejection, overshoot, stability margins, and so on. This often leads to the need of trial and error to define the weight matrices R and Q.

Figure 7.3 shows the block scheme of optimal control (also called linear quadratic regulator). The transfer matrix is given by $G_{LQ} = K(sI - A)^{-1}B$. It can be shown that this matrix has certain robustness margins with respect to delays and gains in the direct chain.

The main features of optimal control are summarized in the following theorem:

Theorem 15 *Given the system* $\dot{\mathbf{x}} = A\mathbf{x} + B\mathbf{u}$ *with initial condition* $\mathbf{x}_0$ *and given the index* $J = \int_0^\infty (\mathbf{x}^T Q\mathbf{x} + \mathbf{u}^T R\mathbf{u})dt$, *if the system is in minimal form the entire state* $\mathbf{x}$ *can be fed back and matrices* Q *and* R *are symmetric and positive semi-definite and definite, respectively, then:*

1. *There is only one linear quadratic regulator* $\mathbf{u} = -K_{opt}x$ *(with* $K_{opt} = R^{-1}B^T\bar{P}$*) which minimizes the index* J*;*

2. $\overline{P}$ *is the only symmetric positive definite solution of the Riccati equation*

$$PA + A^T P - PBR^{-1}B^T P + Q = 0$$

3. *The closed-loop system with state matrix* $(A - BK_{opt})$ *is asymptotically stable;*

4. The minimum value of index J is $J = \frac{1}{2}\mathbf{x}_0^T \bar{P} \mathbf{x}_0$.

MATLAB® Exercise 7.1

Here, we discuss the use of the Kleinman algorithm to solve the Riccati equation. Let us consider the system $G(s) = \frac{s+2}{s^2-2s-3}$ which is stable and minimal. Consider the functional (7.61) with $Q = C^T C$ and $r = 1$. Let us define the system in MATLAB® from its canonical controllability form:

```
>> A=[0 1; 3 2];
>> B=[0; 1];
>> C=[2 1];
>> D=0;
```

Define the weight matrices

```
>> R=1;
>> Q=C'*C;
```

Choose K_0 such that $A_c = A - BK_0$ is asymptotically stable, and in particular that its eigenvalues are $\lambda_1 = -1$ e $\lambda_2 = -0.5$

```
>> K0=acker(A,B,[-1 -0.5]);
>> Ki=K0;
```

Apply the Kleinman method assuming that $K_{i+1} \simeq K_i$, when $\|K_{i+1} - K_i\| < 0.0001$:

```
>> for i=1:100
    P=lyap((A-B*Ki)',Ki'*R*Ki+Q);
    Kii=inv(R)*B'*P;
    if (norm(Kii-Ki)<0.0001), break; end
    Ki=Kii;
  end
```

After seven iterations the algorithm converges on $K_i = \begin{bmatrix} 6.6056 & 6.2674 \end{bmatrix}$ and $P = \begin{bmatrix} 7.3865 & 6.6056 \\ 6.6056 & 6.2674 \end{bmatrix}$. It is easy to verify that P is positive definite. The optimal eigenvalues (`eig(A-B*Ki)`) are $\lambda_{1,opt} = -1.1605$ and $\lambda_{2,opt} = -3.1070$.

Note that from an unstable matrix A_c (e.g., by imposing eigenvalues $\lambda_1 = -1$ and $\lambda_2 = 2$), the algorithm converges to a non-definite matrix P.

The linear quadratic regulator can be also calculated with the MATLAB command:

```
>> [K,P,E]=lqr(A,B,C'*C,1)
```

Finally, the transfer function G_{LQ} is calculated with command:

```
>> system=ss(A,B,K,0)
```

To verify that the closed-loop system eigenvalues are effectively the optimal ones, the closed-loop transfer function can be calculated with the command

```
>> tf(feedback(system,1))
```

Furthermore, the Nyquist diagram can be plotted to verify the robustness of the optimal control in terms of gain and phase margins with the command

```
>> nyquist(system)
```

MATLAB® Exercise 7.2

Consider now the system with transfer function $G(s) = \frac{s+10}{s^2+s-2}$. As in the previous example, $G(s)$ is unstable and in minimal form. We want to calculate the linear quadratic regulator with $Q = C^T C$ for two cases: $r = 0.1$ and $r = 10$.

Define the system in MATLAB by considering its canonical control form:

```
>> A=[0 1; 2 -1];
>> B=[0; 1];
>> C=[10 1];
>> D=0;
```

Then define the weight matrices:

```
>> Q=C'*C;
```

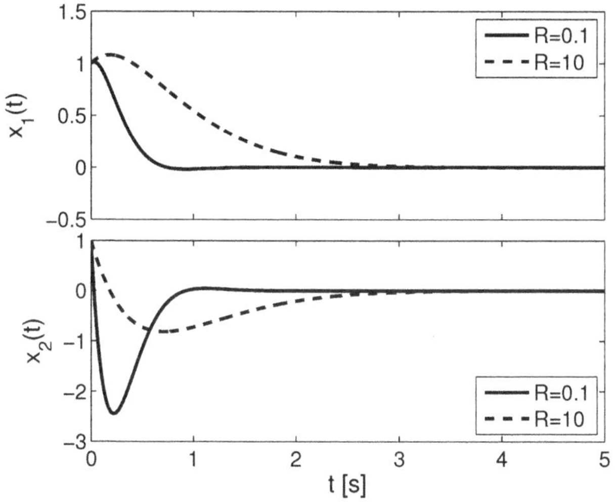

FIGURE 7.4
Zero-input response of the LQR closed-loop system for $r = 0.1$ and $r = 10$.

```
>> r1=0.1;
>> r2=10;
```

The transfer function G_{LQR} of the linear quadratic regulator for the two cases can be calculated through the commands:

```
>> [K1,P1,E1]=lqr(A,B,Q,r1);
>> GLQR1=ss(A,B,K1,0);
>> [K2,P2,E2]=lqr(A,B,Q,r2);
>> GLQR2=ss(A,B,K2,0);
```

and the corresponding closed-loop transfer functions are:

```
>> CLsys1=tf(feedback(GLQR1,1));
>> CLsys2=tf(feedback(GLQR2,1));
```

In order to compare the behavior of the two closed-loop systems, we can calculate and plot the zero-input response for both cases:

```
>> [Y1,T1,X1] = lsim(CLsys1,zeros(5001,1),[0:0.001:5],[1 1]);
>> [Y2,T2,X2] = lsim(CLsys2,zeros(5001,1),[0:0.001:5],[1 1]);
```

In Figure 7.4 the trends of the state variables $x_1(t)$ and $x_2(t)$ are reported for the two cases. When $r = 0.1$ (continuous lines) the energy associated to the input is weighted less than that associated to the states, hence the effect of the linear quadratic regulator, which mostly minimizes the state energy, is a zero-input response with a small time constant which rapidly decreases to zero; on the contrary, when $r = 10$ (dashed lines) the closed-loop system has a larger time constant, as it can be easily observed from its zero-input response.

7.5 Hamiltonian Matrices

We have seen how optimal control is solved by an algebraic Riccati equation (7.62). To each equation of this type a matrix can be associated, a Hamiltonian matrix, which has very particular properties. We can, for example, associate to the Riccati equation (7.62), the Hamiltonian matrix:

$$\mathrm{H} = \begin{bmatrix} \mathrm{A} & -\mathrm{BR}^{-1}\mathrm{B}^T \\ -\mathrm{Q} & -\mathrm{A}^T \end{bmatrix} \tag{7.65}$$

H is a $2n \times 2n$ matrix. In the first line of this block matrix, there are the coefficients (with their sign) of the term which is multiplied to the left by the unknown matrix P (element H_{11}) and the one by the quadratic (component H_{12}). The other two coefficients of the matrix (H_{21} and H_{22}) are given by the known term from the Riccati equation and by the term which is multiplied to the right by P with reverse sign.

The Hamiltonian matrix H is such that, given

$$\mathrm{Y} = \begin{bmatrix} 0 & \mathrm{I} \\ -\mathrm{I} & 0 \end{bmatrix}$$

with I the $n \times n$ identity matrix, then:

$$\mathrm{H}^T\mathrm{Y} = -\mathrm{YH} \tag{7.66}$$

or rather

$$\mathrm{Y}^{-1}\mathrm{H}^T\mathrm{Y} = -\mathrm{H} \tag{7.67}$$

This is the property which defines the Hamiltonian matrix. All matrices with this property are called Hamiltonian.

The matrix associated with the Riccati equation for optimal control has a further important property expressed by the following theorem:

Theorem 16 *Given the optimal eigenvalues* $\lambda_1, \lambda_2, \ldots, \lambda_n$ *(i.e., the eigenvalues of* $\mathrm{A} - \mathrm{BK}_{opt}$*), the eigenvalues of* $\mathrm{H} = \begin{bmatrix} \mathrm{A} & -\mathrm{BR}^{-1}\mathrm{B}^T \\ -\mathrm{Q} & -\mathrm{A}^T \end{bmatrix}$ *are:*

$$\lambda_1, \lambda_2, \ldots, \lambda_n, -\lambda_1, -\lambda_2, \ldots, -\lambda_n$$

Proof *Let us consider the matrix* $\mathrm{T} = \begin{bmatrix} \mathrm{I} & 0 \\ \overline{\mathrm{P}} & \mathrm{I} \end{bmatrix}$ *and calculate its inverse* $\mathrm{T}^{-1} = \begin{bmatrix} \mathrm{T}_{11} & \mathrm{T}_{12} \\ \mathrm{T}_{21} & \mathrm{T}_{22} \end{bmatrix}$ *(remembering that* $\overline{\mathrm{P}}$ *is positive definite, so the inverse of* T *exists).*

Matrix $\mathrm{T}^{-1}\mathrm{T}$ *is given by:*

$$T^{-1}T = \begin{bmatrix} T_{11} & T_{12} \\ T_{21} & T_{22} \end{bmatrix} \begin{bmatrix} I & 0 \\ \overline{P} & I \end{bmatrix} = \begin{bmatrix} T_{11} + T_{12}\overline{P} & T_{12} \\ T_{21} + T_{22}\overline{P} & T_{22} \end{bmatrix}$$

Since $T^{-1}T = I$, *then:*

$$\begin{aligned} &T_{11} + T_{12}\overline{P} = I \Rightarrow T_{11} = I \\ &T_{12} = 0 \\ &T_{21} + T_{22}\overline{P} = 0 \Rightarrow T_{21} = -\overline{P} \\ &T_{22} = I \end{aligned}$$

and so $T^{-1} = \begin{bmatrix} I & 0 \\ -\overline{P} & I \end{bmatrix}$.

Now let us consider the matrix $T^{-1}HT$ *which, as we know, has the same eigenvalues of matrix* H*:*

$$T^{-1}HT = \begin{bmatrix} I & 0 \\ -\overline{P} & I \end{bmatrix} \begin{bmatrix} A & -BR^{-1}B^T \\ -Q & -A^T \end{bmatrix} \begin{bmatrix} I & 0 \\ \overline{P} & I \end{bmatrix} =$$

$$= \begin{bmatrix} A & -BR^{-1}B^T \\ -\overline{P}A - Q & \overline{P}BR^{-1}B^T - A^T \end{bmatrix} \begin{bmatrix} I & 0 \\ \overline{P} & I \end{bmatrix} =$$

$$= \begin{bmatrix} A - BR^{-1}B^T\overline{P} & -BR^{-1}B^T \\ -\overline{P}A - Q + \overline{P}BR^{-1}B^T\overline{P} - A^T\overline{P} & \overline{P}BR^{-1}B^T\overline{P} - A^T \end{bmatrix} =$$

$$= \begin{bmatrix} A - BK_{opt} & -BR^{-1}B^T \\ 0 & -(A - BK_{opt})^T \end{bmatrix}$$

In finding the last expression, the algebraic Riccati equation was used ($\overline{P}A + Q - \overline{P}BR^{-1}B^T\overline{P} + A^T\overline{P} = 0$)*. Notice that the eigenvalues of* H *are given by union of* $\lambda_1, \lambda_2, \ldots, \lambda_n$ *and of* $-\lambda_1, -\lambda_2, \ldots, -\lambda_n$*.*

For SISO systems, Theorem 16 also provides an alternative method for finding the optimal controller with respect to that based on the Riccati equation (7.62). First, the eigenvalues of H are found and those on the right-hand half of the complex plane are discarded. Then, those eigenvalues with negative real part are used to design a control law $\mathbf{u} = -K\mathbf{x}$ by eigenvalue placement (a system of n equations with n unknowns). The method guarantees that no eigenvalues are found on the imaginary axis: if there are eigenvalues of H on the imaginary axis, then the system is uncontrollable. If this were to be verified, the base hypothesis for solving optimal control would be violated. Furthermore in the SISO case, the Lyapunov equation could be used to find $\overline{P}$: $A_c^T\overline{P} + \overline{P}A_c = -K_{opt}^T R K_{opt} - Q$.

The property illustrated is characteristic of all Hamiltonian matrices: they all have symmetric eigenvalues with respect to the imaginary axis, or rather if λ is an eigenvalue, then so is $-\lambda$. However, the opposite is not true as the following example shows.

MATLAB® Exercise 7.3

Consider the matrix $P = \begin{bmatrix} 0 & 1 & 0 & 0 \\ 0 & 0 & 1 & 0 \\ 0 & 0 & 0 & 1 \\ 1 & 0 & 0 & 0 \end{bmatrix}$. Despite its eigenvalues $\lambda_{1,2} = \pm 1$ and $\lambda_{3,4} = \pm j$, it is not a Hamiltonian matrix. In fact, if we calculate $-Y^{-1}P^TY$, we obtain

$$-Y^{-1}P^TY = \begin{bmatrix} 0 & 0 & 0 & 1 \\ -1 & 0 & 0 & 0 \\ 0 & 1 & 0 & 0 \\ 0 & 0 & -1 & 0 \end{bmatrix}$$

i.e., $-Y^{-1}P^TY \neq P$, and P does not satisfy equation (7.67). This can be checked in MATLAB with the following commands:

```
>> H=diag([1 1 1],1)
>> H(4,1)=1
>> eig(H)
>> Y=[zeros(2) eye(2); -eye(2) zeros(2)]
>> -inv(Y)*H'*Y
```

This example demonstrates that it is not true that, if a matrix has eigenvalues $\lambda_1, \lambda_2, \ldots, \lambda_n, -\lambda_1, -\lambda_2, \ldots, -\lambda_n$, it is Hamiltonian.

MATLAB® Exercise 7.4

In this MATLAB® exercise we show an example of the calculation of the optimal eigenvalues from the Hamiltonian matrix. We then show how they can be assigned as closed-loop eigenvalues with the command `acker`.

Consider the continuous-time LTI system given by

$$A = \begin{bmatrix} -1 & 0 & 0 \\ 0 & 2 & 0 \\ 0 & 0 & 3 \end{bmatrix}; B = \begin{bmatrix} 1 \\ 1 \\ 1 \end{bmatrix}; C = \begin{bmatrix} 1 & 1 & 1 \end{bmatrix} \tag{7.68}$$

and the optimal control problem with

$$Q = \begin{bmatrix} 2 & 0 & 0 \\ 0 & 1 & 0 \\ 0 & 0 & 1 \end{bmatrix}; R = 15 \tag{7.69}$$

Let us first define in MATLAB the state-space matrices:

```
>> A=diag([-1 2 3])
>> B=[1; 1; 1]
>> C=[1 1 1]
```

and matrices Q and R:

```
>> R=15
>> Q=[2 0 0; 0 1 0; 0 0 1]
```

We then calculate the Hamiltonian matrix H and its eigenvalues:

```
>> H=[A -B*inv(R)*B'; -Q -A']
>> p=eig(H)
```

One gets: $p_1 = 3.0114$, $p_2 = 2.0171$, $p_3 = 1.0627$, $p_4 = -1.0627$, $p_5 = -2.0171$ and $p_6 = -3.0114$.

The optimal closed-loop eigenvalues are those with negative real part, i.e., p_4, p_5 and p_6. The value of the gain that corresponds to such eigenvalues can be calculated with the command `acker` as follows:

```
>> K=acker(A,B,p(4:6))
```

The result is a closed-loop system having as eigenvalues p_4, p_5 and p_6 as it can be verified with the following command:

```
>> eig(A-B*K)
```

7.6 Solving the Riccati Equation via the Hamiltonian Matrix

The Riccati equation (7.62) can be solved by a method which holds true for all quadratic equations and is based on the associated Hamiltonian matrix. Consider a diagonalization of matrix H, obtained by ordering the Hamiltonian eigenvalues so that those at the top of the diagonal have negative real part:

$$\mathrm{H} = \mathrm{T}\begin{bmatrix} -\Lambda & 0 \\ 0 & \Lambda \end{bmatrix}\mathrm{T}^{-1} \tag{7.70}$$

with $\mathrm{T} = \begin{bmatrix} \mathrm{T}_{11} & \mathrm{T}_{12} \\ \mathrm{T}_{21} & \mathrm{T}_{22} \end{bmatrix}$. At this point P can be calculated:

$$\mathrm{P} = \mathrm{T}_{21}\mathrm{T}_{11}^{-1} \tag{7.71}$$

Note that the non-singularity of T_{11} is assured by the controllability of the system.

The eigenvalue order within the sub-matrix Λ is irrelevant as regards this calculation. Ordering in terms of eigenvalues with negative real part and those with positive real part is possible only if matrix H has no imaginary eigenvalues, a hypothesis required for solving the Riccati equation.

This method based on the Jordan decomposition of matrix H can prove to be computationally demanding. At the end of the 80s a method based on another decomposition (Schur decomposition) proved more advantageous.

In Schur decomposition, matrix H is the product of three matrices:

$$\mathrm{H} = \mathrm{USU}^T \tag{7.72}$$

with $\mathrm{S} = \begin{bmatrix} \mathrm{S}_{11} & \mathrm{S}_{12} \\ 0 & \mathrm{S}_{22} \end{bmatrix}$ and U an orthonormal matrix ($\mathrm{UU}^T = \mathrm{I}$). So, S is a higher triangular matrix or quasi-triangular. Furthermore, it is structured such that S_{11} has eigenvalues with negative real part, whereas S_{22} has eigenvalues with positive real part. The solution to the Riccati equation is found analogously. Since $\mathrm{U} = \begin{bmatrix} \mathrm{U}_{11} & \mathrm{U}_{12} \\ \mathrm{U}_{21} & \mathrm{U}_{22} \end{bmatrix}$, matrix P is given by:

$$\mathrm{P} = \mathrm{U}_{21}\mathrm{U}_{11}^{-1} \tag{7.73}$$

7.7 The Control Algebraic Riccati Equation

Let us now consider a particular case of optimal control. Consider a linear time-invariant linear system:

$$\begin{aligned} \dot{\mathbf{x}} &= \mathrm{A}\mathbf{x} + \mathrm{B}\mathbf{u} \\ \mathbf{y} &= \mathrm{C}\mathbf{x} \end{aligned} \tag{7.74}$$

and a LQR problem with the following index:

$$J = \int_0^\infty (\mathbf{y}^T\mathbf{y} + \mathbf{u}^T\mathbf{u})dt \tag{7.75}$$

The index defined by equation (7.75) is a special case of the more general one defined by equation (7.61), with $\mathrm{Q} = \mathrm{C}^T\mathrm{C}$ and $\mathrm{R} = \mathrm{I}$.

Note that, in this case, (7.75) has two terms: the term $\int_0^\infty \mathbf{y}^T\mathbf{y}dt$ is the energy associated to the system output, whereas $\int_0^\infty \mathbf{u}^T\mathbf{u}dt$ is the energy associated to the system input.

The Riccati equation (7.62) associated with this problem is called the Control Algebraic Riccati Equation (CARE):

$$\mathrm{A}^T\mathrm{P} + \mathrm{P}\mathrm{A} - \mathrm{P}\mathrm{B}\mathrm{B}^T\mathrm{P} + \mathrm{C}^T\mathrm{C} = 0 \tag{7.76}$$

The Hamiltonian matrix associated to the CARE is:

$$\mathrm{H} = \begin{bmatrix} \mathrm{A} & -\mathrm{B}\mathrm{B}^T \\ -\mathrm{C}^T\mathrm{C} & -\mathrm{A}^T \end{bmatrix} \tag{7.77}$$

If the system is not strictly proper

$$\begin{aligned} \dot{\mathbf{x}} &= \mathrm{A}\mathbf{x} + \mathrm{B}\mathbf{u} \\ \mathbf{y} &= \mathrm{C}\mathbf{x} + \mathrm{D}\mathbf{u} \end{aligned} \tag{7.78}$$

the CARE equation becomes:

$$\mathrm{A}^T\mathrm{P} + \mathrm{P}\mathrm{A} - (\mathrm{P}\mathrm{B} + \mathrm{C}^T\mathrm{D})(\mathrm{I} + \mathrm{D}^T\mathrm{D})^{-1}(\mathrm{B}^T\mathrm{P} + \mathrm{D}^T\mathrm{C}) + \mathrm{C}^T\mathrm{C} = 0 \tag{7.79}$$

and the Hamiltonian becomes:

$$\mathrm{H} = \begin{bmatrix} \mathrm{A} - \mathrm{B}(\mathrm{I} + \mathrm{D}^T\mathrm{D})^{-1}\mathrm{D}^T\mathrm{C} & -\mathrm{B}(\mathrm{I} + \mathrm{D}^T\mathrm{D})^{-1}\mathrm{B}^T \\ -\mathrm{C}^T\mathrm{C} + \mathrm{C}^T\mathrm{D}(\mathrm{I} + \mathrm{D}^T\mathrm{D})^{-1}\mathrm{D}^T\mathrm{C} & -\mathrm{A}^T + \mathrm{C}^T\mathrm{D}(\mathrm{I} + \mathrm{D}^T\mathrm{D})^{-1}\mathrm{B}^T \end{bmatrix} \tag{7.80}$$

MATLAB® Exercise 7.5

Consider the continuous-time LTI system with state-space matrices:

$$\mathrm{A} = \begin{bmatrix} -1 & 0 \\ 0 & 3 \end{bmatrix}; \mathrm{B} = \begin{bmatrix} 1 \\ -1 \end{bmatrix}; \mathrm{C} = \begin{bmatrix} 2 & 1 \end{bmatrix}; \mathrm{D} = 1 \tag{7.81}$$

and let us compute the optimal control with index (7.75).

Once the system has been defined

```
>> A=[-1 0; 0 3]
>> B=[1; -1]
>> C=[2 1]
```

```
>> D=1
```
let us compute the Hamiltonian
```
>> H=[A-B*inv(eye(1)+D'*D)*D'*C -B*inv(eye(1)+D'*D)*B';
-C'*C+C'*D*inv(eye(1)+D'*D)*D'*C -A'+C'*D*inv(eye(1)+D'*D)*B']
```
and its eigenvalues
```
>> eig(H)
```
We find the optimal eigenvalues: $\lambda_1 = -3.1794$ and $\lambda_2 = -2.3220$.
It is also possible to use the command `lqr` as follows:
```
>> Q=C'*C-C'*D*inv(eye(1)+D'*D)*D'*C
>> [K,P,E]=lqr(A-B*inv(eye(1)+D'*D)*D'*C,B,Q,(eye(1)+D'*D))
```

7.8 Optimal Control for SISO Systems

Let us now consider the case of linear time-invariant SISO systems. For this class of systems there exist results of significant interest in simplifying optimal control. In particular, let us consider a generalization of index (7.75) so as to weight unequally the energy terms associated with the input and output:

$$J = \int_0^\infty (\mathbf{y}^T\mathbf{y} + \mathbf{u}^T r\mathbf{u})dt = \int_0^\infty (y^2 + ru^2)dt \tag{7.82}$$

This optimal control problem is the same as that associated with (7.61) on condition that $\mathrm{Q} = \mathrm{C}^T\mathrm{C}$ and $\mathrm{R} = r$ (note that because it is a SISO system, $\mathrm{R} \in \mathbb{R}^{1\times 1}$ is a scalar quantity).

Let us suppose that the system has transfer function $G(s) = \frac{b(s)}{a(s)}$ and it is in minimal form.

To resolve optimal control a realization $(\mathrm{A}, \mathrm{B}, \mathrm{C})$ of the system could be adopted and its optimal eigenvalues be found from the Hamiltonian:

$$\mathrm{H} = \begin{bmatrix} \mathrm{A} & -\mathrm{B}r^{-1}\mathrm{B}^T \\ -\mathrm{C}^T\mathrm{C} & -\mathrm{A}^T \end{bmatrix} \tag{7.83}$$

Let us recall that generally, given the weight matrices Q and R (in the SISO case, r), the optimal eigenvalues do not depend on the chosen reference system. In this case, therefore, the optimal eigenvalues do not depend on the adopted system realization $(\mathrm{A}, \mathrm{B}, \mathrm{C})$ but only on the transfer function $G(s)$.

For SISO systems there is another method for finding optimal eigenvalues which is simpler. They are given by the Letov theorem.

Theorem 17 (Letov theorem) *For a minimal linear time-invariant SISO system with transfer function $G(s) = \frac{b(s)}{a(s)}$, the optimal eigenvalues according to index (7.82) are given by the roots with negative real part of the following equation:*

$$a(s)a(-s) + r^{-1}b(s)b(-s) = 0 \tag{7.84}$$

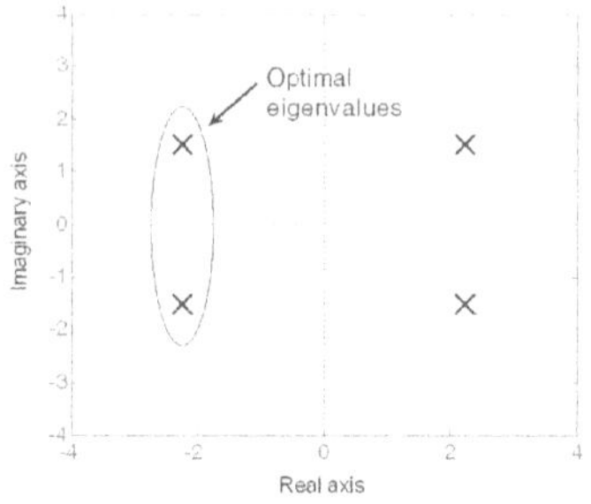

FIGURE 7.5
Solutions for equation (7.87) and optimal eigenvalues (with $r = 2$) for the system $G(s) = \frac{10-s}{(s+1)(s-2)}$.

Example 7.7

Consider a linear time-invariant SISO system with transfer function $G(s) = \frac{1}{s}$. Suppose that one wants to find the optimal eigenvalues according to the functional (7.82) with $r = 1$.
Applying the Letov theorem then

$$-s^2 + 1 = 0 \tag{7.85}$$

and so the optimal eigenvalue is $\lambda_{opt} = -1$.
Alternatively, a realization of the system is given by $\mathrm{A} = 0$, $\mathrm{B} = 1$, $\mathrm{C} = 1$. In this reference system, the CARE equation reads:

$$-\mathrm{P}^2 + 1 = 0 \Rightarrow \bar{\mathrm{P}} = 1 \tag{7.86}$$

So $\mathrm{K}_{opt} = -\mathrm{B}\bar{\mathrm{P}} = -1$. Finally, given that $\mathrm{A}_c = \mathrm{A} - \mathrm{BK}_{opt} = -1$, the optimal eigenvalue is found: $\lambda_{opt} = -1$.

Example 7.8

Let us consider the linear time-invariant SISO system with transfer function $G(s) = \frac{10-s}{(s+1)(s-2)}$ and suppose $r = 2$. In this case, the optimal eigenvalues are given by:

$$(s+1)(s-2)(1-s)(-s-2) + \frac{1}{2}(10-s)(10+s) = 0$$

$$\Rightarrow s^4 - 5.5s^2 + 54 = 0 \tag{7.87}$$

The solutions to equation (7.87) are $s_{1,2} = -2.2471 \pm j1.5163$ and $s_{3,4} = 2.2471 \pm j1.5163$ (see Figure 7.5). The optimal eigenvalues are $\lambda_{1,2} = -2.2471 \pm j1.5163$.

Example 7.9

Let us consider a linear time-invariant SISO system with transfer function $G(s) = \frac{1}{s^2}$ and let $r = 1$. In this case, the Letov formula becomes:

$$s^4 + 1 = 0 \tag{7.88}$$

The solutions to equation (7.88) are $s_{1,2} = -0.7071 \pm j0.7071$ and $s_{3,4} = 0.7071 \pm j0.7071$ and they belong to the circumference of unitary radius (Figure 7.6).

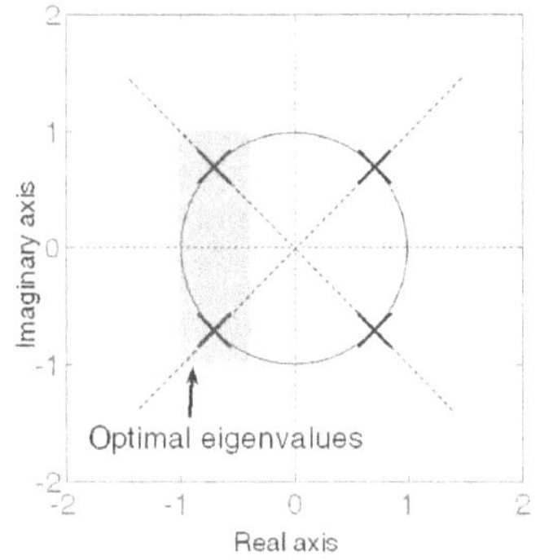

FIGURE 7.6
Solutions for equation (7.88) and the optimal eigenvalues ($r = 1$) for system $G(s) = \frac{1}{s^2}$.

The result in Example 7.9 can be generalized to a system of n integrators in cascade ($G(s) = \frac{1}{s^n}$). In fact, in this case ($r = 1$), the equation for obtaining the optimal eigenvalues becomes:

$$(-1)^n s^{2n} + 1 = 0 \tag{7.89}$$

Example 7.10

Let us consider the all-pass system $G(s) = \frac{(1-s)(2-s)}{(1+s)(2+s)}$. In this case the Letov formula becomes:

$$(1-s)(2-s)(1+s)(2+s) + \frac{1}{r}(1-s)(2-s)(1+s)(2+s) = 0$$

$$\Rightarrow (1+\frac{1}{r})(1-s)(2-s)(1+s)(2+s) = 0$$

The optimal eigenvalues are $\lambda_1 = -1$ and $\lambda_2 = -2\ \forall r$.

Generalizing the result of the previous exercise, it may be intuited that for whatever type of all-pass system the optimal eigenvalues do not depend on r and are given by the eigenvalues (if it is the case, sign changed) of the open-loop system.

Example 7.11

$G(s) = \frac{(1+s)(2-s)}{(1-s)(2+s)}$ is an all-pass system. For $\forall r$ the optimal eigenvalues are $\lambda_1 = -1$ and $\lambda_2 = -2$.

Example 7.12

Consider now a system $G(s) = \frac{b(s)}{a(s)}$ satisfying the following property: $G(s) = -G^T(-s)$. For such a system the optimal eigenvalues for $r = 1$ are given by the roots of $a(s)+b(s) = 0$. In fact, since $\frac{b(s)}{a(s)} = -\frac{b(-s)}{a(-s)}$

$$b(s)a(-s) + b(-s)a(s) = 0 \tag{7.90}$$

and since from the Letov formula with $r = 1$ one has

$$a(s)a(-s) + b(s)b(-s) = 0 \tag{7.91}$$

adding equations (7.90) and (7.91), one gets:

$$a(s)a(-s) + b(s)b(-s) + b(s)a(-s) + b(-s)a(s) = 0$$

and thus

$$(a(-s) + b(-s))\,(a(s) + b(s)) = 0$$

We will see in Chapter 9 that systems satisfying $G(s) = -G^T(-s)$ are *loss-less* systems. For such systems, it can be proven (the theorem is often referred as Chebyshev theorem) that, given that $G(s) = \frac{b(s)}{a(s)}$ is loss-less, then the polynomial $a(s) + b(s)$ has negative real part roots. Therefore, for loss-less systems, the optimal eigenvalues with $r = 1$ are given by:

$$a(s) + b(s) = 0 \tag{7.92}$$

From the Letov theorem, some important considerations can be drawn in the extreme cases where $r \to 0$ or $r \to \infty$.

If r is very large (in the limit case $r \to \infty$), the term $\frac{1}{r}b(s)b(-s)$ in the Letov formula is negligible compared to term $a(s)a(-s)$. The optimal eigenvalues are given by $a(s)a(-s) = 0$. In this case, conclude that if the system has poles with negative real part, the optimal eigenvalues coincide with the open-loop system poles. If, instead, the system also has some poles with positive real part, then the optimal eigenvalues are given by the open-loop poles with negative real part and the sign reversed open-loop poles with positive real part. In conclusion, if the open-loop system is asymptotically stable then the optimal control with $r \to \infty$ is the one which leaves the system eigenvalues unchanged. The case with $r \to \infty$ from a physical viewpoint implies minimizing as much as possible energy at the input (and, at the limit, not acting on the system at all, given that by definition it is asymptotically stable).

When $r \to 0$, the term $a(s)a(-s)$ can be neglected with respect to $\frac{1}{r}b(s)b(-s)$ and the Letov formula can be approximated to $\frac{1}{r}b(s)b(-s) = 0$. This formula can calculate m optimal eigenvalues (where m is the order of polynomial $b(s)$). The other $n - m$ eigenvalues (n is the system order and so also the order of polynomial $a(s)$) can be obtained from the formula $s^{2(n-m)} = (-1)^{n-m+1}b_0^2 r^{-1}$ where b_0 is the coefficient associated to the higher power in the polynomial $b(s)$, that is, $b(s) = b_0 s^m + b_1 s^{m-1} + \ldots + b_m$. The eigenvalues are therefore arranged in a Butterworth configuration in a circle of radius $(\frac{b_0^2}{r})^{\frac{1}{2(n-m)}}$.

MATLAB® Exercise 7.6

Let us reconsider the example in Exercise 7.1 and calculate the optimal eigenvalues from the Letov formula for various values of r.

- If $r = 1$, then $\Delta(s) = (s^2 - 2s - 3)(s^2 + 2 - 3s) + (s+2)(-s+2) = s^4 - 11s + 13$. The solutions of $\Delta(s) = 0$ with negative real part are $\lambda_{1,opt} = -3.1070$ and $\lambda_{2,opt} = -1.1605$ and coincide with the optimal eigenvalues calculated in Exercise 7.1.
- If $r = 10$, then the Letov formula produces $\Delta(s) \simeq (s^2 - 2s - 3)(s^2 + 2 - 3s)$ and so the optimal eigenvalues are given by the open-loop eigenvalues (if it is the

case, sign changed) $\lambda_{1,opt} \simeq -1$ and $\lambda_{2,opt} \simeq -3$. To evaluate the accuracy of the approximation use the MATLAB command:

```
>> [K,P,E]=lqr(A,B,C'*C,10)
```

$\lambda_{1,opt} = -1.0184$ and $\lambda_{2,opt} = -3.0104$ are obtained.

- If $r = 0.01$, then the Letov formula gives a closed-loop eigenvalue equal to the zero of the open-loop system, that is $\lambda_{1,opt} = -2$, whereas the other eigenvalue is given by the formula $s^2 = (-1)^2 \cdot 100$ (in fact, $b_0 = 1$), and so $\lambda_{2,opt} = -10$.
  Let us consider the eigenvalues obtained by applying the MATLAB command

  ```
  >> [K,P,E]=lqr(A,B,C'*C,0.01)
  ```

  that is $\lambda_{1,opt} = -1.9629$ and $\lambda_{2,opt} = -10.3028$. Notice that even here the approximation from the Letov formula in the extreme case of very small r is more than sufficient.

Example 7.13

Given the system $G(s) = \frac{s+20}{(s^2-4s+8)^2}$ find the optimal eigenvalues with respect to r assuming that $J = \int_0^\infty (y^T y + u^T r u)dt$. Then fix r so that one optimal eigenvalue is equal to $\lambda = -5$.

Solution

System $G(s) = \frac{s+20}{(s^2-4s+8)^2}$ is a SISO system, in minimal form, and unstable with poles equal to $s_{1,2} = 2 \pm 2j$. Since the functional J is defined as per equation (7.82), all the assumptions that allow one to use the Letov formula to solve the assigned optimal control problem are respected.
So, we have:

$$a(s)a(-s) + r^{-1}b(s)b(-s) = 0 \Rightarrow$$

$$(s^2 - 4s + 8)^2(s^2 + 4s + 8)^2 + r^{-1}(20 + s)(20 - s) = 0 \Rightarrow$$

$$s^4 - r^{-1}s^2 + 400r^{-1} + 64 = 0 \tag{7.93}$$

Solving the bi-quadratic equation and considering only the two solutions with negative real part, we have:

$$s_{opt,1,2} = -\sqrt{\frac{r^{-1} \pm \sqrt{r^{-2} - 1600r^{-1} - 256}}{2}}$$

To ensure that one of these solutions is equal to $\bar{\lambda}$, we have to impose that (7.93) equals zero for $s = \bar{\lambda}$:

$$\bar{\lambda}^4 - r^{-1}\bar{\lambda}^2 + 400r^{-1} + 64 = 0 \Rightarrow$$

$$r = \frac{\bar{\lambda}^2 - 400}{\bar{\lambda}^4 + 64}$$

Since $r > 0$, only $\bar{\lambda} < -20$ is admissible. So it is not possible to follow the assigned specifications.
The roots of the equation (7.93) can be studied by building an opportune root locus. Consider

$$s^4 - r^{-1}s^2 + 400r^{-1} + 64 = 0 \Rightarrow$$

$$1 + r^{-1}\frac{400 - s^2}{s^4 + 64} = 0$$

By studying the root locus of the fictitious transfer function $\tilde{G}(s) = \frac{400-s^2}{s^4+64}$, all the

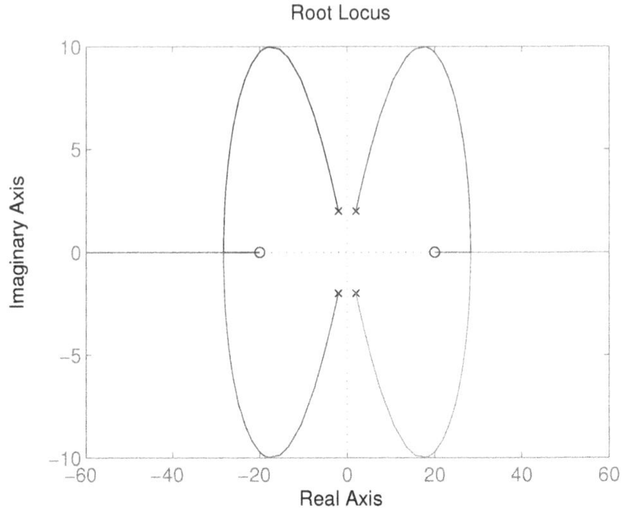

FIGURE 7.7
Root locus of $\tilde{G}(s) = \frac{400-s^2}{s^4+64}$.

solutions of the equation (7.93) can be obtained (Figure 7.7). To know the optimal eigenvalues with respect to r, we have to refer to the left half plane of this root locus. Notice there are no values of r^{-1} (and so of r) which ensure real solutions with $-20 < \bar{\lambda} < 0$.

7.9 Linear Quadratic Regulator with Cross-weighted Cost

There exist several generalizations of the LQR problem presented above. An important one derives from the consideration that the variables $\mathbf{z}(t)$ to keep small may contain a feedthrough term. In fact, if we consider $\mathbf{z}(t) = \mathrm{C}\mathbf{x}(t) + \mathrm{D}\mathbf{u}(t)$, the index to minimize becomes:

$$J = \int_0^\infty (\mathbf{x}^T\mathrm{C}^T\mathrm{C}\mathbf{x} + 2\mathbf{x}^T\mathrm{C}^T\mathrm{D}\mathbf{u} + \mathbf{u}^T(\mathrm{R} + \mathrm{D}^T\mathrm{D})\mathbf{u})dt \tag{7.94}$$

For this reason, in LQR problems the more general definition of the index to minimize

$$J = \int_0^\infty (\mathbf{x}^T\bar{\mathrm{Q}}\mathbf{x} + 2\mathbf{x}^T\bar{\mathrm{N}}\mathbf{u} + \mathbf{u}^T\bar{\mathrm{R}}\mathbf{u})dt \tag{7.95}$$

may be considered with general weight matrices $\bar{Q}$, $\bar{N}$, and $\bar{R}$.

Index (7.95) appears when the variables to keep small contain derivatives of the state variables, but cross-coupled costs also appear when frequency-dependent weighting terms are used.

In this case, the Riccati equation becomes:

$$P(A-B\bar{R}^{-1}\bar{N}^T)+(A-B\bar{R}^{-1}\bar{N}^T)^TP-PB\bar{R}^{-1}B^TP+(\bar{Q}-\bar{N}\bar{R}^{-1}\bar{N}^T)=0 \quad (7.96)$$

with $P \in \mathbb{R}^{n \times n}$. If $\overline{P}$ is the solution, the optimum gains are now given by:

$$K_{opt} = \bar{R}^{-1}(\bar{N}^T + B^T\overline{P}) \quad (7.97)$$

The LQR problem defined by index (7.95) may be also solved in MATLAB® with the command `lqr`.

7.10 Finite-horizon Linear Quadratic Regulator

For system (7.59), let us consider now the following performance index:

$$J = \int_t^T (\mathbf{x}^T(\tau)Q\mathbf{x}(\tau) + \mathbf{u}^T(\tau)R\mathbf{u}(\tau))d\tau \quad (7.98)$$

which assumes that the upper bound of the integral in the performance index is a finite time T rather than infinite and that the lower bound is t rather than zero.

The gains of the optimal control law are now given by:

$$K_{opt} = R^{-1}B^TP(t) \quad (7.99)$$

where P(t) is now a time-varying matrix given by the solution of the *differential Riccati equation*:

$$\dot{P} + Q - PBR^{-1}B^TP + PA + A^TP = 0 \quad (7.100)$$

with $P(t) = 0$.

Equation (7.100) is difficult to analytically solve for systems with order higher than two. It is usually numerically solved with an iterative backward in time procedure.

The solution of the infinite horizon problem is recovered from that of equation (7.100) by taking:

$$\overline{P} = \lim_{t \to -\infty} P(t,T) = \lim_{T \to \infty} P(0,T) \quad (7.101)$$

In the following example the use of MATLAB® symbolic computation to solve a simple differential Riccati equation is illustrated.

MATLAB® Exercise 7.7

Consider system:

$$\dot{x} = x + u \tag{7.102}$$

and the performance index (7.98) with $Q = 1$ and $R = 1$. Since a state-space realization of system (7.102) is given by $A = 1$ and $B = 1$, equation (7.100) becomes:

$$\dot{P} - P^2 + 2P + 1 = 0 \tag{7.103}$$

It can be solved with the MATLAB command:

```
>> dsolve('DP=P^2-2*P-1')
```

obtaining

$$P(t) = 1 - \sqrt{2}\tanh(\sqrt{2}t + \sqrt{2}c_1) \tag{7.104}$$

where the constant c_1 is obtained by imposing $P(0) = 0$.
The solution of the infinite horizon problem is obtained as:

$$\overline{\mathrm{P}} = \lim_{t \to -\infty} \mathrm{P}(t, T) = 1 + \sqrt{2} \tag{7.105}$$

that is the same result that can be obtained by solving the algebraic Riccati equation (7.62) with the command:

```
>> are(1,1,1)
```

7.11 Optimal Control for Discrete-time Linear Systems

To conclude this chapter, let us look briefly at discrete-time systems. Consider a discrete-time system in state-space form:

$$\begin{aligned} \mathbf{x}_{k+1} &= \mathrm{A}\mathbf{x}_k + \mathrm{B}\mathbf{u}_k \\ \mathbf{y}_k &= \mathrm{C}\mathbf{x}_k \end{aligned} \tag{7.106}$$

The quadratic index in this case is defined as:

$$J = \sum_{k=0}^{\infty} (\mathbf{x}_k^T \mathrm{Q} \mathbf{x}_k + \mathbf{u}_k^T \mathrm{R} \mathbf{u}_k) dt \tag{7.107}$$

with Q positive semi-definite and R positive definite. The gain matrix of control law $\mathbf{u}_k = -\mathrm{K}\mathbf{x}_k$ which minimizes index (7.107) is found by solving the algebraic Riccati equation for discrete-time systems:

$$\mathrm{A}^T\mathrm{P}\mathrm{A} - \mathrm{P} - \mathrm{A}^T\mathrm{P}\mathrm{B}(\mathrm{R} + \mathrm{B}^T\mathrm{P}\mathrm{B})^{-1}\mathrm{B}^T\mathrm{P}\mathrm{A} + \mathrm{Q} = 0 \tag{7.108}$$

and by making

$$\mathrm{K} = (\mathrm{B}^T\mathrm{P}\mathrm{B} + \mathrm{R})^{-1}\mathrm{B}^T\mathrm{P}\mathrm{A} \tag{7.109}$$

MATLAB® Exercise 7.8

Consider the discrete-time LTI MIMO system with state-space matrices:

$$\mathrm{A} = \begin{bmatrix} -2 & 0 & 0 \\ 0 & 1 & 0 \\ 0 & 0 & 0.2 \end{bmatrix}; \mathrm{B} = \begin{bmatrix} 1 & 0.1 \\ 2 & 5 \\ 0.3 & 3 \end{bmatrix}; \mathrm{C} = \begin{bmatrix} 1 & 0 & 1 \\ -1 & 2 & 1 \end{bmatrix} \tag{7.110}$$

and the optimal control problem (7.107) with $\mathrm{Q} = \mathrm{C}^T\mathrm{C}$ and $\mathrm{R} = 2\mathrm{I}$.
Define system (7.110) in MATLAB

```
>> A=[-2 0 0; 0 1 0; 0 0 0.2]
>> B=[1 0.1; 2 5; 0.3 3]
>> C=[1 0 1; -1 2 1]
>> D=zeros(2)
```

and then use the command `dlqr` to calculate the linear quadratic regulator for discrete-time LTI systems:

```
>> [K,P,E]=dlqr(A,B,C'*C,2*eye(2))
```

One obtains the closed-loop optimal eigenvalues: $\lambda_1 = -0.3530$, $\lambda_2 = 0.3753$ and $\lambda = 0.0050$ as it can be verified with command:

```
>> eig(A-B*K)
```

7.12 Exercises

1. Given the system $G(s) = \frac{s+2}{s^2+4s+6}$ calculate the optimal controller that minimizes the index defined by $\mathrm{Q} = \mathrm{C}^T\mathrm{C}$ in these three cases: $r = r_1 = 1$, $r = r_2 = 0.01$ and $r = r_3 = 20$. Verify the performance of the control law obtained.
2. Given the system with transfer function $G(s) = \frac{2s-1}{s(s-1)}$ calculate the optimal eigenvalues with respect to r, if the index to optimize is $J = \int_0^\infty (\mathbf{y}^T\mathbf{y} + \mathbf{u}^T r \mathbf{u})dt$. Then calculate the characteristic values of the system fixing r so that the optimal eigenvalue is $\lambda = -\sqrt{2}$.
3. Design the linear quadratic regulator (with $r = 2$) for the system $G(s) = \frac{s^2+2s+1}{s^3-s^2+5s+3}$ and calculate the optimal eigenvalues.
4. Given the system with state-space matrices:

$$\mathrm{A} = \begin{bmatrix} 0 & 1 \\ 3 & -2 \end{bmatrix}; \mathrm{B} = \begin{bmatrix} 0 \\ 1 \end{bmatrix}; \mathrm{C} = \begin{bmatrix} -1 & 1 \end{bmatrix}$$

design the linear quadratic regulator with $\mathrm{Q} = \mathrm{C}^T\mathrm{C}$ and $r = 5$.

5. Given the system with state-space matrices:

$$\mathrm{A} = \begin{bmatrix} 0 & 1 & 0 \\ -1 & 0 & 0 \\ 0 & 0 & 3 \end{bmatrix}; \mathrm{B} = \begin{bmatrix} 0 \\ 1 \\ -1 \end{bmatrix}; \mathrm{C} = \begin{bmatrix} 1 & 1 & 1 \end{bmatrix}$$

design the linear quadratic regulator with $\mathrm{Q} = \mathrm{I}$ and $r = 7$.

8

Closed-loop Balanced Realization

CONTENTS

For the open-loop balanced representation, linear Lyapunov equations related to the controllability gramian and to the obsevability gramian are considered. Similarly, in this chapter, to derive the closed-loop representation, the CARE and the FARE are considered. The chapter is devoted to derive low order controllers with guaranteed closed loop-stability. Another class of invariants for the systems, the LQR characteristic values, are introduced. Moreover, another canonical form of the original system, called closed loop representation, is presented. Robust algorithms to derive it are dealt with, along with the corresponding MATLAB codes.

8.1 Synthesis of a Compensator for High-Order Systems

In this chapter, we will look at synthesizing a compensator for a high order system. One strategy is to consider a lower order model of the process, $G_r(s)$, and design a lower order compensator $C_r(s)$. In this way, the regulator is bound to work well for the lower order system (see Figure 8.1), but not necessarily for the original model $G(s)$ (see Figure 8.2). Therefore, the efficacy of the project needs to be verified (with numerical simulations).

Another strategy is to design $C(s)$ by assigning poles (design of the observer and of the control law) on the basis of system $G(s)$. In this way, an n order compensator is obtained which can be approximated through model order reduction. This case is called *direct approximation of the compensator*,

DOI: 10.1201/9781003196921-8

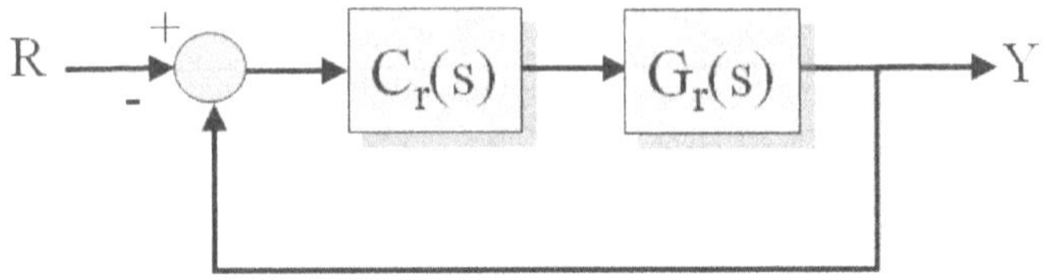

FIGURE 8.1
Control scheme with lower-order compensator applied to a lower order system $G_r(s)$.

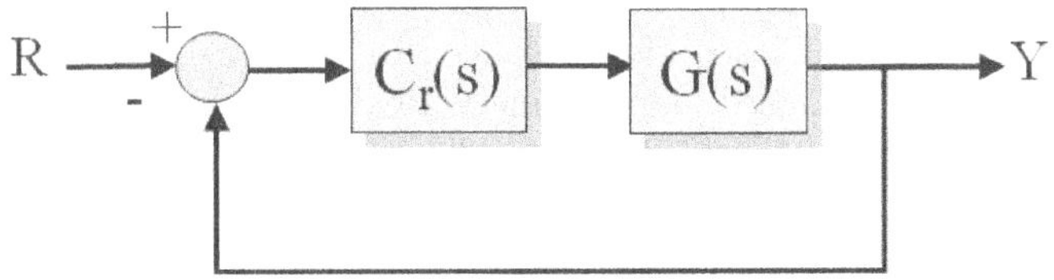

FIGURE 8.2
Control scheme with lower-order compensator applied to a $G(s)$ system.

whereas designing a lower-order compensator is known as *indirect approximation of the compensator.*

In both techniques, approximating the compensator is quite separate from designing it. Below, we will see how closed-loop balancing is performed by using the CARE introduced in the previous chapter and a dual equation, the so-called FARE (solving the dual problem of optimal filtering). With this technique, instead of approximating the controller or the process prior to or after the synthesis, the controller and process are approximated simultaneously and during synthesis.

8.2 Filtering Algebraic Riccati Equation

The dual equation of the CARE is called FARE (Filtering Algebraic Riccati Equation) and is defined as:

$$\mathrm{A}\Pi + \Pi\mathrm{A}^T - \Pi\mathrm{C}^T\mathrm{C}\Pi + \mathrm{B}\mathrm{B}^T = 0 \tag{8.1}$$

Just as the CARE solves an optimal control problem once a certain functional to minimize is given, the FARE is used to design an optimal observer according to a particular criterion.

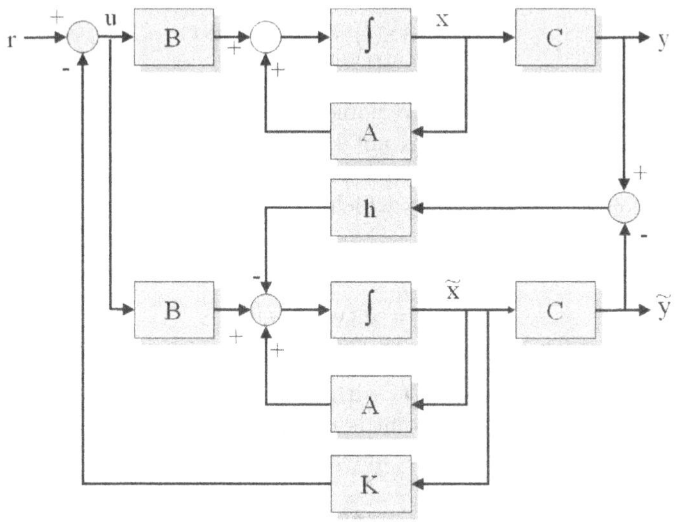

FIGURE 8.3
Control scheme with a state linear regulator and an asymptotic observer.

Referring to Figure 8.3, remember that optimal control based on a linear regulator requires knowing the entire state vector $\mathbf{x}(t)$ or, should all the state variables not be measurable, an estimate $\tilde{\mathbf{x}}(t)$ of the state via an observer. To design an optimal controller, even designing the observer must be based on optimizing a particular criterion. Designing a compensator means finding gains K and h on the basis of established criteria. In the previous chapter, an optimal strategy was found based on minimizing a quadratic functional in choosing the gains K. An optimization criterion can be dually defined for the design of the observer.

Remember that the fundamental condition which must be satisfied in an observer project is to make matrix A_0 a stability matrix. Once the error is defined as $\mathbf{e}(t) = \mathbf{x}(t) - \tilde{\mathbf{x}}(t)$, then the error system dynamic is given by:

$$\dot{\mathbf{e}}(t) = \mathrm{A}_0\mathbf{e}(t) \tag{8.2}$$

so that, to make $\mathbf{x}(t) \simeq \tilde{\mathbf{x}}(t)$, that is, $\mathbf{e}(t) = e^{\mathrm{A}_0 t}\mathbf{e}(0) \to 0$, the gains h must be chosen such that all the eigenvalues of $\mathrm{A}_0 = \mathrm{A} - \mathrm{hC}$ have negative real part.

Generally, noise added to the state variables, that is an uncertainty which makes the variables of state differ from the mathematical model, must also be considered. Another source of noise to be considered is the measurement noise. All this leads to formulating the following state-space model:

$$\begin{aligned}\dot{\mathbf{x}}(t) &= \mathrm{A}\mathbf{x}(t) + \mathrm{B}\mathbf{u}(t) + \mathbf{d}(t) \\ \mathbf{y}(t) &= \mathrm{C}\mathbf{x}(t) + \mathbf{v}(t)\end{aligned} \tag{8.3}$$

where $\mathbf{d}(t)$ (noise on the state variables) and $\mathbf{v}(t)$ (measurement noise) are stochastic signals. Their value is not known exactly but their statistics are. Suppose that $\mathbf{d}(t)$ and $\mathbf{v}(t)$ are random signals with zero mean and suppose we know the covariance matrices which characterize them:

$$\begin{aligned}\mathrm{M}_d &= E\{\mathbf{d}(t)\mathbf{d}(t)^T\} \\ \mathrm{M}_v &= E\{\mathbf{v}(t)\mathbf{v}(t)^T\}\end{aligned}$$

where $\mathrm{M}_d \in \mathbb{R}^{n\times n}$ and $\mathrm{M}_v \in \mathbb{R}^{p\times p}$. $E\{x(t)\} = \int_{-\infty}^{\infty} x p_{x(t)}(x)dt$ is the expected value of the stochastic signal $x(t)$ with probability density function $p_{x(t)}(x)$. Even the error system at this point is non-deterministic. Let us establish the eigenvalues so as to minimize the index defined by:

$$\bar{J} = E\{(\mathbf{x}(t) - \tilde{\mathbf{x}}(t))^T(\mathbf{x}(t) - \tilde{\mathbf{x}}(t))\} \tag{8.4}$$

Optimal filtering is the dual problem of optimal control. Optimal gains are given by $\mathrm{h}_{opt}^T = \mathrm{M}_v^{-1}\mathrm{C}\Pi$ where Π is the solution of the Riccati equation:

$$\mathrm{A}\Pi + \Pi\mathrm{A}^T - \Pi\mathrm{C}^T\mathrm{M}_v^{-1}\mathrm{C}\Pi + \mathrm{M}_d = 0 \tag{8.5}$$

M_v is a $p \times p$ matrix. If $\mathrm{M}_v = \mathrm{I}$ and $\mathrm{M}_d = \mathrm{B}\mathrm{B}^T$ in equation (8.5), the result is the FARE equation (8.1). As regards signal $\mathbf{v}(t)$, this choice means that its covariance matrix is unitary. As for signal $\mathbf{d}(t)$ making $\mathrm{M}_d = \mathrm{B}\mathrm{B}^T$ means hypothesizing that $\mathbf{d}(t) = \mathrm{B}\xi(t)$ where $\xi(t)$ is a stochastic signal which also has a unitary covariance matrix.

8.3 Computing the Closed-loop Balanced Realization

The CARE (7.76) and FARE (8.1) are nonlinear equations, they are also normalized given that $\mathrm{R} = \mathrm{I}$ and $\mathrm{M}_v = \mathrm{I}$. With these two equations and using the same procedure which produces open-loop balanced realizations from gramians equations, we can obtain a closed-loop balanced form. To simplify, we will refer to the closed-loop balanced realization based on these normalized equations even though the procedure is general and can be applied to the corresponding not-normalized equations.

Let us consider the CARE and FARE equations:

$$\mathrm{A}^T\mathrm{P} + \mathrm{P}\mathrm{A} - \mathrm{P}\mathrm{B}\mathrm{B}^T\mathrm{P} + \mathrm{C}^T\mathrm{C} = 0 \tag{8.6}$$

$$\mathrm{A}\Pi + \Pi\mathrm{A}^T - \Pi\mathrm{C}^T\mathrm{C}\Pi + \mathrm{B}\mathrm{B}^T = 0 \tag{8.7}$$

and let us indicate as $\bar{\mathrm{P}}$ and $\bar{\Pi}$ the only positive definite solutions of CARE and FARE. These two matrices depend on the reference system in the same way as the gramians depend on the reference system (this can be proved by following the same steps done for the case of open-loop balanced realization). The eigenvalues of $\bar{\Pi}\bar{\mathrm{P}}$, however, do not depend on the reference system. They allow to define another set of system invariants, called *characteristic values of the system.*

Definition 17 (Characteristic values of the system) *The characteristic values of a system in minimal form $\mu_1 \geq \mu_2 \geq \ldots \geq \mu_n$ are the square roots in descending order of the eigenvalues of $\bar{\Pi}\bar{\mathrm{P}}$. They are invariant quantities, i.e., they do not depend on the state-space representation of the system.*

The characteristic values of a system are also known as LQR characteristic values. They are fundamental to derive the closed-loop balanced realization, that is, the realization where the CARE and FARE produce the same diagonal solution, i.e., $\bar{\Pi} = \bar{\mathrm{P}} = \mathrm{diag}(\mu_1, \mu_2, \ldots, \mu_n)$.

Definition 18 (Closed-loop balanced realization) *Given a linear time-invariant system in minimal form, the closed-loop balanced realization is that realization in which $\bar{\Pi} = \bar{\mathrm{P}} = \mathrm{diag}(\mu_1, \mu_2, \ldots, \mu_n)$, where $\bar{\mathrm{P}}$ and $\bar{\Pi}$ are the only positive definite solutions of the CARE and FARE.*

Note that the hypothesis which makes possible to define the closed-loop balanced realization of a system is the existence of the positive definite solutions for the CARE and FARE, that is, that the system is completely controllable and observable. As opposed to the case of open-loop balancing, it is no longer necessary that the open-loop system is asymptotically stable. Besides, since the system is controllable and optimal control is being designed, the closed-loop system is guaranteed to be asymptotically stable.

This balancing technique arose from the need to balance systems with small damping. Known as large space flexible structures they are used to model antennae or light satellites with solar energy panels. In particular, often distributed parameter models for these structures or concentrated parameter models with a large number of state variables are formulated. The closed-loop balancing of this class of structures arose from the need to design controllers which could stabilize the oscillations of these very flexible structures. Open-loop balancing for these types of structures is numerically difficult because they have small damping. Closed-loop balancing has the advantage of not causing numerical errors in finding the solution and it allows the controller to be designed while effecting the balancing.

Closed-loop balancing facilitates the design of the optimal regulator and observer at the same time by proceeding analogously to the open-loop case to reduce system order. In the next paragraph, we will analyze in detail the issues concerning lower order models based on closed-loop balancing.

8.4 Procedure for Closed-loop Balanced Realization

The procedure for closed-loop balanced realization is analogous to the one for open-loop, on condition that the CARE and FARE equations are used instead of the two Lyapunov equations of the two gramians. The procedure is illustrated below. Note that the FARE solution is calculated first, then singular value decomposition is applied to matrix $\bar{\Pi}$ and from this the transformation matrix P_1 is defined. In the new reference system (where it is easy to verify that $\tilde{\Pi} = I$) the CARE solution needs to be calculated, then proceeding to its singular value decomposition and then defining the transformation matrix T_2.

MATLAB® Exercise 8.1

The procedure for closed-loop balanced realization is reported below.

```
function [system_bal,G,PIGREEK_bal,P_care_bal]=balrealmcc(system)
   % closed-loop balanced realization
   %    [system_bal, G, PIGREEK_bal, P_care_bal]=balrealmcc(system)
   % system is defined in state-space form
   A=system.A;
   B=system.B;
   C=system.C;
   %rank(ctrb(system))
   %rank(obsv(system))
   %controllability and observability hypotheses
   PIGREEK=are(A',C'*C,B*B');  %FARE
   [Uc,Sc2,Vc]=svd(PIGREEK);
   P1=Vc*sqrt(Sc2);
   Atilde=inv(P1)*A*P1;
   Btilde=inv(P1)*B;
   Ctilde=C*P1;
   %%%verify: Wtildec2=I
   PIGREEKtilde=are(Atilde',Ctilde'*Ctilde,Btilde*Btilde'); %FARE
   Pcaretilde=are(Atilde,Btilde*Btilde',Ctilde'*Ctilde);  %CARE
   [Uo,So2,Vo]=svd(Pcaretilde);
   T2=Vo*(So2)^(-1/4);
   Abil=inv(T2)*Atilde*T2;
   Bbil=inv(T2)*Btilde;
   Cbil=Ctilde*T2;
   system_bal=ss(Abil,Bbil,Cbil,0);
   PIGREEK_bal=are(Abil',Cbil'*Cbil,Bbil*Bbil'); %New FARE
   P_care_bal=are(Abil,Bbil*Bbil',Cbil'*Cbil);  %CARE
   %%%%%%%%%%%%Verify: calculate the characteristic values of system
   Pcare=are(A,B*B',C'*C);  %CARE
   G=sqrt(eig(PIGREEK*Pcare));
```

This function can be applied to a system in state-space form. For instance, given the system:

```
>> A=[-0.2 0.5 0 0 0;
   0 -0.5 1.6 0 0;
   0 0 -14.3 87.7 0;
   0 0 0 -25 75;
   0 0 0 0 -10]
```

```
>> B=[0 0 0 0 30]'
>> C=[1 0 0 0 0]
>> system=ss(A,B,C,0)
```

the closed-loop balanced realization is constructed with the commands:

```
>> [system_bal,G,PIGREEK_bal,P_care_bal]=balrealmcc(system)
```

We get the characteristic values $\mu_1 = 3.9859$, $\mu_2 = 0.7727$, $\mu_3 = 0.1276$, $\mu_4 = 0.0101$ and $\mu_5 = 0.0003$, suggesting the choice of $r = 3$ for the order of the reduced order model.

8.5 Reduced Order Models Based on Closed-loop Balanced Realization

Let $\mathbf{X}^*$ represent the reference system for the closed-loop balanced form. In this reference system the CARE solution, $\bar{\mathrm{P}}^*$, is diagonal and the optimal gains are of the form:

$$K^*_{opt} = \mathrm{B}^{*T}\mathrm{diag}(\mu_1, \mu_2, \ldots, \mu_n)$$

Furthermore, $\mathrm{A}^*_c = \mathrm{A}^* - \mathrm{B}^*\mathrm{K}^*_{opt} = \mathrm{A}^* - \mathrm{B}^*\mathrm{B}^{*T}\bar{\mathrm{P}}^*$ certainly has all eigenvalues with negative real part.

If the index r is such that the system characteristic values can be divided into two groups, so that $\mu_1 \geq \mu_2 \geq \ldots \geq \mu_r \gg \mu_{r+1} \geq \ldots \geq \mu_n$, then a lower order model could be constructed which only accounts for the first r state variables of the system. In other words, instead of feeding back the n state variables, only r are fed back.

So, in A^*_c, we will use the matrix $\bar{\mathrm{P}}^*$ defined by

$$\bar{\mathrm{P}}^* = \begin{bmatrix} \mu_1 & & & & & \\ & \ddots & & & & \\ & & \mu_r & & & \\ & & & 0 & & \\ & & & & \ddots & \\ & & & & & 0 \end{bmatrix}$$

Corresponding to this approximation, the control law becomes:

$$\mathbf{u} = -\mathrm{K}^*_{opt}\mathbf{x}^* = -\mathrm{B}^{*T}\bar{\mathrm{P}}^*\mathbf{x}^* = -\mathrm{B}^{*T} \begin{bmatrix} \mu_1 x_1^* \\ \vdots \\ \mu_r x_r^* \\ 0 \\ \vdots \\ 0 \end{bmatrix}$$

If the discarded characteristic values $\mu_{r+1}, \ldots, \mu_n$ are effectively small, the signals $\mu_{r+1}x^*_{r+1}, \ldots, \mu_n x^*_n$ contribute minimally to the feedback and so it should be expected that matrix A^*_c remains a stability matrix. In other words, the characteristic values $\mu_1, \ldots, \mu_n$ represent weights indicating how important the corresponding state variables are in the feedback. When there are two groups of characteristic values of very differing orders ($\mu_r \gg \mu_{r+1}$), some system variables have a small effect in the feedback and the closed-loop system remains stable even when only the first r state variables are used for the feedback. Obviously, in this case, the closed-loop system is stable, but the control system performance could deteriorate. Alternatively, what happens when an open-loop optimum approximation is made is that the closed-loop system destabilizes because an important state variable was discarded to carry out the feedback (i.e., associated with a high characteristic value). Although generally true, it is not unconditional that the strongly controllable and observable parts are the most important for guaranteeing closed-loop stability.

Suppose that the system is closed-loop balanced. We have seen that if $\mu_r \gg \mu_{r+1}$, then the state vector can be partitioned: $\mathbf{x} = \begin{bmatrix} \mathbf{x}_1 \\ \mathbf{x}_2 \end{bmatrix}$, where $\mathbf{x}_1$ represents the first r components and $\mathbf{x}_2$ the remaining $n - r$ components. Correspondingly, the matrices A, B and C can be partitioned as:

$$A = \begin{bmatrix} A_{11} & A_{12} \\ A_{21} & A_{22} \end{bmatrix}; B = \begin{bmatrix} B_1 \\ B_2 \end{bmatrix}; C = \begin{bmatrix} C_1 & C_2 \end{bmatrix}$$

Taking into account that $K_{opt} = B^T\bar{P}$ with $\bar{P} = \begin{bmatrix} \mu_1 & & & \\ & \mu_2 & & \\ & & \ddots & \\ & & & \mu_n \end{bmatrix} = \begin{bmatrix} \bar{P}_1 & \\ & \bar{P}_2 \end{bmatrix}$, the closed-loop matrix is given by:

$$A_c = A - BK_{opt} = A = \begin{bmatrix} A_{11} & A_{12} \\ A_{21} & A_{22} \end{bmatrix} - \begin{bmatrix} B_1 \\ B_2 \end{bmatrix} \begin{bmatrix} B_1^T & B_2^T \end{bmatrix} \begin{bmatrix} \bar{P}_1 & 0 \\ 0 & \bar{P}_2 \end{bmatrix} =$$

$$= \begin{bmatrix} A_{11} - B_1B_1^T\bar{P}_1 & A_{12} - B_1B_2^T\bar{P}_2 \\ A_{21} - B_2B_1^T\bar{P}_1 & A_{22} - B_2B_2^T\bar{P}_2 \end{bmatrix}$$

We are guaranteed that the eigenvalues of those matrices have negative real part, whereas for the lower order model obtained neglecting $n - r$ state variables, or equivalently considering $\bar{P}_2 = 0$, this issue should be critically studied. The closed-loop matrix becomes:

$$\tilde{A}_c = \begin{bmatrix} A_{11} - B_1B_1^T\bar{P}_1 & A_{12} \\ A_{21} - B_2B_1^T\bar{P}_1 & A_{22} \end{bmatrix}$$

such that the state linear regulator is applied on r state variables instead of on all the n state variables and the closed-loop system may also be unstable as shown in the next example.

MATLAB® Exercise 8.2

Consider the continuous-time LTI system with state-space matrices:

$$A = \begin{bmatrix} 1 & 0 & 0 & 0 \\ 0 & 10 & 0 & 0 \\ 0 & 0 & 2 & 0 \\ 0 & 0 & 0 & 3 \end{bmatrix}; B = \begin{bmatrix} 1 \\ 1 \\ 1 \\ 1 \end{bmatrix}; C = \begin{bmatrix} 1 & 1 & 1 & 1 \end{bmatrix} \tag{8.8}$$

Define the system in MATLAB®

```
>> A=diag([1 10 2 3])
>> B=ones(4,1)
>> C=ones(1,4)
>> system=ss(A,B,C,0)
```

and compute the closed-loop balanced realization with the procedure described in Section 8.4:

```
>> [systemb, G, Pgreekb, Pcareb]=balrealmcc(system)
```

Now, let us consider $r = 2$ and impose $\bar{P}_2$

```
>> Pcarebred=Pcareb
>> Pcarebred(3:4,3:4)=zeros(2)
```

We calculate the gains

```
>> Kred=systemb.b'*Pcarebred
```

and the closed-loop eigenvalues

```
>> eig(systemb.a-systemb.b*Kred)
```

One obtains $\lambda_{1,2} = 0.0655 \pm 14.6635j$, $\lambda_3 = -0.7484$ and $\lambda_4 = 0.2539$ which show that the closed-loop system is not stable. Instead, if the full control is applied:

```
>> K=systemb.b'*Pcareb
>> eig(systemb.a-systemb.b*K)
```

The closed-loop eigenvalues are: $\lambda_1 = -10.3287$, $\lambda_2 = -3.9679$, $\lambda_3 = -1.3195$, $\lambda_4 = -2.4151$ and the closed-loop system is stable.

We have already seen that, if $\bar{P}_2$ is made up of small coefficients, the approximation deriving from the fact that we are applying the linear regulator only on r state variables instead of on all the n state variables, is allowed. Let us now find out to what point the approximation is allowed or in other words to what point is $\tilde{A}_c$ a stability matrix. To do this, the contribution $\bar{P}_2$ can be interpreted as a perturbation term

$$\tilde{A}_c = A_c + \Delta A_c$$

with $\Delta A_c = \begin{bmatrix} 0 & +B_1 B_2^T \bar{P}_2 \\ 0 & +B_2 B_2^T \bar{P}_2 \end{bmatrix}$.

Generally, given matrices

$$\tilde{A} = A + \Delta A$$

let us consider what specifications ΔA must have to guarantee that, given that A is a stability matrix, also $\tilde{A}$ is a stability matrix.

In certain particular cases the problem can be simply resolved. Think of the case in which matrix A is an inferior triangular matrix. Any type of perturbation which modifies the coefficients of A below the diagonal means that $\tilde{A}$ is a stability matrix. So, if matrix ΔA has a particular structure, it does not affect the stability of $\tilde{A}$. Therefore, the conditions depend on the fact that the perturbation is localized onto some terms or spread across all the terms of the matrix.

More generally, the only information we have about matrix A is that it is a stability matrix. The conditions on ΔA are of the type

$$\|\Delta \mathrm{A}\| < \beta \tag{8.9}$$

so that if the 2-norm of this matrix is less than β, the perturbed system is bound to remain stable. There are various methods of calculating β. The least conservative (to which the highest value of β is associated) establishes that:

$$\beta_{max} = \frac{1}{\|(s\mathrm{I} - \mathrm{A})^{-1}\|_\infty} \tag{8.10}$$

where $\|(s\mathrm{I} - \mathrm{A})^{-1}\|_\infty$ is the H_∞ norm of matrix $(s\mathrm{I} - \mathrm{A})^{-1}$:

$$\|(s\mathrm{I} - \mathrm{A})^{-1}\|_\infty = \sup_\omega\{\|(j\omega\mathrm{I} - \mathrm{A})^{-1}\|_S : \omega \in \mathbb{R}\} \tag{8.11}$$

Matrix $(s\mathrm{I} - \mathrm{A})$ is a complex matrix, function of ω. As ω varies, the matrix whose spectral norm is greatest should be considered. β_{max} is the greatest value which assures that if relation (8.9) is verified, matrix $\tilde{\mathrm{A}}$ is still stable.

Going back to the more specific case under examination, note that matrix ΔA can be written as:

$$\Delta\mathrm{A} = \tilde{\mathrm{B}}\tilde{\mathrm{B}}^T\Delta\mathrm{P} \tag{8.12}$$

with $\Delta\mathrm{P} = \begin{bmatrix} 0 & 0 \\ 0 & \mathrm{P}_2 \end{bmatrix}$. ΔA can be therefore considered a structured uncertainty.

By using consistent norms we have that:

$$\|\Delta\mathrm{A}\| = \|\tilde{\mathrm{B}}\tilde{\mathrm{B}}^T\Delta\mathrm{P}\| \leq \|\tilde{\mathrm{B}}\tilde{\mathrm{B}}^T\| \cdot \|\Delta\mathrm{P}\| \leq \|\tilde{\mathrm{B}}\tilde{\mathrm{B}}^T\| \cdot \mu_{r+1} \tag{8.13}$$

Since P_2 is a diagonal matrix, its norm will be less than μ_{r+1}. In this way, a relationship between the uncertainty matrix and the characteristic value μ_{r+1} (which corresponds to the first state variable neglected in the closed-loop reduced model) is established. If $f(\mathrm{A}_c) = \beta$, then

$$\mu_{r+1} \leq \frac{f(\mathrm{A}_c)}{\|\tilde{\mathrm{B}}\tilde{\mathrm{B}}^T\|} \tag{8.14}$$

μ_{r+1} is a quantity which depends on A_c and allows us to establish the condition for which $\tilde{\mathrm{A}}_c$ is a matrix of (asymptotic) stability.

From the closed-loop balanced realization, the characteristic values can therefore be calculated as well as that which satisfies (8.14). At this point (having found index r), the number of state variables which must be used in feedback is established. Naturally, performance deteriorates as one may notice from the performance index $J = \mathbf{x}^T(0)\mathrm{P}\mathbf{x}(0)$ which shows an error proportional to the neglected coefficients $\mu_{r+1}, \ldots, \mu_n$ (in fact P does not contain the sub-matrix P_2).

The direct truncation and singular perturbation techniques for model order reduction can be applied to closed-loop balanced realization exactly as in the case of the open-loop balanced realization widely discussed in Chapter 6. We mention here another result on reduced order models based on the closed-loop balanced realization which is important for the design of low-order compensators. Starting from a system in closed-loop balanced realization, the reduced order model obtained with direct truncation is still in a closed-loop balanced form. When A_{22} is stable, the singular perturbation approximation can be applied and the reduced order model is also closed-loop balanced.

8.6 Closed-loop Balanced Realization for Symmetric Systems

As we saw in Chapter 6.3 in the case of open-loop balanced realization, symmetric systems have the advantage of certain properties which simplify the computation of the balancing. Very similar properties also hold for the closed-loop balanced realization.

Let us consider the CARE equation (7.76) and apply the relations characteristic of symmetric systems ($C = B^T T$, $B = T^{-1} C^T$, $A = T^{-1} A^T T$, $A^T = T A T^{-1}$):

$$A^T P + PA - PBB^T P + C^T C = 0 \Rightarrow$$

$$TAT^{-1}P + PA - PBCT^{-1}P + TBC = 0$$

Multiplying left by matrix T^{-1}, we get:

$$AT^{-1}P + T^{-1}PA - T^{-1}PBCT^{-1}P + BC = 0$$

So making $P^* = T^{-1}P$

$$AP^* + P^*A - P^*BCP^* + BC = 0 \tag{8.15}$$

Proceeding analogously with the FARE (8.1):

$$A\Pi + \Pi A^T - \Pi C^T C \Pi + BB^T = 0 \Rightarrow$$

$$A\Pi + \Pi T A^T T^{-1} - \Pi T B C \Pi + BCT^{-1} = 0$$

Multiplying right by T and making $P^* = \Pi T$, we get:

$$\mathrm{AP}^* + \mathrm{P}^*\mathrm{A} - \mathrm{P}^*\mathrm{BCP}^* + \mathrm{BC} = 0 \tag{8.16}$$

Relation (8.16) is equal to equation (8.15). Therefore, analogous to open-loop balancing, CARE and FARE are equal for symmetric systems. The obtained equation is still quadratic but in contrast to the Riccati equation the solution is not symmetric. Furthermore, there is no guarantee that the solution is positive definite. Note however that:

$$\mathrm{P}^* = \Pi\mathrm{T}, \mathrm{P}^* = \mathrm{T}^{-1}\mathrm{P} \Rightarrow \mathrm{P}^* \cdot \mathrm{P}^* = \Pi\mathrm{P} \tag{8.17}$$

where Π and P are solutions for the FARE and CARE. If, out of all the solutions, the positive definite matrices $\bar{\Pi}$ and $\bar{\mathrm{P}}$ are selected, the characteristic values can still be defined as the square roots of the eigenvalues of $\bar{\Pi}\bar{\mathrm{P}}$. So, for symmetric systems and because of relation (8.17), the characteristic values can be calculated as the square roots of the eigenvalues of the matrix $(\mathrm{P}^*)^2$. Since the matrix T can be calculated from the observability and controllability matrix, to determine the characteristic values, instead of solving two Riccati equations, only one equation has to be solved.

So, equation (8.15) is equivalent to the cross-gramian equation and is called the cross-Riccati equation. Analogously to W_{co}, P^* also holds two pieces of information: one is that it can calculate the system characteristic values and, furthermore, the signs of the eigenvalues allow one to determine the signature matrix of the system.

8.7 Exercises

1. Given the matrix

$$\mathrm{A} = \begin{bmatrix} 0 & 1 & 0 & 0 \\ 0 & 0 & 1 & 0 \\ 0 & 0 & 0 & 1 \\ -3 & -8 & -8 & -5 \end{bmatrix} \tag{8.18}$$

   determine $||\Delta A||_{max}$, so that $A + \Delta A$ is a matrix with eigenvalues with negative real part.

2. Given the system with transfer function

$$G(s) = \frac{1}{(1-s)(s+2)^2(s+0.5)^2}$$

   determine a balanced closed-loop realization and a suitable reduced order model. Then verify that the model is closed-loop stable.

3. Calculate the characteristic values for the system with transfer function $G(s) = \frac{2s}{s^2+1}$.
4. Calculate the closed-loop balanced realization for the system with transfer function $G(s) = \frac{s+1}{s^5+7s^2+6s+5}$. Design a reduced order regulator and observer making sure that the closed-loop system is asymptotically stable.
5. Given the system with transfer function $G(s) = \frac{s}{s^2+1} + \frac{s^2+1}{s(s^2+2)}$ calculate the characteristic values and synthesize the system with circuit components.
6. Given the system with transfer function $G(s) = \frac{1}{s-1}$ calculate the linear state regulator, optimal observer and the compensator $C(s)$ (see Figure 8.4), using the CARE and FARE.

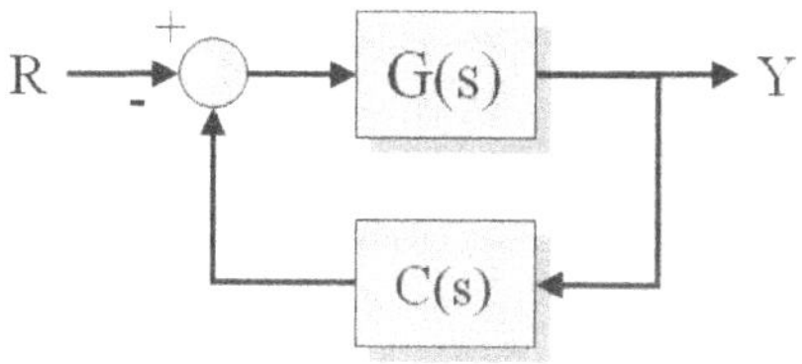

FIGURE 8.4
Block scheme for Exercise 6.

7. Calculate, if possible, the system that has characteristic values μ_1 and μ_2 eigenvalues $\lambda_1 = -5$ and $\lambda_2 = -1$.
8. Verify that the characteristic values of an all-pass system are all equal to the static gain k.
9. Design a reduced order model for system $G(s) = \frac{s+1}{s^3+5s^2+7s-2}$.
10. Given the system

$$G(s) = \frac{1}{(1-s)(s+2)^2(s+0.5)^2}$$

determine a closed-loop balanced realization and a reduced order model. Verify then that the reduced order model is stable in closed-loop.

9

Positive-real, Bounded-real and Negative-imaginary Systems

CONTENTS

In this chapter the main properties of positive-real, bounded-real, and negative-imaginary systems are discussed. Emphasis is also given to the relationships of duality existing among them. These classes of systems are all characterized first in the frequency domain and then in the time domain with the introduction of the positive-real lemma, the bounded-real lemma, and the negative-imaginary lemma. Hamiltonian matrices associated to these lemmas are also dealt with, along with Riccati equations for positive-real and bounded-real systems. The importance of these classes of systems in relation to robust control is also underlined. Several worked examples are included.

DOI: 10.1201/9781003196921-9

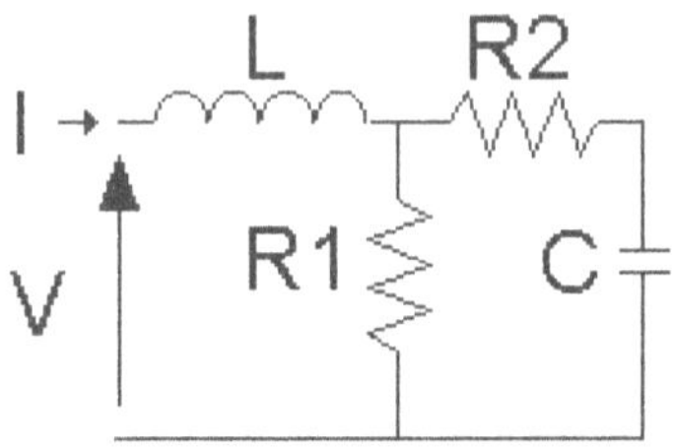

FIGURE 9.1
Example of a passive electrical circuit ($L > 0$, $C > 0$, $R_1 > 0$, $R_2 > 0$).

9.1 Passive Systems

There is perfect duality between dynamical systems and circuits. A dynamical system can be made of circuits which can mimic its behavior. Consider the circuit in Figure 9.1 with parameters $L > 0$, $C > 0$, $R_1 > 0$, $R_2 > 0$. This type of circuit is passive: it dissipates the energy supplied by the generator without ever being able to supply any. By attributing an equivalent impedance to the inductor and capacitor, the circuit can be studied in the Laplace domain (see Figure 9.2). Thus, working with algebraic equations rather than with integral-differential equations, the relationships characterizing the circuit can be derived. For example, consider the impedance of a circuit defined as

$$Z(s) = \frac{V(s)}{I(s)} \tag{9.1}$$

Calculating this characteristic transfer function is straightforward. More generally, when a circuit has more than one input (e.g., the circuit in Figure 9.3 has two independent current generators), the circuit can be characterized by a matrix of impedances, each from each input to each output.

First, let us consider what conditions make a SISO system passive, then we will extend the analysis to systems with multiple inputs and outputs.

9.1.1 Passivity in the Frequency Domain

Deciding whether a SISO system is passive is done by analyzing its impedance characteristics by means of the concept of positive-real function.

Definition 19 (Positive-real function) *A function is positive-real if it satisfies the following properties:*

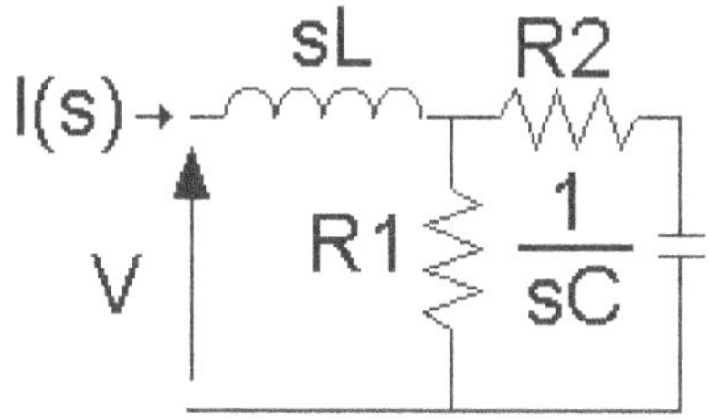

FIGURE 9.2
Example of a passive electrical circuit ($L > 0$, $C > 0$, $R_1 > 0$, $R_2 > 0$) studied in the Laplace domain.

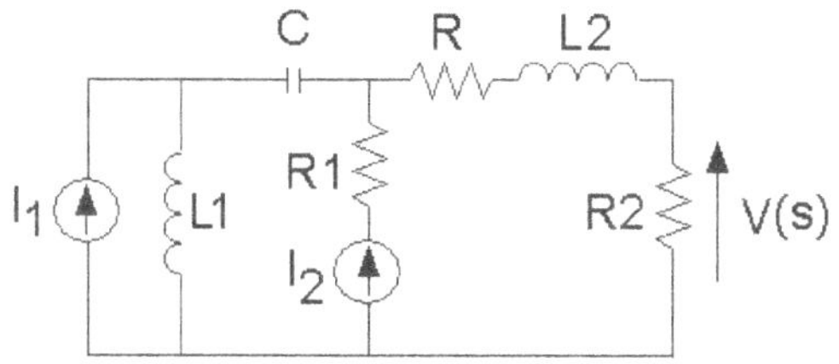

FIGURE 9.3
Example of a passive electrical circuit with two inputs.

1. *Function $Z(s)$ is analytic for $\Re e\ s > 0$ (i.e., there are no poles on the right-hand half of the complex plane);*

2. *Any function poles lying on the imaginary axis are simple and have positive and real residues;*

3. *$\Re e\ (Z(s)) \geq 0\ \forall \Re e\ s \geq 0$.*

Once a positive-real function is defined, we discuss how to determine if the system is passive by analyzing its impedance characteristics.

Theorem 18 *A SISO system is passive if and only if its impedance is positive-real.*

The importance of the last condition of the positive-real definition should be highlighted; the fact that the real part of the impedance is positive is clearly linked to the fact that the system is dissipating energy. Next, we will speak equivalently about positive-real systems and passive ones. Furthermore,

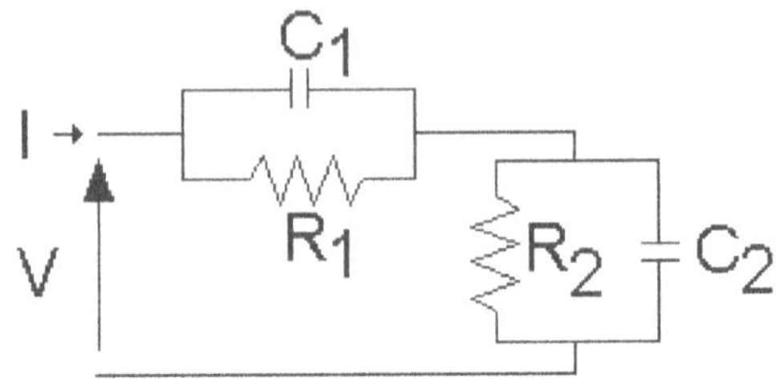

FIGURE 9.4
Example of a relaxation electrical circuit.

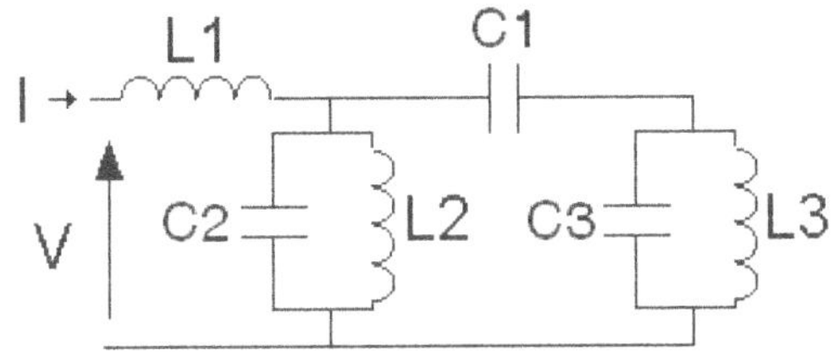

FIGURE 9.5
Example of a *loss-less* circuit.

strictly positive-real systems will refer to those systems strictly satisfying property 3) in Definition 19, i.e., $\Re e\ (Z(s)) > 0\ \forall \Re e\ s \geq 0$. Notice also that the third condition can be equivalently checked by considering the function $\Re e\ (Z(j\omega))$, and in particular requiring that $\Re e\ (Z(j\omega)) \geq 0\ \forall \omega > 0$.

Let us look at some examples of classes of passive systems. Firstly, generic RLC systems (henceforth we assume $R > 0$, $L > 0$ and $C > 0$) have a positive-real impedance which makes them passive.

Another example of passive systems are positive-real functions with simple poles and positive residues whose impedance derives from circuits made up only of resistors and capacitors, the so-called class of *relaxation systems*. An example is shown in Figure 9.4.

In *loss-less* systems, i.e., circuits made up of ideal capacitors and inductors (an example is shown in Figure 9.5) the impedance is always such that $\Re e\ (Z(j\omega)) = 0\ \forall \omega$ (precisely because there are no losses). This impedance is odd positive-real ($Z(s) = -Z(-s)$), its pecularity being that its poles and zeros (all on the imaginary axis, otherwise, it would degenerate or dissipate energy) alternate in frequency.

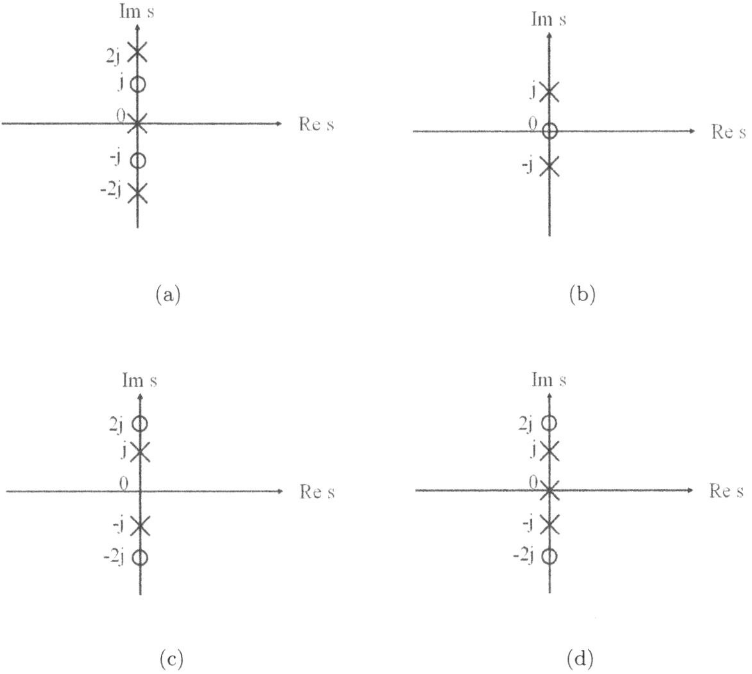

FIGURE 9.6
Examples of systems having (a) and (b) and not having (c) and (d) the property of losslessness.

Example 9.1

Let us consider the systems whose pole-zero maps are shown in Figure 9.6:

- $Z(s) = \frac{s^2+1}{s(s^2+4)}$ (Figure 9.6(a)) is a positive-real loss-less system;
- $Z(s) = \frac{s}{s^2+1}$ (Figure 9.6(b)) is a positive-real loss-less system;
- $Z(s) = \frac{s^2+4}{s^2+1}$ (Figure 9.6(c)) is not a positive-real loss-less system;
- $Z(s) = \frac{s^2+4}{s(s^2+1)}$ (Figure 9.6(d)) is not a positive-real loss-less system.

Note that, relaxation systems are strictly odd positive-real, whereas lossless systems are positive-real but not strictly so.

The characteristic values of an odd positive-real function are all one, whereas the singular values cannot be calculated given that Lyapunov equations have no solution. This means that a lower order model based on closed-loop balancing cannot be obtained, since all the state variables have equal weight in the feedback and so have to be necessarily taken into consideration.

This result also holds for MIMO systems.

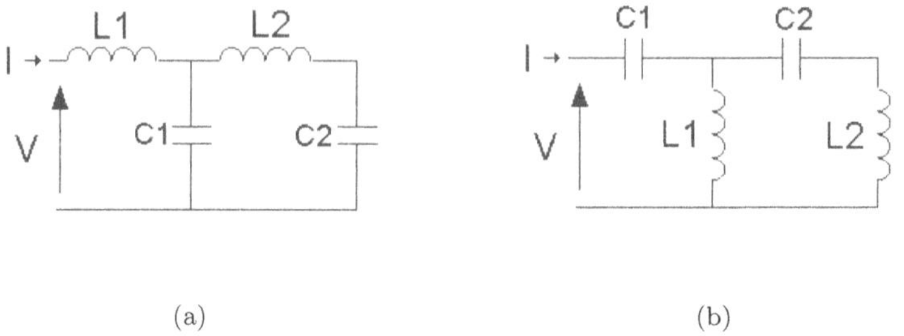

FIGURE 9.7
Examples of loss-less positive-real systems: (a) low-pass filter; (b) high-pass filter.

Example 9.2

Let us calculate the characteristic values of an odd positive-real function $G(s) = \frac{2}{s}$. Consider the realization: A = 0, B = 1, C = 2. The CARE equation is:

$$-P^2 + 4 = 0$$

and so $\bar{P} = 2$. The FARE equation is:

$$-4\Pi^2 + 1 = 0$$

and so $\bar{\Pi} = \frac{1}{2}$.
Therefore,

$$\bar{\Pi}\bar{P} = 1 \Rightarrow \mu = 1$$

Example 9.3

Let us consider the two circuits in Figure 9.7. The one in Figure 9.7(a) is a low-pass filter, whereas the one in Figure 9.7(b) is a high-pass filter. Since both circuits are loss-less, their characteristic values are one.

Note that the term *positive systems* refers to a different class of systems. A linear system is said to be externally positive only if its output (the zero-state response) is non-negative for any non-negative input.

Theorem 19 *A linear system is externally positive if and only if its impulse response is non-negative.*

An as-yet open question is how to verify the non-negativity of $y(t) = \mathrm{C}e^{\mathrm{A}t}\mathrm{B}$, the impulse response of a continuous linear time-invariant system. This requires numerical verification except when there are special pole and residue properties as in the case of relaxation systems which are externally positive (having positive residues).

Analogous to the SISO systems, passive MIMO systems are those whose transfer matrix is positive-real.

Definition 20 (Positive-real transfer matrix) *A transfer matrix is positive-real if and only if:*

1. *Each of its components is analytical for* $\Re e\ s > 0$ *(i.e., there are no poles on the right-hand half of the complex plane);*
2. *For each of the imaginary poles of* $\mathrm{Z}(s)$*, if any, its residue is a positive semi-definite Hermitian matrix;*
3. $\mathrm{Z}(j\omega) + \mathrm{Z}^T(-j\omega)$ *is positive semi-definite for each* ω *such that* $j\omega$ *is not a pole of* $\mathrm{Z}(j\omega)$*.*

Remember that $\mathrm{Z}^\dagger(j\omega)$ is the Hermitian matrix (i.e., the transposed conjugated matrix) of $\mathrm{Z}(j\omega)$, such that condition 3) of the positive-real definition may be formulated in terms of $\mathrm{Z}^\dagger(j\omega) + \mathrm{Z}(j\omega)$. For strictly positive-real systems, the condition is that $\mathrm{Z}^\dagger(j\omega) + \mathrm{Z}(j\omega)$ is positive definite $\forall\omega$. This implies that, as ω varies, the matrix $\mathrm{Z}^\dagger(j\omega) + \mathrm{Z}(j\omega)$ only has positive eigenvalues.

9.1.2 Passivity in the Time Domain

The definition discussed above is in the frequency domain. There are also analogous conditions to characterize passive systems in the time domain. Let us consider a minimal realization $(\mathrm{A}, \mathrm{B}, \mathrm{C}, \mathrm{D})$ of the impedance matrix $\mathrm{Z}(s)$, assumed square (i.e., $m = p$). To characterize a passive system from a realization of it, the following theorem, known as the positive-real lemma, is needed.

Theorem 20 (Positive-real lemma) *A linear time-invariant system in minimal form* $\mathrm{R}(\mathrm{A}, \mathrm{B}, \mathrm{C}, \mathrm{D})$ *is positive-real if there exists a positive definite matrix* P *which satisfies:*

$$\begin{cases} \mathrm{PA} + \mathrm{A}^T\mathrm{P} = -\mathrm{LL}^T \\ \mathrm{PB} = \mathrm{C}^T - \mathrm{LW} \\ \mathrm{D} + \mathrm{D}^T = \mathrm{W}^T\mathrm{W} \end{cases} \tag{9.2}$$

with $\mathrm{L} \in \mathbb{R}^{n \times m}$ *and* $\mathrm{W} \in \mathbb{R}^{m \times m}$*.*

Note in particular that the factorization of equation (9.2) exists if matrix D is positive definite (this in fact allows to obtain a matrix $\mathrm{D} + \mathrm{D}^T$ that can be factorized as the product of two matrices).

Consider, for example, an improper system. It decomposes into a strictly proper part and a remainder. From a circuitry point of view, these two terms correspond to a dynamical circuit block (being a function of s) and a resistance. If the resistance is negative, the circuit is not passive and the function is not positive-real. This corresponds exactly to the case when D is not positive definite.

On the hypothesis that the matrix D is positive definite, L can be calculated from the second equation of (9.2):

$$\mathrm{L} = (-\mathrm{PB} + \mathrm{C}^T)\mathrm{W}^{-1} \tag{9.3}$$

Note that matrix W is invertible since $\mathrm{D}+\mathrm{D}^T \neq 0$, given that D is positive definite.

For this reason, it does not suffice that D is positive semi-definite, but D has to be positive definite.

By substituting into the first equation of (9.2), we obtain:

$$\mathrm{PA} + \mathrm{A}^T\mathrm{P} = -(-\mathrm{PB} + \mathrm{C}^T)(\mathrm{D} + \mathrm{D}^T)^{-1}(-\mathrm{PB} + \mathrm{C}^T)^T \tag{9.4}$$

and by re-ordering

$$\begin{aligned} &\mathrm{P}(\mathrm{A} - \mathrm{B}(\mathrm{D} + \mathrm{D}^T)^{-1}\mathrm{C}) + (\mathrm{A}^T - \mathrm{C}^T(\mathrm{D} + \mathrm{D}^T)^{-1}\mathrm{B}^T)\mathrm{P} + \\ &+\mathrm{PB}(\mathrm{D} + \mathrm{D}^T)^{-1}\mathrm{B}^T\mathrm{P} + \mathrm{C}^T(\mathrm{D} + \mathrm{D}^T)^{-1}\mathrm{C} = 0 \end{aligned} \tag{9.5}$$

Equation (9.5) is a Riccati equation in which the quadratic term is positive so there is no guarantee of a positive definite solution. A Hamiltonian matrix can be associated with the Riccati equation (9.5):

$$\mathrm{H} = \begin{bmatrix} \mathrm{A} - \mathrm{B}(\mathrm{D} + \mathrm{D}^T)^{-1}\mathrm{C} & \mathrm{B}(\mathrm{D} + \mathrm{D}^T)^{-1}\mathrm{B}^T \\ -\mathrm{C}^T(\mathrm{D} + \mathrm{D}^T)^{-1}\mathrm{C} & -\mathrm{A}^T + \mathrm{C}^T(\mathrm{D} + \mathrm{D}^T)^{-1}\mathrm{B}^T \end{bmatrix} \tag{9.6}$$

According to Theorem 20, a system is positive-real if the Riccati equation (9.5) provides a positive definite solution.

The positive-realness of a system can be also studied by examining the Hamiltonian matrix (9.6). The condition can be derived by taking into account the following theorem relating the eigenvalues of the Hamiltonian matrix (9.6) with the singularity of the matrix $\mathrm{Z}(j\omega) + \mathrm{Z}^T(-j\omega)$ at some ω.

Theorem 21 *Assume that* A *has no imaginary eigenvalues,* $(\mathrm{D}+\mathrm{D}^T)$ *is non-singular and* $\omega_0 \in \mathbb{R}$. *Then* $\lambda = 0$ *is an eigenvalue of* $(\mathrm{Z}(j\omega_0) + \mathrm{Z}^T(-j\omega_0))$ *if and only if* $(\mathrm{H} - j\omega_0\mathrm{I})$ *is singular.*

For MIMO systems with $n = 1$ and for SISO systems it can be derived that the system is positive-real if the Hamiltonian matrix (9.6) has no eigenvalues on the imaginary axis.

9.1.3 Factorizing Positive-real Functions

Once obtained P, the matrix L can also be determined, and from these the matrix

$$\mathrm{W}(s) = \mathrm{W} + \mathrm{L}^T(s\mathrm{I} - \mathrm{A})^{-1}\mathrm{B}$$

which factorizes the original positive-real transfer function matrix as follows:

$$\mathrm{Z}(s) + \mathrm{Z}^T(-s) = \mathrm{W}^T(-s)\mathrm{W}(s)$$

Therefore, a positive-real transfer function matrix can be factorized as the product of two matrices which depend on s.

Example 9.4

Consider, for example, $Z(s) = \frac{2}{s}$ that is positive-real.
Since $Z^T(-s) = -\frac{2}{s}$, then $W(s) = 0$ factorizes the original positive-real function.

Example 9.5

Now consider $Z(s) = -\frac{2}{s}$ which is not positive-real. Consider the realization: $A = 0$, $B = 1$ and $C = -2$ (notice that the residue is negative). Consider also the second equation of the positive-real lemma (9.2): $PB = C^T - LW$. Given that $C < 0$, no value of P can satisfy this equation. So, neither L nor $W(s)$ can be calculated.

9.1.4 Passive Reduced Order Models

Now let us consider the issue of finding a reduced order model of a positive-real system. Suppose that the system has no poles on the imaginary axis, and to calculate the open-loop balanced realization from which we can calculate the singular values and derive a lower order model. To evaluate the efficacy of the model, it is not sufficient to consider the error norm between the original system and the lower-order model but, since this latter should accurately describe the original system characteristics, we must also verify that the reduced order model is still positive-real.

There is a method which guarantees this property. As we saw in Chapter 6, it is important to build a lower-order model into a state-space realization which has the desired characteristics. In this case, one may consider the dual equation of the Riccati equation (9.4):

$$\Pi \mathrm{A}^T + \mathrm{A}\Pi = -(-\Pi \mathrm{C}^T + \mathrm{B})(\mathrm{D} + \mathrm{D}^T)^{-1}(-\Pi \mathrm{C}^T + \mathrm{B})^T \tag{9.7}$$

and find a realization in which the solutions to (9.4) and (9.7) are equal and diagonal. Naming these solutions $\bar{\Pi}$ and $\bar{\mathrm{P}}$ note they are positive definite matrices given that the original system is positive-real. Furthermore, since they are diagonal, when direct truncation is applied and the matrix subblock neglected, the subblock obtained is positive definite, so the original characteristics of the system (i.e., passivity) are preserved. When dealing with symmetric systems, the lower-order model can be obtained via one equation.

9.1.5 Energy Considerations Connected to the Positive-real Lemma

The positive-real lemma came about as a result of considerations on circuit energy. Let us consider the impedance in Figure 9.8. Given that the current and voltage are as in Figure 9.8, the energy dissipated by the impedance is:

$$E = \int_0^T v(t)i(t)dt \tag{9.8}$$

The circuit is passive if the dissipated energy is not negative. For a linear time-invariant system:

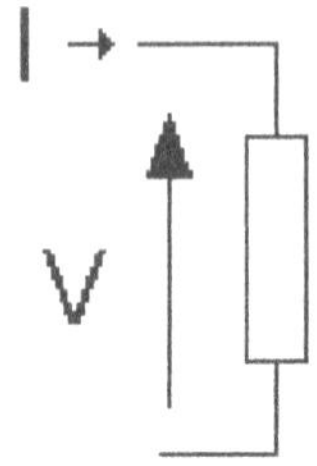

FIGURE 9.8
A generic impedance.

$$\begin{cases} \dot{\mathbf{x}} = \mathrm{A}\mathbf{x} + \mathrm{B}\mathbf{u} \\ \mathbf{y} = \mathrm{C}\mathbf{x} + \mathrm{D}\mathbf{u} \end{cases} \tag{9.9}$$

analogous to equation (9.8) the energy can be defined as:

$$E = \int_0^T \mathbf{u}^T(t)\mathbf{y}(t)dt \tag{9.10}$$

System (9.9) is passive if energy (9.10) is non-negative. This condition on the energy is equivalent to:

$$\begin{gathered} \mathrm{P} > 0; \\ \begin{bmatrix} \mathrm{A}^T\mathrm{P} + \mathrm{PA} & \mathrm{PB} - \mathrm{C}^T \\ \mathrm{B}^T\mathrm{P} - \mathrm{C} & -\mathrm{D}^T - \mathrm{D} \end{bmatrix} \leq 0 \end{gathered} \tag{9.11}$$

where the notation $\mathrm{P} > 0$ indicates that P has to be positive definite. (9.11) is a collection of matrix inequalities; if they have a solution then the system is passive. In Chapter 12, we will discuss the exact meaning of these expressions and show that many control problems can be re-formulated as linear matrix inequalities.

9.1.6 Closed-loop Stability and Positive-real Systems

There is a significant connection between the stability of feedback systems and positive-real systems. Let us consider the feedback system in Figure 9.9, it can be shown that if $G_1(s)$ and $G_2(s)$ are positive-real and at least one of the two systems is strictly positive-real, then the closed-loop system is asymptotically stable.

Theorem 22 *The feedback system shown in Figure 9.9 is asymptotically stable if $G_1(s)$ and $G_2(s)$ are positive-real, at least one of the two systems being strictly positive-real.*

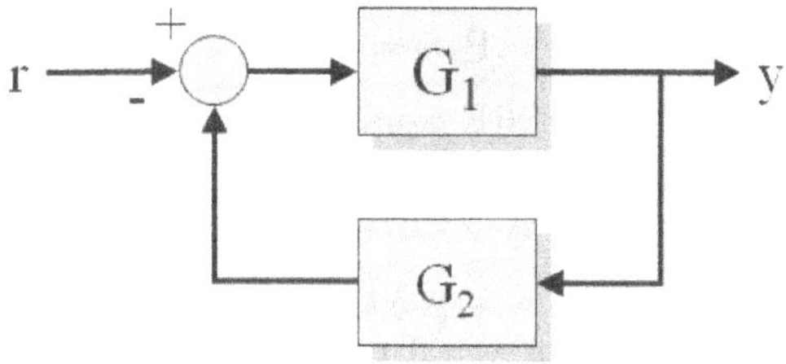

FIGURE 9.9
Closed-loop system with positive-real $G_1(s)$ and $G_2(s)$.

Example 9.6

Consider the system in Figure 9.9 with $G_1(s) = \frac{1}{s+4}$ and $G_2(s) = \frac{s}{s^2+1}$. Since $G_2(s)$ is loss-less (therefore positive-real) and $G_1(s)$ is a relaxation system (therefore strictly positive-real), the closed-loop system is asymptotically stable. It is easy to verify that the closed-loop transfer function equals $W(s) = \frac{G_1(s)}{1+G_1(s)G_2(s)} = \frac{s^2+1}{s^3+4s^2+2s+4}$ and that the closed-loop system is asymptotically stable.

Example 9.7

The condition expressed by Theorem 22 is only sufficient, the opposite not being verified. Let us look at two counter-examples of systems which do not satisfy the theorem conditions, one case showing a closed-loop unstable system, the other asymptotically stable.
First, let us consider $G_1(s) = \frac{1}{s-1}$ and $G_2(s) = \frac{s}{s^2+1}$. The closed-loop transfer function is $W(s) = \frac{s^2+1}{s^3-s^2+2s-1}$ such that the closed-loop system is unstable.
Now let us consider $G_1(s) = \frac{1}{s-0.1}$ and $G_2(s) = \frac{1}{s^2+2s+1}$, here the closed-loop transfer function is $W(s) = \frac{s-0.1}{s^3+1.9s^2+0.8s+0.9}$. By calculating the system poles or by applying the Routh criterion, the closed-loop system can be verified as being asymptotically stable.

9.1.7 Optimal Gain for Loss-less Systems

There is a significant tie-in between the optimal gain and coefficients of C for the class of loss-less systems. Consider the optimal control problem for a loss-less system $S(\mathrm{A}, \mathrm{B}, \mathrm{C})$, with the following optimization index:

$$J = \int_0^\infty (x^T \mathrm{C}^T \mathrm{C} x + u^T u) dt \tag{9.12}$$

The optimal gain turns out to be equal to matrix C, i.e., K = C.

A system $S(\mathrm{A}, \mathrm{B}, \mathrm{C})$ is loss-less if and only if the positive-real lemma holds, in other words if there exists a symmetric and positive definite matrix P_1 such that:

$$\mathrm{P}_1 \mathrm{A} + \mathrm{A}^T \mathrm{P}_1 = 0 \tag{9.13}$$

$$P_1B = C^T \tag{9.14}$$

Now, let us consider the CARE equation for loss-less systems:

$$A^TP + PA - PBB^TP + C^TC = 0 \tag{9.15}$$

The solution of this equation allows the optimal gain to be calculated, but by using relations (9.13) and (9.14), it can be proved that P_1 satisfies CARE and therefore $K = B^TP_1 = C$.

Moreover, it follows another result concerning the optimal eigenvalues for SISO loss-less systems with index (9.12): the optimal eigenvalues are the roots of

$$N(s) + D(s) = 0 \tag{9.16}$$

where $N(s)$ and $D(s)$ are the numerator and denominator polynomials of the transfer function of system $S(A, B, C)$.

MATLAB® Exercise 9.1

Consider the continuous-time LTI system with state-space matrices:

$$A = \begin{bmatrix} 0 & 1 & 0 & 0 \\ 0 & 0 & -1 & 0 \\ 0 & 0 & 0 & 8 \\ 0 & 0 & -8 & 0 \end{bmatrix}; \quad B = \begin{bmatrix} 1 \\ 0 \\ 0 \\ 1 \end{bmatrix}; \quad C = \begin{bmatrix} 2 & 0 & 0 & 4 \end{bmatrix}$$

The transfer function of this system is:

$$G(s) = \frac{6s^3 + 132s}{s^4 + 65s^2 + 64}$$

from which it is immediate to verify that the system is loss-less ($G(s) = -G^T(-s)$). Now, with the help of MATLAB, we solve the Riccati equation for optimal control and verify that $K = C$.

First, let us define the state-space matrices:

```
>> A=[0 1 0 0; -1 0 0 0; 0 0 0 8; 0 0 -8 0]
>> B=[1 0 0 1]'
>> C=[2 0 0 4]
```

and calculate its transfer function:

```
>> system=ss(A,B,C,0)
>> tf(system)
```

Let us now solve the Riccati equation $A^TP + PA - PBB^TP + C^TC$ associated with index (9.12). To do this, use the command:

```
>> [K,P,E]=lqr(A,B,C'*C,1)
```

One gets:

$$K = \begin{bmatrix} 2 & 0 & 0 & 4 \end{bmatrix}$$

i.e., $K = C$.

The corresponding optimal eigenvalues are given by: $\lambda_{1,2} = -1.8456 \pm i7.2052$, $\lambda_3 = -1.5738$ and $\lambda_4 = -0.7351$. They can be also calculated by equation (7.92) through the command:

```
>> roots([1 6 65 132 64])
```

Also note that equation (9.16) leads to the same result of equation (7.92).

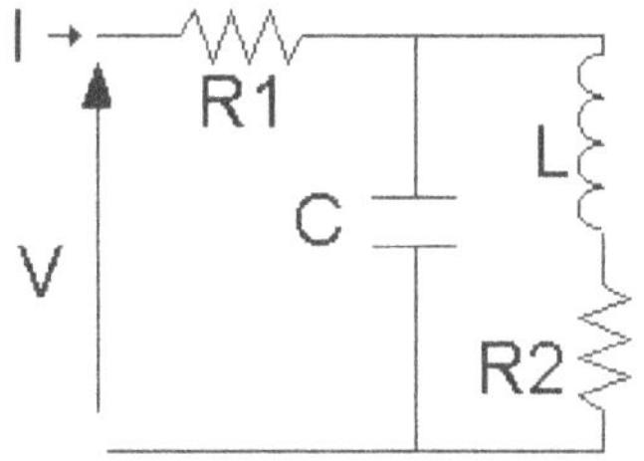

FIGURE 9.10
For $R1 = 1k\Omega$, $C = 2\mu F$, $L = 1mH$ and $R2 = 5k\Omega$ the passive circuit implements the transfer function $G(s) = \frac{2\cdot 10^{-6}s^2+(10+10^{-3})s+6\cdot 10^3}{2\cdot 10^{-9}s^2+10^{-2}s+1}$.

9.2 Circuit Implementation of Positive-real Systems

In this section, an example of the implementation of positive-real systems is illustrated.

Example 9.8

Given the system with transfer function $G(s) = \frac{2\cdot 10^{-6}s^2+(10+10^{-3})s+6\cdot 10^3}{2\cdot 10^{-9}s^2+10^{-2}s+1}$ propose, if possible, a passive electrical circuit implementation.

Solution

First verify if the function $G(s)$ is positive-real, i.e., if it is possible to obtain a passive electric circuit which is analogous to the given system. To do this, we verify if conditions 1-3 of Definition 19 are true. Points 1-2 can be checked immediately. For point 3, the real part of $G(j\omega)$ needs to be calculated:

$$\Re e(G(j\omega)) = \frac{6\cdot 10^3 + (10^{-1} - 4\cdot 10^{-6})\omega^2 + 4\cdot 10^{-15}\omega^4}{(1 - 2\cdot 10^{-9}\omega^2)^2 + 10^{-4}\omega^2}$$

Note that $\Re e\ (G(j\omega)) \geq 0\ \forall\omega$, so the system is positive-real.
Proceed by rewriting the transfer function as the sum of two terms, one strictly proper and one improper. To do this, we divide the polynomial numerator by the denominator, obtaining:

$$G(s) = 1000 + \frac{10^{-3}s + 5\cdot 10^3}{2\cdot 10^{-9}s^2 + 10^{-2}s + 1}$$

The first term can be interpreted as a resistor (R1) of value $1k\Omega$, while the second term is a generic impedance in series with R1. Rewrite the second term as follows:

$$G(s) = 1000 + \frac{1}{\frac{2\cdot 10^{-9}s^2+10^{-2}s+1}{10^{-3}s+5\cdot 10^3}}$$

and iterate the procedure with the term $\frac{2\cdot 10^{-9}s^2+10^{-2}s+1}{10^{-3}s+5\cdot 10^3}$. Note that this term should be interpreted as the admittance associated with the impedance in series with R1. Divide the polynomial numerator by the denominator to obtain:

$$G(s) = 1000 + \frac{1}{2 \cdot 10^{-6} s + \frac{1}{10^{-3} s + 5 \cdot 10^3}}$$

This form of $G(s)$ yields a direct interpretation in terms of series and parallel impedances or elementary admittances. In fact the term $10^{-3}s + 5 \cdot 10^3$ can be interpreted as the series impedance of an inductor with a resistance (with $L = 1mH$ and $R2 = 5k\Omega$). The series of these two components should be then connected in parallel with an admittance equal to $2 \cdot 10^{-6}s$ (and so with a capacitor $C = 2\mu F$). Finally, we have to consider the resistor R1 in series to this circuit block. The equivalent circuit is shown in Figure 9.10.

9.3 Bounded-real Systems

Before defining bounded-real systems in the more general case, consider a SISO system. A SISO system $G(s)$ is bounded-real if $G(s)$ is stable and if the magnitude plot of the Bode diagram is below the value 0dB for any ω, that is, if the maximum magnitude is less than or equal to one.

From this definition it follows that $G(s) = \frac{1}{s}$ and $G(s) = \frac{4}{s+1}$ are not bounded-real systems, while $G(s) = \frac{1}{s+2}$ is.

Generally, the following defines bounded-real systems.

Definition 21 *A system* $\mathrm{S}(s)$ *is bounded-real if these conditions apply:*

1. *All the elements of the transfer function matrix, i.e.,* $S_{ij}(s)$*, are analytic in* $\Re e\ s > 0$ *(i.e., the polynomials in the denominator of* $S_{ij}(s)$ *are Hurwitz polynomials);*
2. *Matrix* $\mathrm{I} - \mathrm{S}^T(-s)\mathrm{S}(s)$ *is a non-negative Hermitian matrix for* $\Re e\ s > 0$ *or matrix* $\mathrm{I} - \mathrm{S}^T(-j\omega)\mathrm{S}(j\omega)$ *is non-negative for all the values of* ω.

Condition 2) can also be formulated according to the standard definition of the H_∞ norm of a system.

Remember that H_∞ norm of a system $\mathrm{S}(s)$ is defined as

$$\|\mathrm{S}(s)\|_\infty = \max_\omega[\sigma_{\max}(\mathrm{S}(j\omega))]$$

Condition 2) can be thus written as

$$\|\mathrm{S}(s)\|_\infty \leq 1$$

Moreover a system is said to be strictly bounded-real if $\|\mathrm{S}(s)\|_\infty < 1$.

Example 9.9

Consider system $G(s) = \frac{1}{s+2}$; condition 1) is clearly satisfied. Condition 2) implies that

$$1 - \frac{1}{2-j\omega}\frac{1}{2+j\omega} < 1$$

i.e., that

$$\frac{1}{4+\omega^2} < 1$$

which is the same condition obtained by applying the definition of bounded-real systems to SISO systems.

As in the case of positive-real systems, conditions can be given directly in the time domain through the bounded-real lemma.

Theorem 23 (Bounded-real lemma) *A system* $\mathrm{S}(s) = \left[\begin{array}{c|c} \mathrm{A} & \mathrm{B} \\ \hline \mathrm{C} & \mathrm{D} \end{array}\right]$ *in minimal form is bounded-real if* $\exists$ P *symmetric and positive definite that satisfies*

$$\mathrm{PA} + \mathrm{A}^T\mathrm{P} = -\mathrm{C}^T\mathrm{C} - \mathrm{LL}^T \tag{9.17}$$

$$-\mathrm{PB} = \mathrm{C}^T\mathrm{D} + \mathrm{LW} \tag{9.18}$$

$$\mathrm{I} - \mathrm{D}^T\mathrm{D} = \mathrm{W}^T\mathrm{W} \tag{9.19}$$

with $\mathrm{L} \in \mathbb{R}^{n\times m}$ *and* $\mathrm{W} \in \mathbb{R}^{m\times m}$.

Provided that W is invertible, we can write

$$\mathrm{L} = (-\mathrm{PB} - \mathrm{C}^T\mathrm{D})\mathrm{W}^{-1}$$

and substituting in the condition (9.17), we obtain the quadratic equation associated with the bounded-real condition of a system:

$$\begin{gathered}\mathrm{P}(\mathrm{A} + \mathrm{B}(\mathrm{I} - \mathrm{D}^T\mathrm{D})^{-1}\mathrm{D}^T\mathrm{C}) + (\mathrm{A}^T + \mathrm{C}^T\mathrm{D}(\mathrm{I} - \mathrm{D}^T\mathrm{D})^{-1}\mathrm{B}^T)\mathrm{P} + \\ +\mathrm{PB}(\mathrm{I} - \mathrm{D}^T\mathrm{D})^{-1}\mathrm{B}^T\mathrm{P} + \mathrm{C}^T\mathrm{D}(\mathrm{I} - \mathrm{D}^T\mathrm{D})^{-1}\mathrm{D}^T\mathrm{C} + \mathrm{C}^T\mathrm{C} = 0\end{gathered} \tag{9.20}$$

If the system is strictly proper, (D = 0), then equation (9.20) becomes

$$\mathrm{PA} + \mathrm{A}^T\mathrm{P} + \mathrm{PBB}^T\mathrm{P} + \mathrm{C}^T\mathrm{C} = 0 \tag{9.21}$$

Equation (9.20) can be associated with the following Hamiltonian matrix:

$$\mathrm{H} = \begin{bmatrix} \mathrm{A} + \mathrm{B}(\mathrm{I} - \mathrm{D}^T\mathrm{D})^{-1}\mathrm{D}^T\mathrm{C} & \mathrm{B}(\mathrm{I} - \mathrm{D}^T\mathrm{D})^{-1}\mathrm{B}^T \\ -\mathrm{C}^T\mathrm{D}(\mathrm{I} - \mathrm{D}^T\mathrm{D})^{-1}\mathrm{D}^T\mathrm{C} - \mathrm{C}^T\mathrm{C} & -\mathrm{A}^T - \mathrm{C}^T\mathrm{D}(\mathrm{I} - \mathrm{D}^T\mathrm{D})^{-1}\mathrm{B}^T \end{bmatrix} \tag{9.22}$$

If there exists a positive definite solution of the Riccati equation (9.20), then the system is bounded-real, otherwise not. Analogous to positive-real systems, the real boundness of a system can be checked from the eigenvalues of H. For strictly proper MIMO systems, for MIMO systems with $n = 1$, and for SISO systems, the condition is that H has no eigenvalues on the imaginary axis.

In strictly proper systems (D = 0) the Hamiltonian matrix (9.22) can be written as

$$\mathrm{H} = \begin{bmatrix} \mathrm{A} & \mathrm{BB}^T \\ -\mathrm{C}^T\mathrm{C} & -\mathrm{A}^T \end{bmatrix} \tag{9.23}$$

Example 9.10

Consider the system $G(s) = \frac{k}{s+1}$, taking into account its Bode diagram, the system is bounded-real if $k \leq 1$, and strictly bounded-real if $k < 1$.
Applying the bounded-real lemma, we obtain the same result. In fact, a realization in state form of the system is given by A = −1, B = 1, C = k, for which the Hamiltonian is:

$$\mathrm{H} = \begin{bmatrix} -1 & 1 \\ -k^2 & 1 \end{bmatrix}$$

The characteristic polynomial is given by:

$$\det(\lambda \mathrm{I} - \mathrm{H}) = \lambda^2 - 1 + k^2$$

The condition that the eigenvalues are not on the imaginary axis is $|k| \leq 1$. For these values the system is bounded-real.
Note that if we consider the CARE associated with $G(s)$

$$\mathrm{H} = \begin{bmatrix} -1 & -1 \\ -k^2 & 1 \end{bmatrix}$$

we obtain

$$\det(\lambda \mathrm{I} - \mathrm{H}) = \lambda^2 - 1 - k^2$$

In this case, the eigenvalues of H are never purely imaginary (independently of the value assumed by k), so the issue of the optimal regulator is always solvable (since the system is always controllable).

If the system S(s) is bounded-real, i.e., if we are able to determine the positive definite solution of the Riccati equation (9.20), then $\mathrm{W}(s) = \mathrm{W} + \mathrm{L}^T(s\mathrm{I} - \mathrm{A})^{-1}\mathrm{B}$ can be defined, leading to the following factorization:

$$\mathrm{I} - \mathrm{S}^T(-s)\mathrm{S}(s) = \mathrm{W}^T(-s)\mathrm{W}(s)$$

The bounded-real lemma derives from a more general definition which can be given for dynamical systems (not necessarily linear). A system is bounded-real if $\forall u$, $\forall T$ it holds that

$$\int_0^T y^T(t)y(t)dt \leq \int_0^T u^T(t)u(t)dt$$

The conditions expressed by the bounded-real lemma are limit conditions of this more general definition.

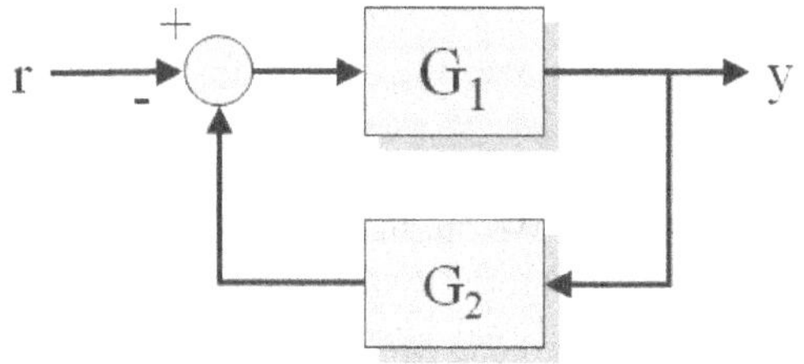

FIGURE 9.11
System constituted by the feedback of $G_1(s)$ and $G_2(s)$. The system is asymptotically stable if $G_1(s) \cdot G_2(s)$ is a strictly bounded-real system.

Finally, the problem can be reformulated in terms of linear matrix inequalities. A system is bounded-real if a solutions to the inequalities

$$\begin{aligned} &\mathrm{P} > 0 \\ &\begin{bmatrix} \mathrm{A}^T\mathrm{P} + \mathrm{PA} + \mathrm{C}^T\mathrm{C} & \mathrm{PB} + \mathrm{C}^T\mathrm{D} \\ \mathrm{B}^T\mathrm{P} + \mathrm{D}^T\mathrm{C} & \mathrm{D}^T\mathrm{D} - \mathrm{I} \end{bmatrix} \leq 0 \end{aligned} \tag{9.24}$$

can be found.

9.3.1 Properties of Bounded-real Systems

There is an important result linking the properties of bounded-real systems with closed-loop asymptotic stability as expressed by the following theorem:

Theorem 24 (Small gain theorem) *The system shown in Figure 9.11 is asymptotically stable if* $\mathrm{L}(s) = \mathrm{G}_1(s)\mathrm{G}_2(s)$ *is a strictly bounded-real system.*

The proof of this theorem is based on the fact that, injecting a signal into the summing node, it will recur attenuated in input. In fact, since the system is bounded-real, each frequency component of the signal will be attenuated. Note also that in the SISO case the theorem implies that the Nyquist plot of the equation $L(s) = G_1(s)G_2(s)$ does not surround point $(-1, 0)$.

The theorem expresses a very important result. Since the H_∞ norm is a consistent norm, we can obtain information about the stability of the feedback system from the value of the norm of the single systems. For example, if $\|G_1(s)\|_\infty < \gamma$ we can deduce that the feedback system will be asymptotically stable for any system $G_2(s)$ such that $\|G_2(s)\|_\infty < \frac{1}{\gamma}$. As we will see below, this result is of utmost importance for characterizing the robustness of a control system.

9.3.2 Bounded-real Reduced Order Models

The Riccati equation (9.20) is also important for obtaining a reduced-order model with particular properties. If we approximate a bounded-real system using the open-loop balancing, in fact, we have no guarantee that the approximated system is bounded-real. We ought to have an approximated model that faithfully represents this characteristic of the initial system.

To approximate a bounded-real system through a reduced-order model which is also bounded-real, the Riccati equation (9.20) and its dual can be used for balancing and then proceed similarly to closed-loop balancing. In addition, if the system is symmetric, the two equations can be replaced by a single Riccati equation.

9.4 Relationship Between Passive and Bounded-real Systems

There is a relationship between passive systems and bounded-real systems. Before clarifying this relationship, we will briefly show why we need to find a link between passive systems and bounded-real systems. The theory of passive filter design often requires a specification in the frequency domain which the designed filter must meet. This mask is defined by two curves in the Bode diagram which represent the upper and lower limits within which the transfer function of the designed filter must lie. Given transfer function $G(j\omega)$, it has then to be implemented through components R, L and C. If the transfer function is positive-real, as we have seen, this implementation is always possible.

Once assigned a mask, the problem of designing a filter is to find a transfer function which adheres to the specifications defined by the mask and at the same time is positive-real. This problem is difficult to solve, whereas it is easier to find a bounded-real function that respects the specifications defined by the mask. The importance of a link between passive and bounded-real systems now becomes evident, because it permits us, once the bounded-real function is found, to obtain a passive system which satisfies the assigned specifics. This relation is expressed through a matrix called a "scattering matrix".

If $\mathrm{S}(s)$ is a transfer matrix of a bounded-real system, a scattering matrix can be defined as:

$$\mathrm{G}(s) = [\mathrm{I} + \mathrm{S}(s)][\mathrm{I} - \mathrm{S}(s)]^{-1}$$

which has the positiveness property.

Moreover, if $S(s) = \left[\begin{array}{c|c} \tilde{A} & \tilde{B} \\ \hline \tilde{C} & \tilde{D} \end{array}\right]$, then a realization of $G(s)$ is given by:

$$G(s) = \left[\begin{array}{c|c} (\tilde{A}+I)(\tilde{A}-I)^{-1} & \sqrt{2}(\tilde{A}-I)^{-1}\tilde{B} \\ \hline -\sqrt{2}(\tilde{A}^T-I)^{-1}\tilde{C} & \tilde{D}-\tilde{C}^T(\tilde{A}-I)^{-1}\tilde{B} \end{array}\right]$$

9.5 Negative-imaginary Systems

Another important class of systems is that of negative-imaginary systems. Let us begin with by doing some considerations related to the SISO case. We have seen that in the case of positive-real systems the transfer function is such that $\Re e\ (Z(j\omega)) \geq 0\ \forall \omega$. Consequently, the Nyquist plot entirely lies in the half-plane with positive x-axis. In the case of negative-imaginary systems, a dual property holds, with the Nyquist plot entirely found in the half-plane with negative y-axis. Quite interestingly, as we have seen that positive-real systems represent the model of given physical systems, also for the case of negative-imaginary systems, we will see that there are classes of physical systems whose model is negative-imaginary.

9.5.1 Characterization of Negative-imaginary Systems in the Frequency Domain

Let us now formally introduce the notion of a negative-imaginary transfer function matrix. Here, in particular, we consider different definitions, which correspond to the fact that negative-imaginary systems have been first introduced with the assumption of poles in the open left half plane, and then extended to include simple poles on the imaginary axis and one or two poles in the origin.

Definition 22 *A square real-rational proper transfer function matrix* $G(s)$ *is negative-imaginary if it has no pole in the closed right half of the complex plane and for all* $\omega \geq 0$ $j(G(j\omega) - G^{\dagger}(j\omega)) \geq 0$.

Definition 23 *A square real-rational proper transfer function matrix* $G(s)$ *is negative-imaginary if:*

1. $G(s)$ *has no poles in the origin and in the open right half of the complex plane;*
2. *For all* $\omega \geq 0$ *such that* $j\omega$ *is not a pole of* $G(s)$, $j(G(j\omega) - G^{\dagger}(j\omega)) \geq 0$;
3. *If* $s = j\omega_0$ *is a pole of* $G(s)$, *then it is simple and the residue matrix* $K = \lim_{s \to j\omega_0} (s - j\omega_0) jG(s)$ *is Hermitian and positive semidefinite.*

Definition 24 *A square real-rational proper transfer function matrix* $\mathrm{G}(s)$ *is negative-imaginary if:*

1. $\mathrm{G}(s)$ *has no poles in the open right half of the complex plane;*

2. *for all* $\omega > 0$ *such that* $j\omega$ *is not a pole of* $\mathrm{G}(s)$, $j(\mathrm{G}(j\omega) - \mathrm{G}^\dagger(j\omega)) \geq 0$*;*

3. *If* $s = j\omega_0$ $(\omega_0 > 0)$ *is a pole of* $\mathrm{G}(s)$*, then it is simple and the residue matrix* $\mathrm{K} = \lim_{s \to j\omega_0} (s - j\omega_0) j\mathrm{G}(s)$ *is Hermitian and positive semidefinite;*

4. *If* $s = 0$ *is a pole of* $\mathrm{G}(s)$*, then* $\lim_{s \to 0} s^k \mathrm{G}(s) = 0$ *for all* $k \geq 3$ *and* $\lim_{s \to 0} s^2 \mathrm{G}(s)$ *is Hermitian and positive definite.*

Finally, the next definition formalizes the concept of a strictly negative-imaginary transfer matrix.

Definition 25 (Strictly negative-imaginary transfer matrix) *A square real-rational proper transfer function matrix* $\mathrm{G}(s)$ *is negative-imaginary if it has no pole in the closed right half of the complex plane and for all* $\omega > 0$ $j(\mathrm{G}(j\omega) - \mathrm{G}^\dagger(j\omega)) > 0$.

Negative-imaginary systems are those systems having a negative-imaginary transfer function matrix.

Definition 26 *A linear system is said to be (strictly) negative-imaginary if and only if its transfer function matrix is (strictly) negative-imaginary.*

Example 9.11

Let us consider the SISO system with transfer function equal to $G(s) = \frac{1}{s+1}$. This system is stable. Then, since $G(j\omega) + G^\dagger(j\omega) = \frac{2}{1+\omega^2} > 0$, the system is strictly positive-real. In addition, since $j(G(j\omega) - G^\dagger(j\omega)) = \frac{2\omega}{1+\omega^2} > 0$ for all $\omega > 0$, then the system is also strictly negative-imaginary.
The same conclusion is obtained considering that $G(j\omega) = \frac{1}{1+\omega^2} + j\frac{-\omega}{1+\omega^2}$, which shows that $\Re e(G(j\omega)) > 0$ and $\Im m(G(j\omega)) < 0$ for all $\omega > 0$, and, consequently, the Nyquist plot entirely lies in the quadrant $\Re e(G(j\omega)) > 0$ and $\Im m(G(j\omega)) < 0$.
Therefore, $G(s) = \frac{1}{s+1}$ constitutes an example of a system that is both strictly positive-real and strictly negative-imaginary.

Example 9.12

Consider now the mass-spring-damper system shown in Figure 9.12 with $M, k, b > 0$, and consider two scenarios. In the first one, let us assume that the output of the system is the speed of the mass; in the second one, let us assume that the output is the position of the mass. We will see that in the first case one gets a positive-real system, whereas in the second case a negative-imaginary system. This shows that using a different sensor, and so measuring a different variable (position/speed), yields a model of a different class. This is important also for the type of control that could be implemented in the system.
By applying the Newton's law to the system in Figure 9.12, one gets:

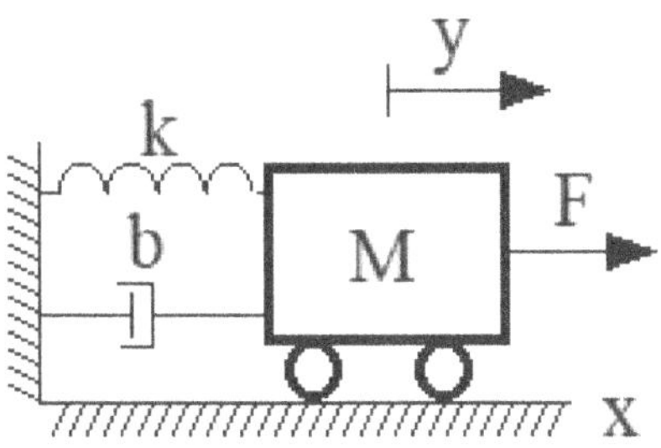

FIGURE 9.12
A mass-spring-damper system that depending on the choice of y may be positive-real or negative-imaginary.

$$M\ddot{x} + b\dot{x} + kx = F \tag{9.25}$$

Then, considering the first scenario, the output corresponds to the derivative of x, that is, $y = \dot{x}$. Hence, the transfer function for case 1 is $G_1(s) = \frac{Y(s)}{U(s)} = \frac{s}{Ms^2+sb+k}$. Instead, in the second case, the output is the position, that is, $y = x$, and the transfer function is given by $G_2(s) = \frac{Y(s)}{U(s)} = \frac{1}{Ms^2+sb+k}$.
The system is stable. Now, notice that in the first case one has:

$$G_1(j\omega) + G_1^\dagger(j\omega) = \frac{2\omega^2 B^2}{(k-\omega^2 M)^2 + \omega^2 B^2} > 0, \forall\omega > 0 \tag{9.26}$$

and in the second case one gets:

$$j(G_2(j\omega) - G_2^\dagger(j\omega)) = \frac{2\omega B}{(k-\omega^2 M)^2 + \omega^2 B^2} > 0, \forall\omega > 0 \tag{9.27}$$

We conclude that $G_1(s)$ is positive-real and $G_2(s)$ is negative-imaginary, such that the choice of the sensor is fundamental to determine the type of system that is obtained.

The result in Example 9.12 is more general than a specific example. We discuss now how it applies to a larger class of systems. Consider a lightly damped flexible structure with a single actuator and a single sensor, located in the same place of the actuator (in this case one says that the sensor is *colocated*). Such a structure can be modeled using the so-called *modal analysis*, which leads to the following transfer function model:

$$G(s) = \sum_{i=1}^{\infty} \frac{\phi_i(s)}{s^2 + \kappa_i s + \omega_i^2} \tag{9.28}$$

where, for each mode i, ω_i is a modal frequency, $\phi_i(s)$ is a first-order polynomial, κ_i is the viscous damping constant associated to mode i, and $\omega_i \neq \omega_j$ for $i \neq j$.

Suppose now to consider a force actuator and a velocity measurement, then the general model (9.28) becomes:

$$G(s) = \sum_{i=1}^{\infty} \frac{\psi_i^2 s}{s^2 + k_i s + \omega_i^2} \tag{9.29}$$

where ψ_i is a real number. It can be easily verified that this system is positive-real.

Now, consider a force actuator and a colocated position sensor. In this case, the general model (9.28) particularizes into:

$$G(s) = \sum_{i=1}^{\infty} \frac{\psi_i^2}{s^2 + k_i s + \omega_i^2} \tag{9.30}$$

that is negative-imaginary.

So, depending on the type of sensor used for the control of a flexible structure one can face a control problem involving either a positive-real system or a negative-imaginary one.

A similar result also holds for the case of m colocated position/velocity sensors and actuators. In the case of velocity measurements, the transfer function matrix has the following form:

$$\mathrm{G}(s) = \sum_{i=1}^{\infty} \frac{s}{s^2 + k_i s + \omega_i^2} \psi_i \psi_i^T \tag{9.31}$$

where, for each mode i, ψ_i is a column vector of m elements. In the case of position measurements, instead, the transfer function matrix reads:

$$\mathrm{G}(s) = \sum_{i=1}^{\infty} \frac{1}{s^2 + k_i s + \omega_i^2} \psi_i \psi_i^T \tag{9.32}$$

The transfer function matrix (9.31) is positive-real, whereas (9.32) is negative-imaginary.

Negative-imaginary systems may also represent the model of an electrical circuit composed of resistors, capacitors, and inductors. Consider the circuit shown in Figure 9.13 where the input is the voltage generator u and the output is the voltage y across the capacitor C. Then, it can be demonstrated that the transfer function $G(s) = \frac{Y(s)}{U(s)}$ is negative-imaginary. The same results holds for the dual case that the input is provided by a current generator and the output by the current in an inductor that is connected in parallel with the generator. Multi-port configurations can be considered as well.

Example 9.13

Consider the electrical circuit of Figure 9.14, with input u and output y. By applying Kirchhoff's circuit laws, one derives the following transfer function:

$$G(s) = \frac{Y(s)}{U(s)} = \frac{sC_1 R_{eq} + 1}{s^2 C_1 C_2 R_2 R_{eq} + sC_1 R_{eq} + 1} \tag{9.33}$$

where $R_{eq} = R_1 + R_2$. Clearly, for positive values of the components this system is stable. Let us now calculate $j(G(j\omega) - G^\dagger(j\omega))$:

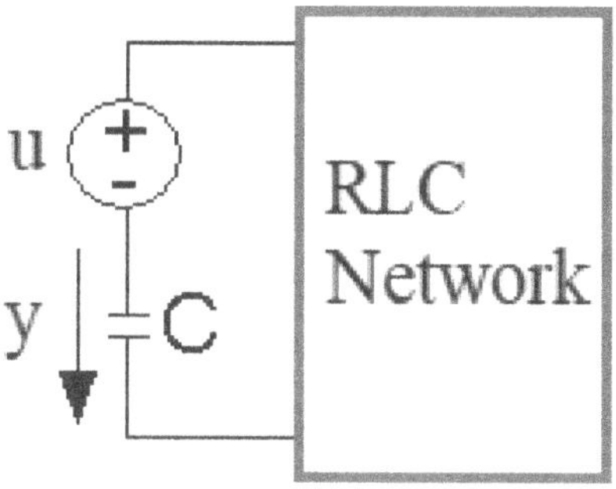

FIGURE 9.13
Electrical circuits where the voltage output is in series with the voltage input are negative-imaginary systems. Here, the gray box encapsulates an arbitrary network of passive components (capacitors, resistors and inductors).

$$j(G(j\omega) - G^{\dagger}(j\omega)) = \frac{2\omega^3 C_1^2 C_2 R_2 R_{eq}}{(1-\omega^2 C_1 C_2 R_2 R_{eq})^2 + \omega^2 C_1 R_{eq}} \tag{9.34}$$

Since $j(G(j\omega) - G^{\dagger}(j\omega)) > 0$ for all $\omega > 0$, the system is strictly negative-imaginary.

9.5.2 Characterization of Negative-imaginary Systems in the Time Domain

Similarly to what observed for positive-real and bounded-real systems, also for negative-imaginary systems a characterization in the time domain can be given. In particular, there exists a result analogous to the positive-real lemma and known as the negative-imaginary lemma.

Theorem 25 (Negative-imaginary lemma) *A linear time-invariant system* $\mathrm{R(A,B,C,D)}$ *with* m *inputs and outpus is negative-imaginary if and only if* A *has no eigenvalues on the imaginary axis,* D *is symmetric, and there exists a positive definite matrix* P *which satisfies:*

$$\begin{cases} \mathrm{AP} + \mathrm{PA}^T \leq 0 \\ \mathrm{B} + \mathrm{APC}^T = 0 \end{cases} \tag{9.35}$$

The following theorem further characterizes negative-imaginary systems in the time domain, providing in particular a necessary and sufficient condition for strictly negative-imaginary systems.

Theorem 26 *A linear time-invariant system* $\mathrm{R(A,B,C,D)}$ *with* m *inputs and outpus is strictly negative-imaginary if and only if i)* A *has no eigenvalues on the imaginary axis; ii)* D *is symmetric; iii) there exists a positive definite matrix* P *which satisfies (9.35); iv) the transfer function matrix*

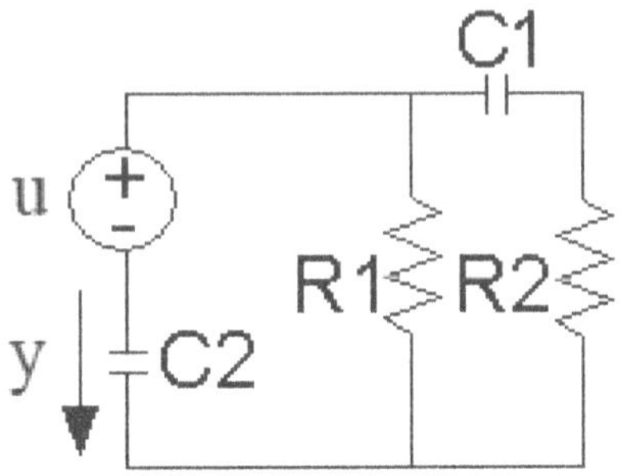

FIGURE 9.14
Example of an electrical circuit ($C_1 > 0$, $C_2 > 0$, $R_1 > 0$, $R_2 > 0$) whose transfer function is negative-imaginary.

$G(s) = C(sI - A)^{-1}B + D$ *is such that* $G(s) - G^T(-s)$ *has no zeros on the imaginary axis* $s = j\omega$ *with* $\omega \neq 0$.

At variance with the positive-real lemma, in the negative-imaginary lemma we find a linear matrix inequality. Linear matrix inequalities are dealt with in detail in Chapter 12, and, therefore, we do not study here the problem of solving (9.35), but limit the discussion to the illustration of a scalar example.

Example 9.14

Consider the following system

$$\begin{aligned} \dot{x} &= -x + u \\ y &= x \end{aligned} \tag{9.36}$$

Here, $A = -1$, $B = 1$, $C = 1$ and $D = 0$. The conditions (9.35) become:

$$\begin{cases} -2P \leq 0 \\ 1 - P = 0 \end{cases} \tag{9.37}$$

which hold for $P = 1$. As $P > 0$, the system is negative-imaginary. In addition, for this system $G(s) = \frac{1}{s+1}$ and $G(s) - G^T(-s) = \frac{2s}{s^2-1}$. As $G(s) - G^T(-s)$ has a zero in the origin, but no zeros on the imaginary axis $s = j\omega$ with $\omega \neq 0$, then also the hypotheses of Theorem 26 hold and we conclude that the system is strictly negative-imaginary.

9.5.3 Closed-loop Stability and Negative-imaginary Systems

We have seen that the negative feedback connection of two positive-real systems (one of which is strictly positive-real) guarantees the stability of the closed-loop system. For negative-imaginary systems there is an analogous result, but in this case a positive-feedback connection has to be used. This difference is particularly important in practice. Consider for instance the case of colocated sensor and actuator pairs in a flexible structure, then, if a velocity

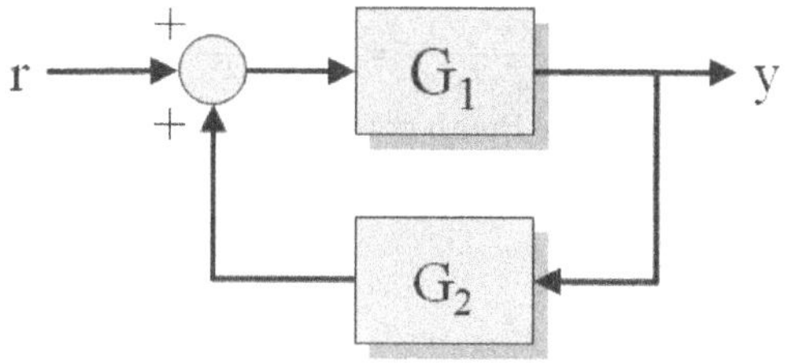

FIGURE 9.15
Positive-feedback configuration with negative-imaginary $G_1(s)$ and $G_2(s)$.

sensor is used, control has to operate in a negative feedback configuration. On the contrary, if position measurements are used for control, then a positive feedback configuration has to be implemented.

The relationship between closed-loop stability and negative-imaginary systems is formally expressed in the following theorem.

Theorem 27 *Consider the positive feedback configuration shown in Figure 9.15 with* $\mathrm{G}_1(s)$ *negative-imaginary,* $\mathrm{G}_2(s)$ *strictly negative-imaginary,* $\bar{G}_1 = \lim_{s\to\infty} \mathrm{G}_1(s)$, $\bar{G}_2 = \lim_{s\to\infty} \mathrm{G}_2(s) \geq 0$ *and* $\bar{G}_1\bar{G}_2 = 0$. *Then, the closed-loop system is asymptotically stable if and only if* $\rho(\mathrm{G}_1(0)\mathrm{G}_2(0)) < 1$.

Notice that this theorem guarantees closed-loop stability through phase stabilization, whereas the small gain theorem is based on gain stabilization. For this reason, it explicitly requires a static gain condition $\rho(\mathrm{G}_1(0)\mathrm{G}_2(0)) < 1$ to hold.

Example 9.15

Consider the system in Figure 9.15 with $G_1(s) = \frac{1}{s+1}$ and $G_2(s) = \frac{1}{s^2+3s+4}$. $G_1(s)$ and $G_2(s)$ are both strictly negative-imaginary. In addition, as $\bar{G}_1 = \bar{G}_2 = 0$ they also satisfy the hypothesys $\bar{G}_1\bar{G}_2 = 0$. Then, since $\rho(G_1(0)G_2(0)) = \frac{1}{4} < 1$, the closed-loop system is asymptotically stable. This can be verified by noticing that the closed-loop transfer function is given by

$$W(s) = \frac{G_1(s)}{1 - G_1(s)G_2(s)} = \frac{s^2 + 3s + 4}{s^3 + 4s^2 + 7s + 3} \tag{9.38}$$

and has the following poles: $s_{1,2} = -1.6963 \pm j1.4359$, and $s_3 = -0.6074$.

Example 9.16

Let us now generalize the previous example, by considering $G_1(s) = \frac{k}{s+1}$ with $k > 0$. $G_1(s)$ is strictly negative-imaginary for any $k > 0$, and, once again, we have that $\bar{G}_1 = \bar{G}_2 = 0$, such that $\bar{G}_1\bar{G}_2 = 0$. Then, applying Theorem 27, since $\rho(G_1(0)G_2(0)) = \frac{k}{4} < 1$ if and only if $k < 4$, we conclude that for $k > 0$ the closed-loop system is asymptotically stable if and only if $k < 4$. This can be also verified by calculating the closed-loop transfer function

$$W(s) = \frac{G_1(s)}{1 - G_1(s)G_2(s)} = \frac{s^2 + 3s + 4}{s^3 + 4s^2 + 7s + 4 - k} \tag{9.39}$$

and studying the stability with the Routh criterion restricted to the case $k > 0$. Notice, in fact, that Theorem 27 expresses a necessary and sufficient condition provided that G_1 and G_2 are negative-imaginary, and so this analysis applies only for $k > 0$.

9.6 Exercises

1. Given the continuous-time system with transfer function $G(s) = \frac{\alpha}{s^3+s^2+4s+4}$ calculate for which values of α the system is bounded-real.

2. Given the continuous-time system with transfer function $G(s) = \frac{4}{s^2+\alpha s+4}$ calculate for which values of α the system is bounded-real.

3. Given the continuous-time system with state-space matrices:

$$A = \begin{bmatrix} 0 & 1 & 0 \\ 0 & 0 & 1 \\ -5 & -3 & -2 \end{bmatrix}; \quad B = \begin{bmatrix} 0 \\ 0 \\ \alpha \end{bmatrix} \quad C = \begin{bmatrix} 1 & 0 & 0 \end{bmatrix}$$

   calculate for which values of α the system is bounded-real.

4. Given the continuous-time system with state-space matrices:

$$A = \begin{bmatrix} -1 & 0 & 0 \\ 0 & -2 & 0 \\ 0 & 0 & -0.5 \end{bmatrix}; \quad B = C^T = \begin{bmatrix} \alpha \\ \alpha \\ \alpha \end{bmatrix}$$

   calculate for which values of α the system is strictly bounded-real.

5. For the feedback system shown in Figure 9.16 with $G(s) = \frac{1}{s^2}$ prove, if possible, that there is a stable first order compensator which makes the system closed-loop passive. Note: do not make any cancellation.

6. Given the system with transfer function $G(s) = \frac{s+1}{(s+2)(s+3)}$ calculate the Cauchy index of the system with two different analytical methods. Then calculate the energy associated with the impulse response. Finally, using the bounded-real lemma, verify analytically if $G(s)$ is bounded-real.

7. Determine for which values of α the system with transfer function $G(s) = \frac{\alpha}{2s^3+s^2+4s+5}$ is bounded-real using two different methods.

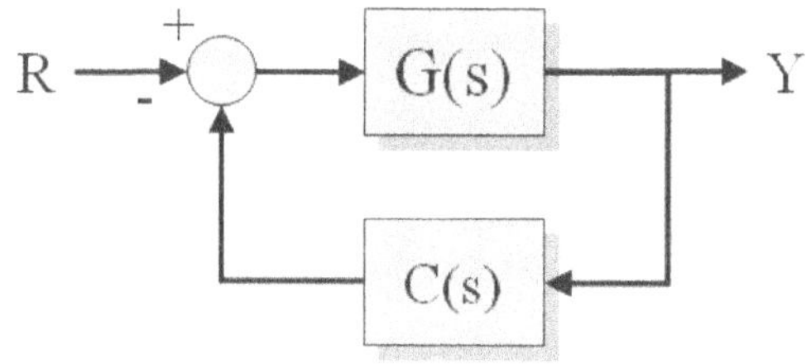

FIGURE 9.16
Block scheme for exercise 5.

8. Given the system with transfer function $G(s) = \frac{s(s^2+2)}{(s^2+1)(s^2+4)}$, propose, if possible, a passive electric circuit realization.
9. Given the system $G(s) = \frac{1}{(s+1)^2}$ determine if there is an R such that the optimal closed-loop system (with functional $J = \int_0^\infty (y^T y + u^T R u)dt$) is passive.
10. Write down an example of a third-order loss-less system and verify that its characteristic values are equal to one.
11. Determine for which values of z_1 and p_1 the continuous-time system with transfer function $G(s) = \frac{s+z_1}{s^2(s+p_1)}$ is negative-imaginary.
12. Given the system in positive feedback configuration (as in Figure 9.15) with $G_1(s) = \frac{40}{s+40}$ and $G(s) = \frac{5}{s^2+5s+\alpha}$, find for which values of $\alpha > 0$ the closed-loop system is stable.
13. Consider the continuous-time system with transfer function $G(s) = \frac{1}{s^2+3s+4}$. If possible, derive an electric circuit realization and a mechanical one.

10

Enforcing the Positive-real or the Negative-imaginary Property in a Linear Model

CONTENTS

This chapter deals with the problem of enforcing either the positive-real or the negative-imaginary property in a linear model through a forward action. Given a linear model, which is for example not positive-real, it is shown how to design a forward action able to make the system passive. In an analogous way, it is discussed how to design a forward action to make a system negative-imaginary. MATLAB problems are included.

10.1 Why to Enforce the Positive-real and Negative-Imaginary Property in a Linear Model

Suppose that we are modeling a system through an identification technique based for instance on the input-output response. In general, even if we know that the system is passive or negative-imaginary, there is no guarantee that the model that we obtain is such. We have already seen a similar problem arising in the context of reduced order model in Section 9.1.4, where starting from a passive system the reduced order model may not be such, unless a specific technique is adopted. In this case, we face the problem of deriving a model of a system that is already known to be positive-real or negative-imaginary, for example because it is a flexible structure with colocated sensors and actuators,

DOI: 10.1201/9781003196921-10

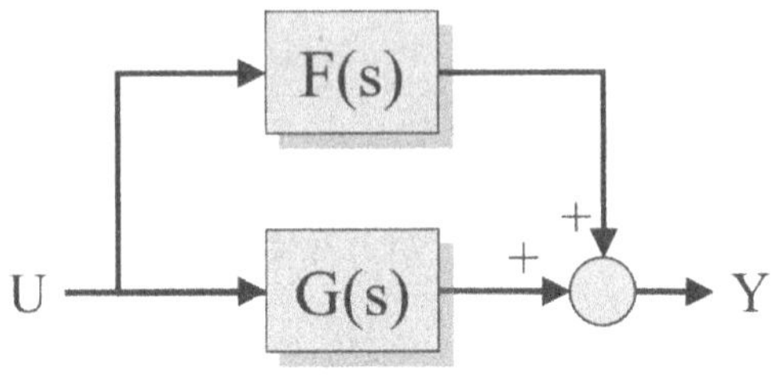

FIGURE 10.1
Block scheme of the feedforward action used to enforce the positive-real or negative-imaginary property.

and having no guarantee that the approximated model produced through identification satisfies the known property of the physical system. Such a scenario requires to compensate in some way the positive-real or negative-imaginary violations.

Another motivation comes from the results that concern closed-loop stability of positive-real and negative-imaginary systems (Chapter 9). In fact, one can think to apply the techniques illustrated in this chapter to make a system positive-real or negative-imaginary using a feedforward control and then to control the system in a feedback scheme exploting the stability results for closed-loop configurations made of positive-real or negative-imaginary (in particular, adopting a negative feedback scheme for positive-real systems and a positive feedback scheme for negative-imaginary systems).

Here, we briefly mention that there are several techniques to enforce the positive-real and negative-imaginary property in a linear model (in particular, there are successful approaches based on perturbing the state-space matrices or using feedback of the state variables), but in this chapter we limit the discussion to illustrate a technique based on the use of a feedforward control action, according to the scheme of Figure 10.1. The technique requires the tuning of a small number of parameters (one or two, as discussed in more detail below).

10.2 Passification

This section deals with the following problem: given a not passive system, can a feedforwad action make it passive? This problem is referred to as the problem of *passification*. It is worth noticing that this can be done for MIMO stable systems and only requires a single feedforward parameter as proven in the following theorem. Essentially, the theorem states that a MIMO stable system $G(s) = \left[\begin{array}{c|c} A & B \\ \hline C & D \end{array}\right]$, in general not passive, can be made passive through the addition of αI to its feedforward matrix D.

Theorem 28 *Let* $G(s) = \left[\begin{array}{c|c} A & B \\ \hline C & D \end{array}\right]$ *be a minimal realization of an asymptotically stable system, then there exists* $\alpha \geq 0$ *such that the system* $G_\alpha(s) = \left[\begin{array}{c|c} A & B \\ \hline C & D+\alpha I \end{array}\right]$ *is passive.*

If $G(s)$ is passive, trivially $\alpha = 0$ and $G_\alpha(s)$ is also passive, so that we have to prove the theorem for the case that $G(s)$ is not passive.

Let us define the following quantity:

$$\bar{\lambda} := \inf_{\omega} \lambda_{min}(G(j\omega) + G^T(-j\omega)) \tag{10.1}$$

Since $G(s)$ is not passive, $\bar{\lambda} < 0$. Moreover, since $G(s)$ is a proper asymptotically stable transfer matrix, it has no poles on the imaginary axis and, therefore, $\bar{\lambda}$ is a real finite quantity.

Let us define as $\bar{\omega}$ the point at which $\lambda_{min}(G(j\bar{\omega}) + G^T(-j\bar{\omega})) = \bar{\lambda}$ (in the limit case, it can be also $\bar{\omega} = +\infty$).

Let us now consider the matrix $(G(j\bar{\omega}) + G^T(-j\bar{\omega}))$. This is an Hermitian (and therefore diagonalizable) matrix:

$$G(j\bar{\omega}) + G^T(-j\bar{\omega}) = T\Lambda T^T \tag{10.2}$$

where T is an unit matrix containing the orthogonal eigenvector of $G(j\bar{\omega}) + G^T(-j\bar{\omega})$ and $\Lambda = \text{diag}(\lambda_1, \lambda_2, \ldots, \lambda_n)$. The smallest eigenvalue of $G(j\bar{\omega}) + G^T(-j\bar{\omega})$ is $\bar{\lambda}$.

If the quantity αI is added to the D matrix of the original system $G(s)$, one obtains:

$$\begin{aligned} G_\alpha(j\bar{\omega}) + G_\alpha^T(-j\bar{\omega}) &= G(j\bar{\omega}) + G^T(-j\bar{\omega}) + 2\alpha I = \\ &= T\Lambda T^T + 2\alpha I = T(\Lambda + 2\alpha I)T^T \end{aligned} \tag{10.3}$$

Therefore, if α is chosen as $\alpha \geq -\bar{\lambda}/2$, all the eigenvalues of $G_\alpha(j\bar{\omega}) + G_\alpha^T(-j\bar{\omega})$ are non-negative, from which it can be concluded that the original system may be made passive by adding αI to the original D matrix. Moreover, if $\alpha > -\bar{\lambda}/2$, the system G_α is strictly passive.

The result can be also extended to Lyapunov stable systems.

The minimum value of α, say $\bar{\alpha}$, that makes the system passive can be found checking the eigenvalues of the Hamiltonian matrix:

$$H_\alpha = \begin{bmatrix} A - B(D + D^T + 2\alpha I)^{-1}C & B(D + D^T + 2\alpha I)^{-1}B^T \\ -C^T(D + D^T + 2\alpha I)^{-1}C & -A^T + C^T(D + D^T + 2\alpha I)^{-1}B^T \end{bmatrix} \tag{10.4}$$

associated to the system having in place of matrix D matrix $\tilde{\mathrm{D}} = \mathrm{D} + \alpha \mathrm{I}$. This is rigorously defined for all α such that $\alpha \neq -\lambda_i((\mathrm{D}+\mathrm{D}^T)/2)$, i.e., $-\alpha$ should not be an eigenvalue of the symmetric part of D. $\bar{\alpha}$ is that value that guarantees that the minimum eigenvalue $\lambda_{min}(\omega)$ of $\mathrm{G}_\alpha(j\omega) + \mathrm{G}_\alpha^T(-j\omega)$ is positive.

Let us consider a value of α, say $\alpha*$, such that $\lambda_{min}(\omega)$ crosses the real axis (i.e., the corresponding H_α has imaginary eigenvalues). For $\alpha \geq \alpha*$, $\mathrm{G}_\alpha(s)$ is passive if and only if H_α has no eigenvalues on the imaginary axis. Therefore, the following steps can be adopted to find the minimum value of α such that $\mathrm{G}_\alpha(s)$ is passive. Pick a random value of ω, say ω_0, and let $\alpha_l = -\lambda_{min}(\mathrm{G}(j\omega_0) + \mathrm{G}^T(-j\omega_0))/2$. Let then α_h be a real number large enough that H_{α_h} has no eigenvalues on the imaginary axis (i.e., such that $G_{\alpha_h}(s)$ is passive). In summary, the following bisection algorithm has been used to find the minimum value of α such that $G_\alpha(s)$ is passive:

- Fix $\alpha = (\alpha_l + \alpha_h)/2$;
- Calculate H_α;
- If H_α has no imaginary eigenvalues, then fix $\alpha_h = \alpha$, otherwise $\alpha_l = \alpha$;
- Stop the procedure when $\alpha_h - \alpha_l < \epsilon$, where $\epsilon > 0$ is the required precision for the calculation of the minimum value of α.

Example 10.1

Let us consider a SISO system defined by the following state-space matrices:

$$\mathrm{A} = \begin{bmatrix} 0 & 1 \\ -5 & -5 \end{bmatrix}; \mathrm{B} = \begin{bmatrix} 0 \\ 1 \end{bmatrix}; \mathrm{C} = \begin{bmatrix} 7 & -9 \end{bmatrix}; \mathrm{D} = 1 \tag{10.5}$$

The transfer function of system (10.5) is $G(s) = \frac{s^2-4s+12}{s^2+5s+5}$. Since, for SISO systems, $G(j\omega)+G^T(-j\omega) = 2\mathrm{Re}\,[G(j\omega)]$, the real part of $G(j\omega)$ has to be studied. In particular, the values at which $\mathrm{Re}\,[G(j\omega)] = 0$ can be calculated from the characteristic polynomial of the Hamiltonian matrix H. This is given by:

$$\psi(s) = \det(s\mathrm{I} - \mathrm{H}) = s^4 + 37s^2 + 60 \tag{10.6}$$

Solving for $\psi(j\omega) = 0$ yields two positive solutions: $\omega_1 = 1.3037$ and $\omega_2 = 5.9414$. For these values, we thus have $\mathrm{Re}\,[G(j\omega)] = 0$. In fact, $\mathrm{Re}\,[G(j\omega)]$ is given by:

$$\mathrm{Re}\,[G(j\omega)] = \frac{\omega^4 - 37\omega^2 + 60}{(5-\omega^2)^2 + 25\omega^2} \tag{10.7}$$

The plot of this function is shown in Figure 10.2.
Solving for $\frac{d\mathrm{Re}\,[G(j\omega)]}{d\omega} = 0$ (i.e., $\frac{\omega(104\omega^4-140\omega^2-3650)}{((5-\omega^2)^2+25\omega^2)^2} = 0$) yields $\bar{\omega} = 2.5759$. Since $\mathrm{Re}\,[G(j\bar{\omega})] = -0.8394$, then $\bar{\lambda} = -1.6788$. Therefore, α has to be chosen so that $\alpha \geq \frac{\bar{\lambda}}{2} = 0.8394$. For instance, the system

$$\begin{array}{ll} \mathrm{A} = \begin{bmatrix} 0 & 1 \\ -5 & -5 \end{bmatrix}; & \mathrm{B} = \begin{bmatrix} 0 \\ 1 \end{bmatrix}; \\ \mathrm{C} = \begin{bmatrix} 7 & -9 \end{bmatrix}; & \mathrm{D} = 1 + \alpha; \\ \alpha = 0.8394 & \end{array} \tag{10.8}$$

is passive, since $\mathrm{Re}\,[G_\alpha(j\omega)] \geq 0\ \forall\omega$ as shown in Figure 10.3.

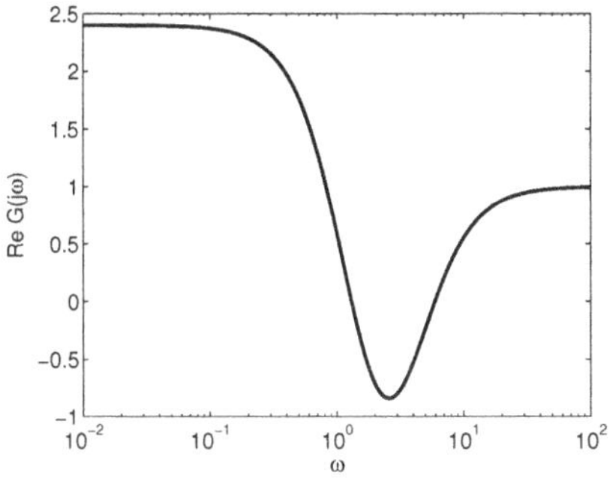

FIGURE 10.2
Plot of Re $[G(j\omega)]$ vs. ω for example 10.1.

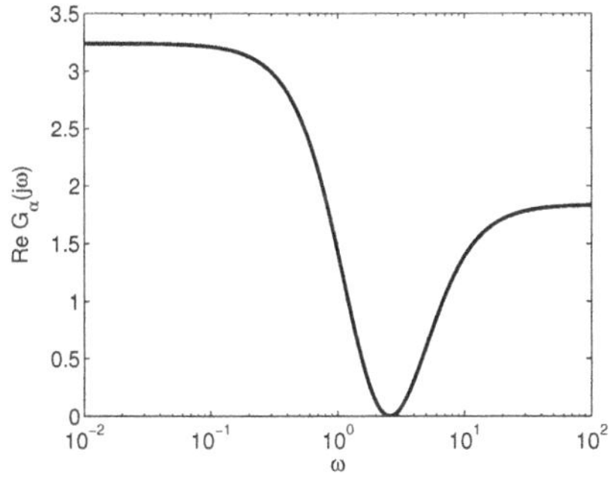

FIGURE 10.3
Plot of Re $[G_\alpha(j\omega)]$ for system (10.8) vs. ω.

MATLAB® Exercise 10.1

Let us consider the MIMO system defined by the following state-space matrices:

$$\mathrm{A} = \begin{bmatrix} 0 & 1 & 0 & 0 \\ 0 & 0 & 1 & 0 \\ 0 & 0 & 0 & 1 \\ -1 & -2 & -5 & -2 \end{bmatrix}; \quad \mathrm{B} = \begin{bmatrix} 1 & 2 \\ -1 & 1 \\ 0 & 1 \\ -1 & 1 \end{bmatrix};$$
$$\mathrm{C} = \begin{bmatrix} 1 & -1 & 1 & -1 \\ 1 & 1 & 1 & 1 \end{bmatrix}; \quad \mathrm{D} = \begin{bmatrix} 1.5 & 2 \\ 1 & 4 \end{bmatrix}. \tag{10.9}$$

The eigenvalues of $\mathrm{G}(j\omega) + \mathrm{G}^T(-j\omega)$ are shown in Figure 10.4. The corresponding eigenvalues of H are: $\lambda_{1,2}(H) = \pm 3.4219$, $\lambda_{3,4}(H) = \pm j0.5945$, $\lambda_{5,6}(H) = \pm j1.0602$, $\lambda_{7,8}(H) = \pm j1.6629$. These values indicate that the eigenvalues of $\mathrm{G}(j\omega) + \mathrm{G}^T(-j\omega)$ cross the real axis three times, i.e., in correspondence of $\omega_1 = 0.5945$, $\omega_2 = 1.0602$ and $\omega_3 = 1.6629$. In this case, it is the minimum eigenvalue of $\mathrm{G}(j\omega) + \mathrm{G}^T(-j\omega)$ that crosses the real axis at ω_1, ω_2 and ω_3.
By applying the bisection algorithm with $\epsilon = 0.0001$, it can be found that the minimum value of α such that $G(s)$ is passive is $\alpha = 3.2674$.
In correspondence of this value, the eigenvalues of H are: $\lambda_{1,2,3,4}(H) = \pm 0.7699 \pm$

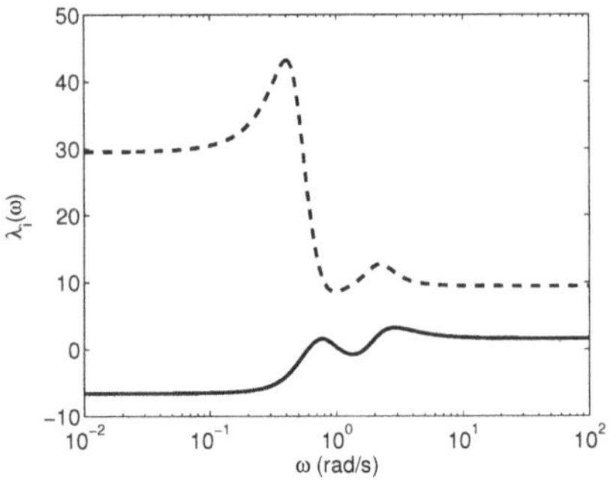

FIGURE 10.4
Eigenvalues of $G(j\omega) + G^T(-j\omega)$ for system (10.9).

$j1.5088$, $\lambda_{5,6}(H) = \pm 0.4873$ and $\lambda_{7,8}(H) = \pm 0.0040$. The eigenvalues of $G_\alpha(j\omega) + G_\alpha^T(-j\omega)$ for $\alpha = 3.2674$ are shown in Figure 10.5.

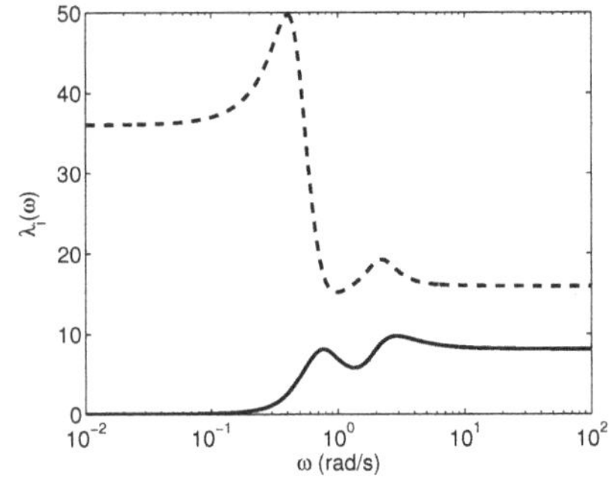

FIGURE 10.5
Eigenvalues of $G_\alpha(j\omega) + G_\alpha^T(-j\omega)$ for system (10.9) with $\tilde{D} = D + \alpha I$ and $\alpha = 3.2674$.

The MATLAB® code used is here reported:

```
A=[0 1 0 0; 0  0 1 0; 0 0 0 1; -1 -2 -5 -2];
B=[1 2; -1 1; 0 1; -1 1];
C=[1 -1 1 -1; 1 1 1 1];
D=[1.5 2; 1 4];
system=ss(A,B,C,D)
MH=[A-B*inv(D+D')*C B*inv(D+D')*B';
  -C'*inv(D+D')*C -A'+C'*inv(D+D')*B'];
eig(MH)
w=logspace(-2,2,1000);
autovaloreminvsomega=w;
E=[];
Z=freqresp(system,w);
ZZ=Z;
for i=1:length(w)
```

```
    ZZ=Z(:,:,i)+ctranspose(Z(:,:,i));
    autovaloreminvsomega(i)=min(eig(ZZ));
    E=[E'; eig(ZZ)']';
end
figure,semilogx(w,E)
alpha=3.2674;
D=alpha*eye(2)+system.D;
systemalpha=ss(A,B,C,D);
MH=[A-B*inv(D+D')*C B*inv(D+D')*B';
  -C'*inv(D+D')*C -A'+C'*inv(D+D')*B'];
eig(MH)
w=logspace(-2,2,1000);
autovaloreminvsomega=w;
E=[];
Z=freqresp(systemalpha,w);
ZZ=Z;
for i=1:length(w)
    ZZ=Z(:,:,i)+ctranspose(Z(:,:,i));
    autovaloreminvsomega(i)=min(eig(ZZ));
    E=[E'; eig(ZZ)']';
end
figure,semilogx(w,E)
```

10.3 Forward Action to make a System Negative-Imaginary

In this section, the problem of designing a forward action to enforce the negative-imaginary property into a system is dealt with, first for the SISO case and then for the more general MIMO case.

10.3.1 The SISO Case

In the SISO case, it is convenient to distinguish the results according to the different definitions of negative-imaginary systems given in Section 9.5.1. Let us begin with the case dealt with in Definition 22, where the original system is asymptotically stable.

Theorem 29 *Let us consider an asymptotically stable SISO system with transfer function $G(s) = \frac{N(s)}{D(s)}$. The forward action $F(s) = \frac{k}{s+\alpha}$ for suitable values of $k > 0$ and $\alpha > 0$ makes the system $G(s)$ negative imaginary.*

If $F(s) = \frac{k}{s+\alpha}$ is used, then the poles of $G(s) + F(s)$ are those of $G(s)$ plus the one of $F(s)$ and so they are all in the open left half of the complex plane. This $F(s)$ also ensures that for all $\omega \geq 0$, $j(G(j\omega) + F(j\omega) - G^*(j\omega) - F^*(j\omega)) \geq 0$.

To see this, let us rewrite $G(j\omega)$ as the sum of its real and imaginary part, that is, $G(j\omega) = G_R(\omega) + jG_I(\omega)$, and do the same for $F(j\omega) = F_R(\omega) + jF_I(\omega) = \frac{k\alpha}{\alpha^2+\omega^2} - j\frac{k\omega}{\alpha^2+\omega^2}$. Now, we notice that, since all the coefficients of $N(s)$ and $D(s)$ are real quantities and $G(s)$ is asymptotically stable, then $\lim_{\omega\to 0} G_I(\omega) = 0$ and $\lim_{\omega\to+\infty} G_I(\omega) = 0$. The system $\tilde{G}(s) = G(s) + F(s)$ is negative-imaginary if the imaginary part of $\tilde{G}(j\omega)$ is negative. Since

$$\tilde{G}(j\omega) = \tilde{G}_R(\omega) + j\tilde{G}_I(\omega) = G_R(\omega) + F_R(\omega) + j\left[G_I(\omega) - \frac{k\omega}{\alpha^2+\omega^2}\right] \quad (10.10)$$

if there exist $k > 0$ and $\alpha > 0$ such that $\forall\omega$:

$$k > \alpha^2\frac{G_I(\omega)}{\omega} + G_I(\omega)\omega \quad (10.11)$$

then the imaginary part of $\tilde{G}(j\omega)$ is negative.

Taking into account that $\lim_{\omega\to 0} G_I(\omega)\omega = 0$ and $\lim_{\omega\to+\infty} \frac{G_I(\omega)}{\omega} = 0$, one can apply the De L'Hopital's theorem to obtain that the limits $\lim_{\omega\to+\infty} G_I(\omega)\omega$ and $\lim_{\omega\to 0} \frac{G_I(\omega)}{\omega}$ are finite. This yields that the right hand term of (10.11) is uniformly bounded from above $\forall\omega \geq 0$ and, for any choice of $\alpha > 0$, there exists $k > 0$ such that (10.11) is satisfied.

It is interesting to note that Theorem 29 also applies with another choice of $F(s)$, namely $F(s) = k\frac{\alpha-s}{\alpha+s}$. The imaginary part of $F(j\omega)$ is $F_I(j\omega) = -\frac{2k\omega}{\alpha^2+\omega^2}$, and in this case a smaller gain k is obtained.

Also notice that α in $F(s) = \frac{k}{s+\alpha}$ is a free parameter. Once a value is set, then, consequently, the gain k has to be chosen to satisfy (10.11). α can be selected, for instance, by taking into account that the function $\frac{k\omega}{\alpha^2+\omega^2}$, representing the imaginary part of $-F(j\omega)$, is non-negative for any value of ω and has a maximum, equal to $\frac{k}{2\alpha}$, at $\omega = \alpha$.

Let us now discuss a design strategy for $F(s)$. Consider a system with the positive-frequency Nyquist plot shown in Figure 10.6(a). Note that $G_I(\omega)$ has only one local maximum at ω_1. Let M be this maximum, that is, $M = \max G_I(\omega)$. The first step is to select α, for instance as $\alpha = \omega_1$. Once fixed α, to find k, an iterative procedure may be used. Let k_1 be $k_1 = M(\alpha^2 + \omega_1^2)/\omega_1$ (if $\alpha = \omega_1$, then $k_1 = 2M\alpha$). Given k_i, if $\frac{k_i\omega}{\alpha^2+\omega^2} < G_I(\omega)$ for any $\omega \in (0, +\infty)$, then the gain has to be increased, that is, $k_{i+1} = k_i + \Delta k$, where $\Delta k > 0$ represents the increment at each iteration. Otherwise, if $\frac{k_i\omega}{\alpha^2+\omega^2} > G_I(\omega)$ in the whole interval $\omega \in (0, +\infty)$, then the procedure may be stopped and the value $k = k_i$ has to be selected. In virtue of Theorem 29 the procedure reaches convergence.

Given k_i, testing if $\frac{k_i\omega}{\alpha^2+\omega^2} < G_I(\omega)$ over a continuum of frequencies could be impractical. This step can be performed more efficiently by using the negative-imaginary lemma given in Section 9.5.2.

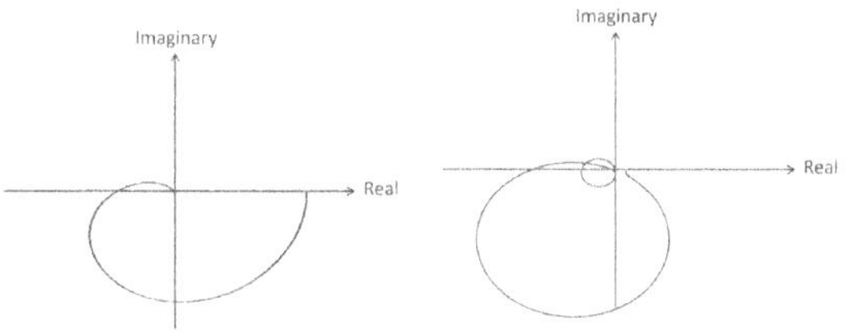

(a) (b)

FIGURE 10.6
Positive-frequency Nyquist plot for two generic systems used to illustrate the design strategy for the forward action making negative-imaginary the original system: (a) $G_I(\omega)$ has only one maximum; (a) $G_I(\omega)$ has more than one maximum.

Let us now consider the more general case, illustrated in Figure 10.6(b), where $G_I(s)$ displays several local maxima (say that their number is n_H) and let us indicate with ω_h, with $h = 1, \ldots, n_H$, these points. Two cases are distinguished: unconstrained gain k and constrained gain k. In the first case, the procedure previously described may be adopted.

The second case, that is, when the value of gain k is constrained, is faced by using several forward blocks in parallel: $F(s) = \frac{k_h}{s+\alpha_h}$ with $h = 1, \ldots, n_H$, where α_h are selected as $\alpha_i = \omega_i$ and the parameter k_h of each block are sequentially obtained in accordance with the following procedure. Given the local maximum M_h of $G_I(\omega)$, fix $k_{h,i} = M_h$ and then check if $\frac{k_{h,i}\omega}{\alpha^2+\omega^2} > G_I(\omega)$ in the interval $[\omega_{h-1}, \omega_{h+1}[$ (for $h = 1$ $\omega_{h-1} = 0$ has to be considered and for $h = n_H$ $\omega_{h+1} = +\infty$). If this is the case, then $k_h = k_{h,i}$, otherwise $k_{h,i+1} = k_{h,i} + \Delta k$, and so on. As each of the blocks $\frac{k_h}{s+\alpha_h}$ contributes with an imaginary part that is negative $\forall \omega$, $F(s)$ makes the original system negative-imaginary. In any case, the procedure seeks to minimize the values of the gains k_h that allow us to achieve the negative-imaginary condition.

Example 10.2

Consider the mass-springer-damper system of Example 9.12 with the mass velocity as output and with the following parameters: $M = 1$, $b = 3$, $k = 4$. In this case the system is not imaginary-negative. Its transfer function is given by:

$$G(s) = \frac{s}{s^2 + 3s + 4} \tag{10.12}$$

Let us now apply Theorem 29 with $F(s) = \frac{k}{s+1}$, that is, having fixed $\alpha = 1$. Since $G_I(\omega) = \frac{\omega(4-\omega^2)}{(4-\omega^2)^2+9\omega^2}$, condition (10.11) becomes:

$$k > \frac{(4-\omega^2)(1+\omega^2)}{(4-\omega^2)^2 + 9\omega^2} \tag{10.13}$$

The maximum of the function $\frac{G_I(\omega)}{\omega} + G_I(\omega)\omega = \frac{(4-\omega^2)(1+\omega^2)}{(4-\omega^2)^2+9\omega^2}$ occurs at $\omega = 1$ where

$2G_I(1) = 1/3$. Hence, selecting $k > 1/3$ guarantees that the system $G(s) + F(s)$ is negative-imaginary.

Consider now the case of a system having simple poles on the imaginary axis, which is dealt with taking into account Definition 23. Suppose that the system is not negative-imaginary because condition 3 of Definition 23 does not hold. In more detail, suppose that $\omega_{0,1}, \ldots, \omega_{0,n_p}$ are the (simple) poles for which the residue $K_l = \lim_{s \to j\omega_{0,l}} (s - j\omega_{0,l})jG(s)$ $(l = 1, \ldots, n_p)$ is negative. Then, the forward action $F(s) = \sum_{l=1}^{n_p} \frac{k_l}{s^2+\omega_{0,l}^2}$ for suitable values of $k_l > 0$ may be used to enforce the negative-imaginary property. In fact, $F(s)$ is able to change the sign of the residues of $\tilde{G}(s) = G(s) + F(s)$ associated to $\omega_{0,1}, \ldots, \omega_{0,n_p}$ as $\lim_{s \to j\omega_{0,l}} (s - j\omega_{0,l})j\tilde{G}(s) = \lim_{s \to j\omega_{0,l}} (s - j\omega_{0,l})jG(s) + \frac{k_l}{\omega_{0,l}}$. Hence, $k_l > -\omega_{0,l} \lim_{s \to j\omega_{0,l}} (s - j\omega_{0,l})jG(s)$ may be used for the purpose. Also note that, on the contrary, adding $F(s)$ does not change the residues associated to the other poles on the imaginary axis as $\lim_{s \to j\omega_0} (s - j\omega_0)jF(s) = 0$ for $\omega_0 \neq \omega_{0,1}, \ldots, \omega_{0,n_p}$.

Example 10.3

Consider the system $G(s) = \frac{1}{(s^2+4)(s^2+20)}$. This system is not negative-imaginary. In fact, we have that $\lim_{s \to j\sqrt{20}} (s - j\sqrt{20})jG(s) = -\frac{1}{32\sqrt{20}} < 0$, which violates condition 3 of Definition 23.
Consider then $F(s) = \frac{k}{s^2+20}$ and select the parameter k to satisfy:

$$\lim_{s \to j\sqrt{20}} (s - j\sqrt{20})jG(s) + \frac{k}{2\sqrt{20}} > 0 \tag{10.14}$$

that yields $k > 1/16$. With this choice of the parameter, then $G(s) + F(s) = \frac{1}{(s^2+4)(s^2+20)} + \frac{k}{s^2+20}$ is negative-imaginary, according to Definition 23.

Finally, consider the case where the original system has one or two poles in the origin, which is dealt with by considering Definition 24. Let us first suppose that the original system $G(s)$ has no poles in the open right half of the complex plane, at most one or two poles in the origin and that $\lim_{s \to 0} s^2 G(s)$ is non-negative. We can also assume that there are no other poles on the imaginary axis as we already have discussed how to deal with such poles. In this case, the forward action $F(s) = \frac{k}{s}$ with a proper choice of k may be used. In fact, conditions 1), 3) and 4) of Definition 24 are trivially satisfied. Concerning condition 2), it has to be proved that the imaginary part of $\tilde{G}(j\omega)$ is non-positive, this yields:

$$k > G_I(\omega)\omega \tag{10.15}$$

The right hand side of equation (10.15) is bounded from above for $\forall \omega \geq 0$. In fact, given $G(s)$ proper and with at most two poles in the origin, then $G_I(\omega)$ is strictly proper and its denominator has at most one root in $\omega = 0$. It follows that the limits $\lim_{\omega\to 0} G_I(\omega)\omega$ and $\lim_{\omega\to+\infty} G_I(\omega)\omega$ are finite and so there exists k satisfying equation (10.15). In particular, k is selected as $k > \max G_I(\omega)\omega$.

If the original system is such that $\lim_{s\to 0} s^2 G(s)$ is negative, $F(s) = \frac{k}{s}$ is not a proper choice as $\lim_{s\to 0} s^2(G(s) + \frac{k}{s}) = \lim_{s\to 0} s^2 G(s)$. A forward action with two poles in the origin, say $F(s) = \frac{k}{s^2}$, is suitable to make non-negative the limit $\lim_{s\to 0} s^2(G(s) + F(s))$, but gives no contribution to the imaginary part. Hence, two terms have to be used: first, $F_1(s) = \frac{k_1}{s^2}$ has to be selected so that $\lim_{s\to 0} s^2(G(s) + F_1(s))$ becomes non-negative; at this point, $F_2(s) = \frac{k_2}{s}$ is applied to system $G(s) + F_1(s)$ that satisfies the hypotheses of the previous case.

Example 10.4

Consider now the system $G(s) = \frac{1}{s^2(s+1)}$ that is not negative-imaginary. In fact, conditions 1), 3) and 4) of Definition 24 hold, but not condition 2) as $G_I(\omega) = \frac{1}{\omega(1+\omega^2)} > 0$ for $\omega > 0$.

Now, since $\lim_{s\to 0} s^2 \frac{1}{s^2(s+1)} = 1 > 0$, we can select $k_2 = 0$. Hence, we use $F(s) = \frac{k_1}{s}$. To derive k_1, we consider the condition

$$\frac{1}{\omega(1+\omega^2)} - \frac{k_1}{\omega} < 0 \tag{10.16}$$

As $\max G_I(\omega)\omega = 1$, then $F(s) = \frac{k}{s}$ with $k > 1$ makes negative-imaginary the system $G(s) + F(s)$.

10.3.2 The MIMO Case

In this section the problem of enforcing the negative-imaginary property in a MIMO system is dealt with. To address this problem, first we need to discuss a fundamental relationship existing between positive-real and negative-imaginary systems is exploited.

Theorem 30 *Given a square proper positive real transfer function matrix* $\mathrm{Z}(s)$, *then* $\mathrm{R}(s) \triangleq \frac{\mathrm{Z}(s)}{s}$ *is negative-imaginary.*

Taking into account this relationship we can reformulate the problem of enforcing the negative-imaginary property in terms of the equivalent problem of enforcing the positive-real property for which we already know that a solution exists. In particular, we have the following result.

Theorem 31 *Given a* $m \times m$ *square proper real-rational transfer function matrix* $\mathrm{G}(s)$ *that has no poles in the open right half plane, has at most two poles in the origin and for each of the poles on the imaginary axis, if any, the residue is a Hermitian matrix, then the forward action* $\mathrm{F}(s) = \frac{k}{s}\mathrm{I}_m$ *for suitable values of* $k > 0$ *makes the system* $\mathrm{G}(s)$ *negative-imaginary.*

In practice, one has to consider the transfer function matrix $s\mathrm{G}(s)$ and find k such that $s\mathrm{G}(s) + k\mathrm{I}_m$ is positive-real. Then, in virtue of Theorem 30, $\mathrm{G}(s) + \frac{k}{s}\mathrm{I}_m$ is negative-imaginary.

Hence, the value of k may be found by using the procedure of Section 10.2, where one has to consider $\lambda_{min}(j\omega[G(j\omega) - G^*(j\omega)])$ vs. ω for $\omega \in (0, +\infty)$ and select $k > \min \lambda_{min}(j\omega[G(j\omega) - G^*(j\omega)])/2$.

Notice that Theorem 31 can be applied to the SISO case, as well. This means to set $F(s) = \frac{k}{s}$ rather than $F(s) = \frac{k}{s+\alpha}$. Using $F(s) = \frac{k}{s}$ yields a system that is negative-imaginary according to Definition 24, while the result of Theorem 29 is stronger as it shows how to obtain negative-imaginary systems that are also asymptotically stable.

MATLAB® Exercise 10.2

Consider the MIMO system

$$\mathrm{A} = \begin{bmatrix} 0 & 1 \\ 0 & -3 \end{bmatrix}; \mathrm{B} = \begin{bmatrix} 1 & 0 \\ 0 & 5 \end{bmatrix}; \mathrm{C} = \begin{bmatrix} 1 & 0.2 \\ 0 & 1 \end{bmatrix}; \mathrm{D} = 0 \tag{10.17}$$

This system is not negative-imaginary as it can be checked by calculating its transfer function matrix

$$G(s) = \begin{bmatrix} \frac{1}{s} & \frac{s+5}{s^2+3s} \\ 0 & \frac{5}{s+3} \end{bmatrix} \tag{10.18}$$

and then $j[G(j\omega) - G^*(j\omega)]$. This matrix, in fact, is not positive semi-definite for any $\omega \in (0, \infty)$ as the smallest eigenvalue of this matrix, namely $\lambda_{min} = \lambda_{min}(j[G(j\omega) - G^*(j\omega)])$, is negative for all $\omega > 0$ (Figure 10.7(a)).

Let us now proceed by considering the procedure described above and study $\lambda_{min}(j\omega[G(j\omega) - G^*(j\omega)])$ (Figure 10.7(b)). Since $\min \lambda_{min}(j\omega[G(j\omega) - G^*(j\omega)]) \simeq 1$, we select $k = 0.5$. This yields $F(s) = \frac{0.5}{s} I_2$ that makes $G(s)$ negative-imaginary (Figure 10.7(c)).

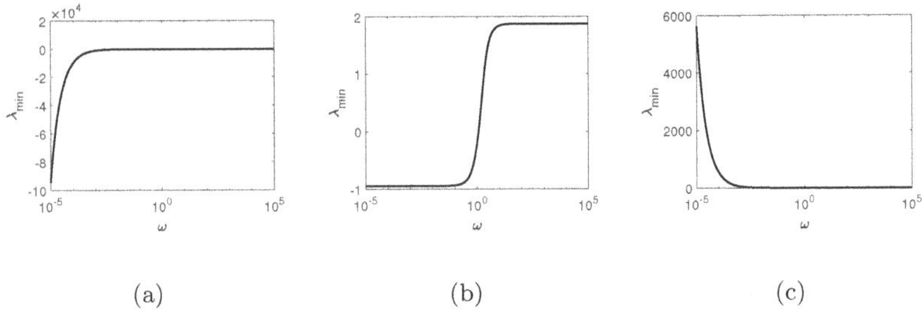

FIGURE 10.7
Enforcing the negative-imaginary property in system (10.17). (a) $\lambda_{min}(j[G(j\omega) - G^*(j\omega)])$; (b) $\lambda_{min}(j\omega[G(j\omega) - G^*(j\omega)])$; (c) $\lambda_{min}(j[\tilde{G}(j\omega) - \tilde{G}^*(j\omega)])$ where $\tilde{G}(s) = G(s) + F(s)$ with $F(s) = \frac{k_2}{s} I_2$ and $k_2 = 0.5$.

The following MATLAB® commands can be used to solve the problem:

```
s=tf('s');
A=[0 1; 0 -3];
B=[1 0; 0 5];
C=[1 0.2; 0 1];
D=0;
system1=ss(A,B,C,D);

w=logspace(-5,5,1000);
mineigvsomega=w;
E=[];
E2=[];
Z=freqresp(system1,w);
ZZ=Z;
for i=1:length(w)
    ZZ=1i*(Z(:,:,i)-ctranspose(Z(:,:,i)));
    mineigvsomega(i)=min(eig(ZZ));
    E=[E'; eig(ZZ)']';
end
figure,semilogx(w,mineigvsomega,'k')
xlabel('\omega'),ylabel('\lambda_m_i_n')

E=[];
for i=1:length(w)
    ZZ=1i*w(i)*(Z(:,:,i)-ctranspose(Z(:,:,i)));
    mineigvsomega(i)=min(eig(ZZ));
    E=[E'; eig(ZZ)']';
end
figure,semilogx(w,mineigvsomega,'k')
xlabel('\omega'),ylabel('\lambda_m_i_n')

k1=0.5; %gain of the forward action
nn=2; %size of the original square tf matrix
feedforwardcontrol=k1/s*eye(nn);
controlledsystem=parallel(system1,feedforwardcontrol)
Z2=freqresp(controlledsystem,w);
ZZ2=Z2;
for i=1:length(w)
    ZZ2=1i*(Z2(:,:,i)-ctranspose(Z2(:,:,i)));
    mineigvsomega(i)=min(eig(ZZ));
    E2=[E2'; eig(ZZ2)']';
end
figure,semilogx(w,autovaloreminvsomega2,'k','LineWidth',2)
xlabel('\omega'),ylabel('\lambda_m_i_n')
```

10.4 Exercises

1. Given the continuous-time system with transfer function $G(s) = \frac{7-s}{s^2+7s+10}$, find $F(s)$ such that $G(s)+F(s)$ is positive-real.
2. Given the continuous-time system with transfer function $G(s) = \frac{7-s}{s^2+7s+10}$, find $F(s)$ such that $G(s)+F(s)$ is negative-imaginary.

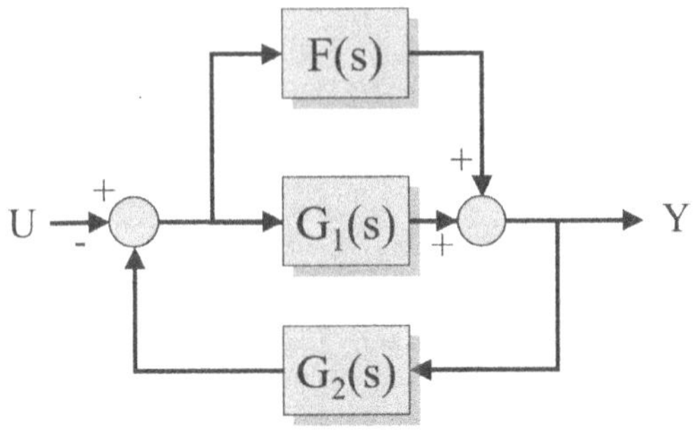

FIGURE 10.8
Block scheme for Exercise 4.

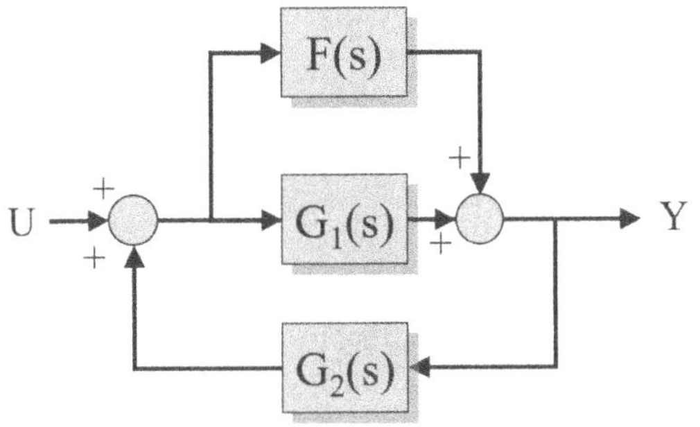

FIGURE 10.9
Block scheme for Exercise 5.

3. Given a system $G(s) = \frac{s}{(s+5)^2}$, find a forward action $F(s)$ such that the system $G(s) + F(s)$ is negative-imaginary. Use $F(s) = F_1(s) = \frac{k}{s+2}$ and repeat the exercise with $F(s) = F_2(s) = \frac{k}{s}$. Then, compare the results.

4. Consider the system in Figure 10.8 with $G_1(s) = \frac{s-2}{s^2+2s+1}$ and $G_2(s) = \frac{1}{s+5}$. Find a transfer function $F(s)$ such that the closed-loop system is stable.

5. Consider the system in Figure 10.9 with $G_1(s) = \frac{s-2}{s^2+2s+1}$ and $G_2(s) = \frac{1}{s+5}$. Find a transfer function $F(s)$ such that the closed-loop system is stable.

11

H_∞ Linear Control

CONTENTS

In this chapter, the H_∞ norm of a system and the tools to calculate it are introduced. Moreover a new representation of the system, useful for deriving a robust controller, is studied. The optimization problem in term of H_∞ norm is presented. The guidelines to derive the H_∞ compensator that results from coupling an observer with the optimal control gains are presented. Even if the H_∞ compensator must be solved by using MATLAB tools, the procedure is described in detail in deriving simple controllers by using a sequence of simple computations. In the chapter the multiobjective control problem is also studied.

11.1 Introduction

Let us consider the classic control scheme for SISO systems shown in Figure 11.1. The transfer functions that characterize the feedback scheme are:

$$\frac{Y(s)}{D(s)} = \frac{1}{1 + C(s)P(s)H(s)} \tag{11.1}$$

$$\frac{Y(s)}{R(s)} = \frac{C(s)P(s)}{1 + C(s)P(s)H(s)} \tag{11.2}$$

$$\frac{Y(s)}{N(s)} = -\frac{C(s)P(s)H(s)}{1 + C(s)P(s)H(s)} \tag{11.3}$$

The analysis of these transfer functions reveals that some of the typical requirements of a control system involve conflicting specifications. For example, the requirement of input tracking requires a large bandwidth, but this implies

DOI: 10.1201/9781003196921-11

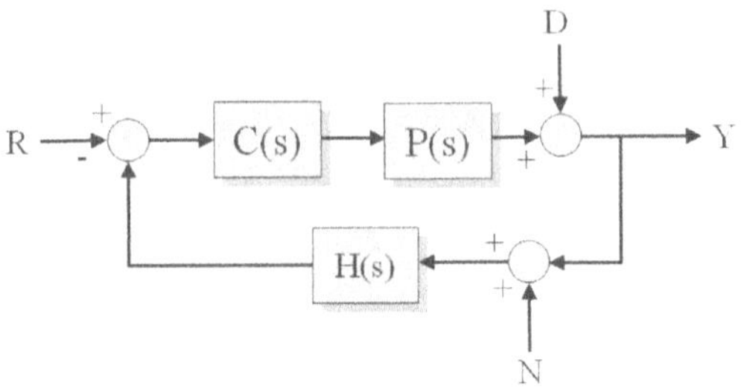

FIGURE 11.1
Scheme of classical control.

a deterioration in performance in terms of rejection of the measurement noise $n(t)$ within the bandwidth. Conversely, the requirements of rejection of noise $d(t)$, of parametric insensitivity and input tracking are not in conflict with each other.

Given these considerations, in relation to the control system discussed above, it makes sense to consider the problem of finding the parameters of $C(s)$ which first assure the stability of the closed-loop system while minimizing the sensitivity of the transfer function $S(s) = \frac{Y(s)}{D(s)}$, thus assuring good disturbance attenuation for all the bandwidth frequencies. Since the index to minimize must account for disturbance varying with frequency, it can be represented by the H_∞ norm of the sensitivity function.

The H_∞ control allows us to solve this kind of problems and not only: it provides a general control scheme which allows us to deal also with MIMO systems and more general problems.

Before illustrating H_∞ control, let us first discuss how to calculate the H_∞ norm of a system. The definition of H_∞ norm, in previous chapters, does not allow the calculation, so we need to use an iterative algorithm. In fact, there is no closed formula to calculate this norm.

Consider the system $G(s)$ and notice that the H_∞ norm is less than γ ($\|G(s)\|_\infty < \gamma$) if and only if the system $G(s)/\gamma$ is bounded-real. This fact can be used to calculate the H_∞ norm of the system. Since it is possible to determine whether a system is bounded-real or not through the bounded-real lemma, calculating the H_∞ norm can be done by performing a bounded-real lemma test.

Hence, given the system $G(s) = (\mathrm{A}, \mathrm{B}, \mathrm{C})$, the H_∞ norm can be found, by verifying for which values of γ the system $G(s)/\gamma = (\mathrm{A}, \frac{\mathrm{B}}{\gamma}, \mathrm{C})$ is bounded-real. So we have to consider the Hamiltonian matrix

$$\mathrm{H} = \begin{bmatrix} \mathrm{A} & \frac{\mathrm{BB}^T}{\gamma^2} \\ -\mathrm{C}^T\mathrm{C} & -\mathrm{A}^T \end{bmatrix} \tag{11.4}$$

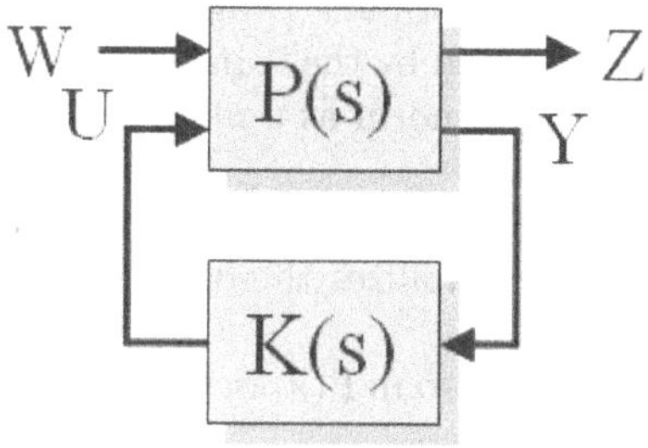

FIGURE 11.2
Scheme of robust control.

fix a value of γ and verify if the matrix (11.4) has no eigenvalues on the imaginary axis. If it does, system $G(s)/\gamma$ is not bounded-real and the norm of $G(s)$ is not less than γ. At this point we have to choose a larger value of γ and run the test again. The H_∞ norm value is given by the smallest value of γ which assures that the system $G(s)/\gamma$ is still bounded-real, that is, the value for which the Hamiltonian matrix (11.4) has no eigenvalues on the imaginary axis.

Example 11.1

Consider the system with transfer function $G(s) = \frac{10}{s+5}$. From the Bode diagram we see that the H_∞ norm of this system is $\|G(s)\|_\infty = 2$. If we consider a realization of the system with $A = -5$; $B = 10$; $C = 1$ we can calculate the H_∞ norm using the procedure based on the Hamiltonian (11.4). In this case we obtain:

$$\mathrm{H} = \begin{bmatrix} -5 & \frac{100}{\gamma^2} \\ -1 & 5 \end{bmatrix} \tag{11.5}$$

The characteristic polynomial of the Hamiltonian is $p(\lambda) = \lambda^2 - 25 + \frac{100}{\gamma^2}$, from which it can be derived that the limit condition for which the Hamiltonian has no eigenvalues on the imaginary axis is $\gamma = 2$.

11.2 Solution of the H_∞ Linear Control Problem

The scheme used for H_∞ control is shown in Figure 11.2. $\mathrm{P}(s)$ represents the system to be controlled, $\mathrm{K}(s)$ is the compensator to be determined. The system outputs are divided in two groups: the output variables available for the control (variables $\mathbf{y}$) and the interest variables (variables $\mathbf{z}$). The objective of the control is to minimize the effect of the exogenous inputs (variables $\mathbf{w}$) on the interest variables, acting through the manipulable inputs (or control signals), represented by the variables $\mathbf{u}$.

The effect of the exogenous inputs (which include disturbances) on the variables of interest is represented by the transfer matrix from $\mathbf{w}$ to $\mathbf{z}$, denoted by $\mathrm{T}_{zw}(s)$. We refer to the H_∞ norm as a measure of the size of the transfer matrix from $\mathbf{w}$ to $\mathbf{z}$.

Once $\gamma > 0$ is assigned, the control problem H_∞ consists in determining the compensator $\mathrm{K}(s)$ which stabilizes the closed-loop system of Figure 11.2 and guarantees $\|\mathrm{T}_{zw}\|_\infty \leq \gamma$.

The state equations of the system $\mathrm{P}(s)$ in the H_∞ control scheme are given by:

$$\begin{cases} \dot{\mathbf{x}} = \mathrm{A}\mathbf{x} + \mathrm{B}_1\mathbf{w} + \mathrm{B}_2\mathbf{u} \\ \mathbf{z} = \mathrm{C}_1\mathbf{x} + \mathrm{D}_{11}\mathbf{w} + \mathrm{D}_{12}\mathbf{u} \\ \mathbf{y} = \mathrm{C}_2\mathbf{x} + \mathrm{D}_{21}\mathbf{w} + \mathrm{D}_{22}\mathbf{u} \end{cases} \tag{11.6}$$

where the dimensions of the vectors $\mathbf{x}$, $\mathbf{w}$, $\mathbf{u}$, $\mathbf{z}$ and $\mathbf{y}$ are respectively, n, m_1, m_2, p_1 and p_2.

With the compact notation given by the realization matrix the system to be controlled can be written as:

$$\mathrm{P}(s) = \left[\begin{array}{c|cc} \mathrm{A} & \mathrm{B}_1 & \mathrm{B}_2 \\ \hline \mathrm{C}_1 & \mathrm{D}_{11} & \mathrm{D}_{12} \\ \mathrm{C}_2 & \mathrm{D}_{21} & \mathrm{D}_{22} \end{array}\right]$$

The problem of the H_∞ control can be solved in the time domain or frequency domain. Originally, it was solved in 1988 in the frequency domain. Two years later, the problem was solved in the time domain, which permits correlating the results to those for optimal control and the Kalman filter.

Throught this section, controllability and observability of the system $\mathrm{P}(s)$ is assumed and it also occurs that $\begin{bmatrix} \mathrm{B}_1 \\ \mathrm{D}_{21} \end{bmatrix} \mathrm{D}_{21}^T = \begin{bmatrix} 0 \\ \mathrm{I} \end{bmatrix}$ and $\mathrm{D}_{12}^T \begin{bmatrix} \mathrm{C}_1 & \mathrm{D}_{12} \end{bmatrix} \mathrm{D}_{21}^T = \begin{bmatrix} 0 & \mathrm{I} \end{bmatrix}$. These latter hypotheses ensure the orthogonality of the variables $\mathbf{y}$ and $\mathbf{z}$, as well as of $\mathbf{w}$ and $\mathbf{u}$, leading to the compact controller formulas discussed herein.

Let us analyze in detail the time domain technique to solve the problem. This allows us to obtain a compensator of order n, defined by the two gain matrices, K_c for the regulator and K_e for the observer. Let us consider $\mathrm{D}_{11} = \mathrm{D}_{22} = 0$ and let us define as X_∞ and Y_∞ the two solutions of the Riccati equations expressed in compact form:

$$\mathrm{X}_\infty = \mathtt{RIC}\begin{bmatrix} \mathrm{A} - \mathrm{B}_2\tilde{\mathrm{D}}_{12}\mathrm{D}_{12}^T\mathrm{C}_1 & \gamma^{-2}\mathrm{B}_1\mathrm{B}_1^T - \mathrm{B}_2\tilde{\mathrm{D}}_{12}\mathrm{B}_2^T \\ -\tilde{\mathrm{C}}_1^T\tilde{\mathrm{C}}_1 & -(\mathrm{A} - \mathrm{B}_2\tilde{\mathrm{D}}_{12}\mathrm{D}_{12}^T\mathrm{C}_1)^T \end{bmatrix} \tag{11.7}$$

and

$$\mathrm{Y}_\infty = \mathtt{RIC}\begin{bmatrix} (\mathrm{A} - \mathrm{B}_1\mathrm{D}_{21}^T\tilde{\mathrm{D}}_{21}\mathrm{C}_2)^T & \gamma^{-2}\mathrm{C}_1^T\mathrm{C}_1 - \mathrm{C}_2^T\tilde{\mathrm{D}}_{21}\mathrm{C}_2 \\ -\tilde{\mathrm{B}}_1\tilde{\mathrm{B}}_1^T & -(\mathrm{A} - \mathrm{B}_1\mathrm{D}_{21}^T\tilde{\mathrm{D}}_{21}\mathrm{C}_2) \end{bmatrix} \tag{11.8}$$

with $\tilde{C}_1 = (I - D_{12}\tilde{D}_{12}D_{12}^T)C_1$ and $\tilde{B}_1 = B_1(I - D_{21}^T\tilde{D}_{21}D_{21})$.

Equation (11.7) is a compact way of indicating the solution of the Riccati equation:

$$A^{*T}X_\infty + X_\infty A^* + X_\infty R^* X_\infty + Q^* = 0$$

with $A^* = A - B_2\tilde{D}_{12}D_{12}^T C_1$, $R^* = \gamma^{-2}B_1B_1^T - B_2\tilde{D}_{12}B_2^T$ and $Q^* = \tilde{C}_1^T\tilde{C}_1$. Similarly, equation (11.8) defines the solution of an analogue Riccati equation.

The solutions of these two Riccati equations depend on γ, which represents the objective reached by the H_∞ control. Starting from the matrices X_∞ and Y_∞ the gains characterizing the compensator can be derived:

$$\begin{cases} \mathbf{K}_c = \tilde{D}_{12}(B_2^T X_\infty + D_{12}^T C_1) \\ \mathbf{K}_e = (Y_\infty C_2^T + B_1 D_{21}^T)\tilde{D}_{21} \end{cases} \tag{11.9}$$

where

$$\begin{aligned} \tilde{D}_{12} &= (D_{12}^T D_{12})^{-1} \\ \tilde{D}_{21} &= (D_{21}D_{21}^T)^{-1} \end{aligned}$$

Note that the Hamiltonian associated with the Riccati equation (11.7), if γ is very large, coincides with the Hamiltonian for optimal control (so a solution can be found). A solution is not guaranteed to exist when γ is small, that is, in the presence of strict requirements for the control system. In this case, the matrix $R^* = \gamma^{-2}B_1B_1^T - B_2\tilde{D}_{12}^T B_2^T$ may be non-negative definite.

The existence of a stabilizing compensator which guarantees $\|T_{zw}\|_\infty < \gamma$ is assured if there are two positive definite solutions to the Riccati equations (11.7) and (11.8) and if the maximum eigenvalue of the product matrix $X_\infty \cdot Y_\infty$ is less than γ^2, that is if $\rho(X_\infty \cdot Y_\infty) < \gamma^2$.

The procedure for solving the H_∞ control problem can be summarized in the following steps:

1. Determine a state-space representation for the process $P(s)$;
2. Verify the existence conditions (invertibility of $D_{12}^T D_{12}$ and $D_{21}D_{21}^T$);
3. Fix a positive value of γ large enough (to solve the two Riccati equations);
4. Solve the two Riccati equations, obtaining the two positive definite solutions;
5. Verify if the condition $\rho(X_\infty \cdot Y_\infty) < \gamma^2$ is met;
6. If the steps 4) and 5) are verified, it is possible to repeat the procedure, lowering γ to point 3).

The obtained compensator consists of a control law expressed as $\mathbf{u} = -K_c\hat{\mathbf{x}}$ and an observer which dynamics is given by the following equation:

$$\dot{\hat{x}} = A\hat{\mathbf{x}} + B_2\mathbf{u} + B_1\hat{\mathbf{w}} + Z_\infty K_e(\mathbf{y} - \hat{\mathbf{y}}) \tag{11.10}$$

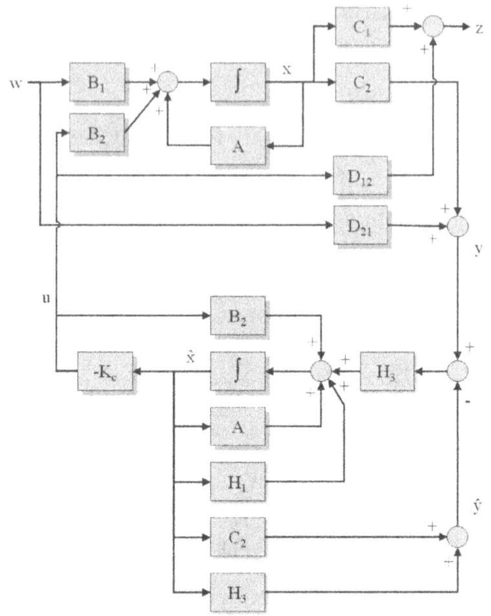

FIGURE 11.3
Block scheme of the H_∞ control.

with $\hat{\mathbf{w}} = \gamma^{-2}\mathrm{B}_1^T\mathrm{X}_\infty$ and $\hat{\mathbf{y}} = \mathrm{C}_2\hat{\mathbf{x}} + \gamma^{-2}\mathrm{D}_{21}\mathrm{B}_1^T\mathrm{X}_\infty\hat{\mathbf{x}}$.

In compact form the state-space equations of the compensator are:

$$\mathrm{K}(s) = \left[\begin{array}{c|c} \mathrm{A} - \mathrm{B}_2\mathrm{K}_c - \mathrm{Z}_\infty\mathrm{K}_e\mathrm{C}_2 + \gamma^{-2}(\mathrm{B}_1\mathrm{B}_1^T - \mathrm{Z}_\infty\mathrm{K}_e\mathrm{D}_{21}\mathrm{B}_1^T)\mathrm{X}_\infty & \mathrm{Z}_\infty\mathrm{K}_e \\ \hline -\mathrm{K}_c & 0 \end{array}\right]$$

with $\mathrm{Z}_\infty = (\mathrm{I} - \gamma^{-2}\mathrm{X}_\infty \cdot \mathrm{Y}_\infty)^{-1}$.

The control scheme relative to the H_∞ control is summarized in Figure 11.3 with $\mathrm{H}_1 = \gamma^{-2}\mathrm{B}_1\mathrm{B}_1^T\mathrm{X}_\infty$, $\mathrm{H}_2 = \gamma^{-2}\mathrm{D}_{21}\mathrm{B}_1^T\mathrm{X}_\infty$, and $\mathrm{H}_3 = \mathrm{Z}_\infty\mathrm{K}_e$.

Relaxing the assumptions on D_{11} and D_{22} leads to more complex formulas. Here we focused on this scenario since it contains all the essential features of the general problem.

Example 11.2

As an example of H_∞ control consider the system in Figure 11.4. Suppose u is the torque applied by the motor to the axis and $J = 1$ is the moment of inertia of the axis. θ is the angle between the motor axis and the x-axis. Assuming the engine is frictionless, the system model is represented by a double integrator. If x_1 is the angular velocity variable and x_2 the state variable representing the angular position, the model in state-space form is expressed by the following equations:

$$\begin{cases} \dot{x}_1 = d + u \\ \dot{x}_2 = x_1 \end{cases} \tag{11.11}$$

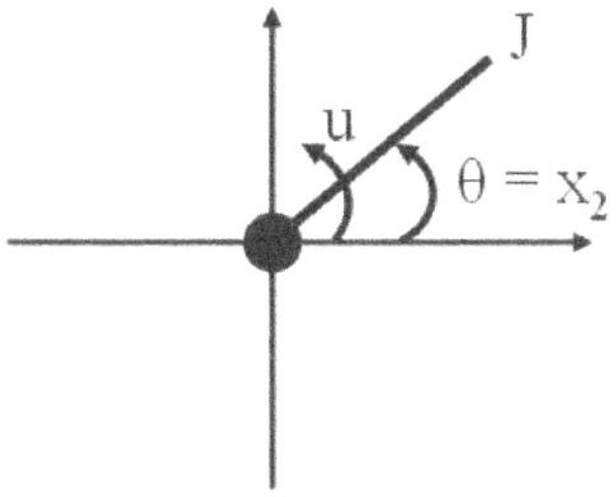

FIGURE 11.4
Example of system to be controlled with the H_∞ control.

where we highlighted the presence of a disturbance d on the engine torque. The output to be regulated is defined as $z = \begin{bmatrix} x_2 \\ u \end{bmatrix}$, so as to take into account the effect of the disturbance on variable x_2, but also on input u. The output is given by variable x_2 affected by noise n ($y = x_2 + n$). Finally, the exogenous input vector is given by $w = \begin{bmatrix} d \\ n \end{bmatrix}$.
Summarizing, the complete model in state-space form is given by:

$$\begin{cases} \dot{x}_1 = w_1 + u \\ \dot{x}_2 = x_1 \\ z_1 = x_2 \\ z_2 = u \\ y = x_2 + w_2 \end{cases} \tag{11.12}$$

Recalling the formulation of state-space equations (11.6) for the H_∞ control scheme:

$$\begin{cases} \mathbf{\dot{x}} = \mathrm{A}\mathbf{x} + \mathrm{B}_1\mathbf{w} + \mathrm{B}_2\mathbf{u} \\ \mathbf{z} = \mathrm{C}_1\mathbf{x} + \mathrm{D}_{11}\mathbf{w} + \mathrm{D}_{12}\mathbf{u} \\ \mathbf{y} = \mathrm{C}_2\mathbf{x} + \mathrm{D}_{21}\mathbf{w} + \mathrm{D}_{22}\mathbf{u} \end{cases}$$

the realization matrix of the system to be controlled can be easily derived:

$$\mathrm{P}(s) = \left[\begin{array}{c|cc} \mathrm{A} & \mathrm{B}_1 & \mathrm{B}_2 \\ \hline \mathrm{C}_1 & \mathrm{D}_{11} & \mathrm{D}_{12} \\ \mathrm{C}_2 & \mathrm{D}_{21} & \mathrm{D}_{22} \end{array}\right] = \left[\begin{array}{cc|ccc} 0 & 0 & 1 & 0 & 1 \\ 1 & 0 & 0 & 0 & 0 \\ \hline 0 & 1 & 0 & 0 & 0 \\ 0 & 0 & 0 & 0 & 1 \\ 0 & 1 & 0 & 1 & 0 \end{array}\right]$$

Applying the H_∞ control procedure (shown in the next chapter), the transfer function of the compensator

$$K(s) = \frac{-977.4(s + 0.40)}{(s + 2.33)(s + 373.4)}$$

and the optimum value $\gamma_{ott} = 2.62$ are obtained. Finally, the solution matrices of the Riccati equations are $\mathrm{X}_\infty = \begin{bmatrix} 1.59 & 1.08 \\ 1.08 & 1.47 \end{bmatrix}$ and $\mathrm{Y}_\infty = \begin{bmatrix} 1.47 & 1.08 \\ 1.08 & 1.59 \end{bmatrix}$.

Example 11.3

In this exercise a further example of H_∞ control is given. Let us consider a Rapid Thermal Processing (RTP) system in semiconductor wafer manufacturing, in which the

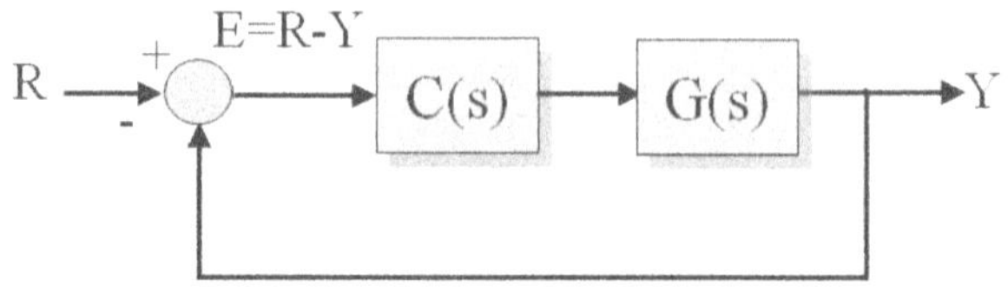

FIGURE 11.5
Block scheme for the control of the RTP system.

ability of rapidly changing the temperature is important for the fabrication of devices with very small crystal length.
Let us consider the following linear model:

$$\begin{aligned} \dot{\mathbf{x}} &= \mathrm{A}\mathbf{x} + \mathrm{B}u \\ y &= \mathrm{C}\mathbf{x} \end{aligned} \tag{11.13}$$

with

$$\mathrm{A} = \begin{bmatrix} -0.0682 & 0.0149 & 0 \\ 0.0458 & -0.1181 & 0.0218 \\ 0 & 0.04683 & -0.1008 \end{bmatrix}; \mathrm{B} = \begin{bmatrix} 0.5122 \\ 0.5226 \\ 0.4185 \end{bmatrix}; \mathrm{C} = \begin{bmatrix} 0 & 1 & 0 \end{bmatrix}; \mathrm{D} = 0$$

The state vector $\mathbf{x}$ represents the temperatures of three tungsten halogen lamps which are tied into one actuator. Only the central one is used for feedback. The reader is referred to the book by Franklin, Powell and Emami-Naeini for a deeper discussion on the linear model of the RTP system and its control by classical methods.
Consider now a H_∞ control problem. Let us start from the control scheme reported in Figure 11.5. The objective of the control (beyond closed-loop stability) is to track the input, i.e., to minimize the error $e(t) = r(t) - y(t)$. So, the first step is to rewrite the problem in terms of this block scheme. To do this, let us write the state-space equations by introducing the variable $\tilde{y} = r - y$ (that is the input of the controller):

$$\begin{aligned} \dot{\mathbf{x}} &= \mathrm{A}\mathbf{x} + \mathrm{B}u \\ \tilde{y} &= w - \mathrm{C}\mathbf{x} \end{aligned} \tag{11.14}$$

To satisfy the objective of input tracking, we choose $z_\infty = e$ and $w = r$. This is the worst case for the error, since for a SISO system the H_∞ norm represents the maximum value of the frequency response, or in other words the maximum amplification for any sinusoidal input.
The whole equations of the system to be controlled are rewritten as:

$$\begin{aligned} \dot{\mathbf{x}} &= \mathrm{A}\mathbf{x} + \mathrm{B}u \\ z_\infty &= w - \mathrm{C}\mathbf{x} \\ \tilde{y} &= w - \mathrm{C}\mathbf{x} \end{aligned} \tag{11.15}$$

from which it is possible to define $\mathrm{B}_1 = \begin{bmatrix} 0 & 0 & 0 \end{bmatrix}^T$, $\mathrm{B}_2 = \mathrm{B}$, $\mathrm{C}_1 = \mathrm{C}_2 = -\mathrm{C}$, $D_{11} = D_{21} = 1$, and $D_{12} = D_{22} = 0$.
The realization matrix of the process can be written as:

$$P(s) = \left[\begin{array}{ccc|cc} -0.0682 & 0.0149 & 0 & 0 & 0.5122 \\ 0.0458 & -0.1181 & 0.0218 & 0 & 0.5226 \\ 0 & 0.04683 & -0.1008 & 0 & 0.4185 \\ \hline 0 & -1 & 0 & 1 & 0 \\ 0 & -1 & 0 & 1 & 0 \end{array}\right];$$

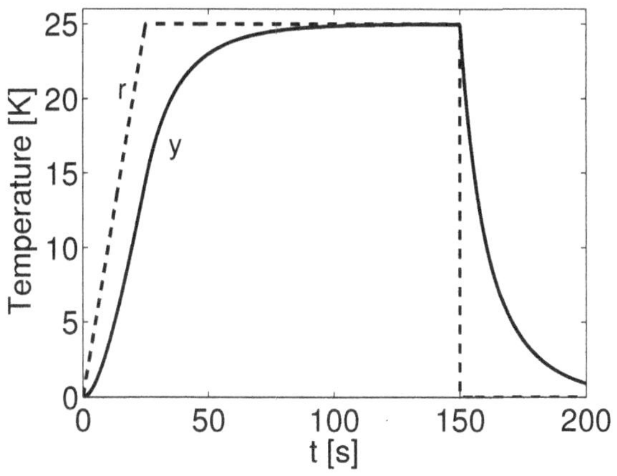

FIGURE 11.6
Response of the closed-loop system with the designed H_∞ controller $K(s)$ as in equation (11.16).

Applying the procedure for the design of the H_∞ control by using the MATLAB® command `hinfric`, we obtain a performance index $\gamma_{opt} = 1.01$ and the following compensator:

$$K(s) = \frac{1.1285(s+0.099)(s+0.071)}{(s+1908)(s+0.14)(s+0.088)} \tag{11.16}$$

The temperature tracking response of the closed-loop system is reported in Figure 11.6 where the input signal (dashed line) is followed by the response (continuous line).

Example 11.4

Let us now discuss a further example related to the design of an H_∞ controller. In this case we will determine the controller and the optimal value of γ through the direct application of the procedure outlined in this section.
Consider the following first order system:

$$\begin{aligned} \dot{x} &= x + u + d \\ y &= x + n \end{aligned} \tag{11.17}$$

Our task is to control the state variable x and the control signal u in presence of a disturbance d and a measurement noise n. Therefore, the equations must be rewritten, considering $\mathrm{z} = \begin{bmatrix} x \\ u \end{bmatrix}$ and $\mathrm{w} = \begin{bmatrix} d \\ n \end{bmatrix}$, as follows:

$$\begin{aligned} \dot{x} &= x + w_1 + u \\ z_1 &= x \\ z_2 &= u \\ y &= x + w_2 \end{aligned} \tag{11.18}$$

According to equations (11.6), the system is characterized by $A = 1$, $\mathrm{B}_1 = \begin{bmatrix} 1 & 0 \end{bmatrix}$, $B_2 = 1$, $\mathrm{C}_1 = \begin{bmatrix} 1 & 0 \end{bmatrix}^T$, $C_2 = 1$, $\mathrm{D}_{11} = \begin{bmatrix} 0 & 0 \\ 0 & 0 \end{bmatrix}$, $\mathrm{D}_{12} = \mathrm{D}_{21}^T = \begin{bmatrix} 0 & 1 \end{bmatrix}^T$, and $D_{22} = 0$.
We then calculate $\tilde{D}_{12} = \left(\mathrm{D}_{12}^T \mathrm{D}_{12}\right)^{-1} = 1$, $\tilde{D}_{21} = \left(\mathrm{D}_{21}\mathrm{D}_{21}^T\right)^{-1} = 1$, $A^* =$

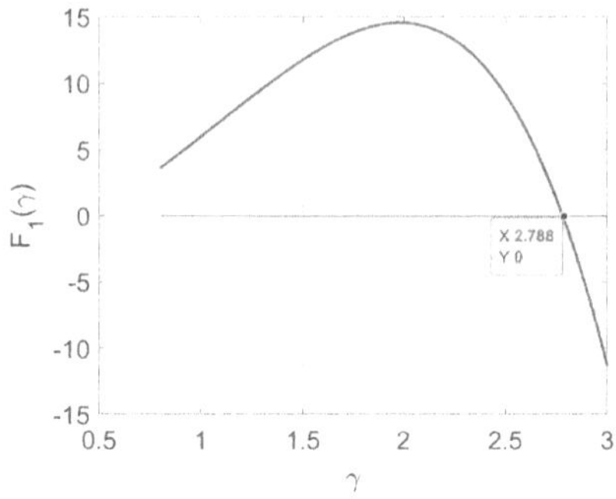

FIGURE 11.7
Graphical solution of the inequality in (11.21) to find γ in Example 11.4.

$A - B_2\tilde{D}_{12}D_{12}^T C_1 = 1$, $\tilde{C}_1 = \left(I - D_{12}\tilde{D}_{12}D_{12}^T\right) C_1 = \begin{bmatrix} 1 & 0 \end{bmatrix}^T$, $R^* = \gamma^{-2}B_1B_1^T - B_2\tilde{D}_{12}B_2^T = \gamma^-2 - 1$, $Q^* = \tilde{C}_1^T\tilde{C}_1 = 1$ and derive the Riccati equation:

$$2X_\infty + \left(\gamma^{-2} - 1\right) X_\infty^2 + 1 = 0 \tag{11.19}$$

whose solution is $X_\infty = \frac{-\gamma^2 - \gamma\sqrt{2\gamma^2-1}}{1-\gamma^2}$. It can be verified that $Y_\infty = X_\infty$.
If $\gamma \gg 0$, the solutions are positive, while for $\gamma = \frac{1}{\sqrt{2}}$ the solutions are negative. In general, $X_\infty > 0$ for $\gamma > 1$.
Let us now calculate the product $X_\infty Y_\infty$ as:

$$\begin{aligned} X_\infty Y_\infty &= \frac{1}{(1-\gamma^2)^2}\left(-\gamma^2 - \gamma\sqrt{2\gamma^2-1}\right)^2 = \\ &= \frac{1}{(1-\gamma^2)^2}\left[\gamma^4 + \gamma^2\left(2\gamma^2-1\right) + 2\gamma^3\sqrt{2\gamma^2-1}\right] \end{aligned} \tag{11.20}$$

According to step 5), we must have that $X_\infty Y_\infty < \gamma^2$. This holds true when:

$$\begin{aligned} &\frac{1}{(1-\gamma^2)^2}\left[\gamma^4 + \gamma^2\left(2\gamma^2-1\right) + 2\gamma^3\sqrt{2\gamma^2-1}\right] < \gamma^2 \\ &\frac{1}{(1-\gamma^2)^2}\left[\gamma^2 + \left(2\gamma^2-1\right) + 2\gamma\sqrt{2\gamma^2-1}\right] < 1 \\ &\left[3\gamma^2 - 1 + 2\gamma\sqrt{2\gamma^2-1}\right] < 1 + \gamma^4 - 2\gamma^2 \\ &-\gamma^4 + 5\gamma^2 + 2\gamma\sqrt{2\gamma^2-1} - 2 < 0 \end{aligned} \tag{11.21}$$

The last inequality in (11.21) can be graphically solved, as in Figure 11.7. We find that $\gamma > 2.67$ represents the limit case where $X_\infty = Y_\infty = 2.729$. The H_∞ controller can be derived calculating $K_c = -K_e = -2.729$, $Z_\infty = (1 - \gamma^{-2}X_\infty Y_\infty)^-1 = 132.44$ and $A_K = A - B_2K_c - Z_\infty K_e C_2 + \gamma^{-2}\left(B_1B_1^T - Z_\infty K_e D_{21}B_1^T\right) X_\infty = -362.9185$. This yields the following H_∞ controller: $K(s) = \frac{-986.99}{s+362.9185}$.

Other examples of H_∞ control are addressed in the following chapters where a very general technique is used to solve them, based on the so-called Linear Matrix Inequalities (LMI).

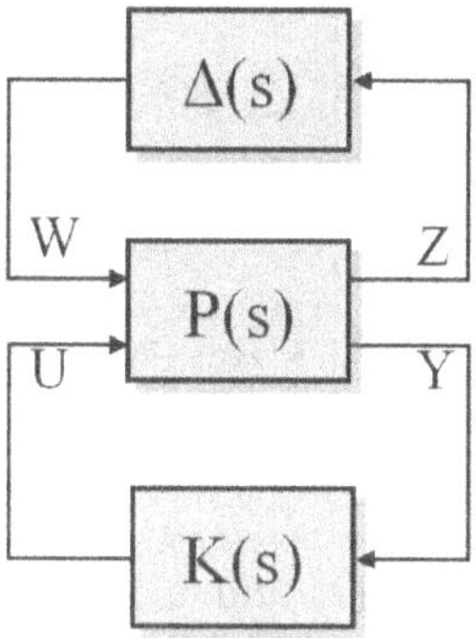

FIGURE 11.8
Scheme of robust control with uncertainty $\Delta(s)$.

11.3 The H_∞ Linear Control and the Uncertainty Problem

As we saw in the introduction, uncertainty is a problem of fundamental importance in control systems. In the case of H_∞ control, we have a significant result to check the robustness of the control system to uncertainties.

Consider again the control scheme shown in Figure 11.2. As discussed above, the objective of the H_∞ control is to determine a stabilizing controller that ensures that $\|T_{zw}(s)\|_\infty < \gamma$. In fact, the transfer matrix depends on the characteristics of the process $P(s)$ to be controlled and on the choice of the compensator, i.e., $T_{zw} = T_{zw}(P, K)$. Thus, for this reason, one can act on K so as to obtain $\|T_{zw}(s)\|_\infty < \gamma$.

Now, let us consider the uncertainty $\Delta(s)$ defined as in Figure 11.8. The uncertainty is modeled through a transfer function from z to w. We will see below that this model can represent various types of uncertainty (multiplicative and additive, for example).

The scheme in Figure 11.8 can be simplified as in Figure 11.9, which highlights that the connection between $\Delta(s)$ and $T_{zw}(s)$ is a feedback connection. At this point, system stability can be assessed by applying Theorem 24 (the small-gain theorem). If system $\Delta(s)T_{zw}(s)$ is bounded-real, then the feedback system is asymptotically stable. Since the H_∞ norm is a consistent norm, then $\|\Delta(s)T_{zw}(s)\|_\infty \leq \|\Delta(s)\|_\infty \|T_{zw}(s)\|_\infty$, so if $\|T_{zw}(s)\|_\infty < \gamma$ then for any uncertainty that $\|\Delta(s)\|_\infty < \frac{1}{\gamma}$ the feedback system is guaranteed to be asymptotically stable.

Therefore, the H_∞ control system is guaranteed to be robust to any stable perturbation having H_∞ norm less than $\frac{1}{\gamma}$.

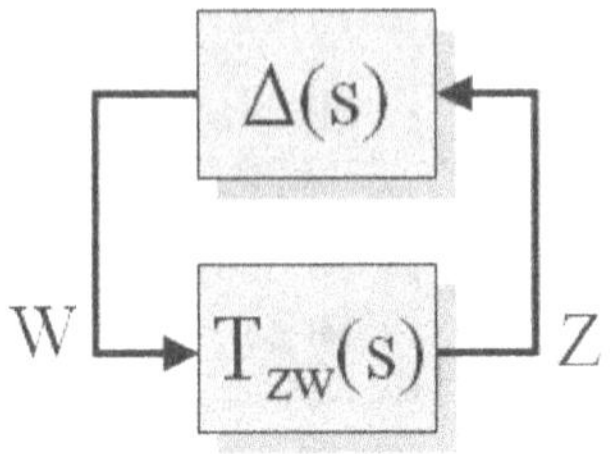

FIGURE 11.9
Feedback connection between $\Delta(s)$ and $T_{zw}(s)$.

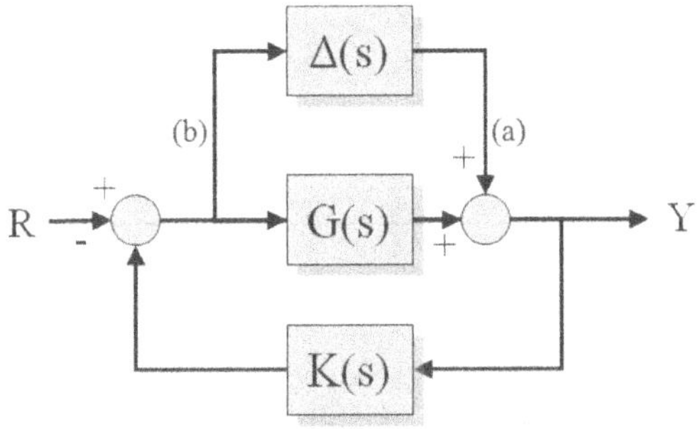

FIGURE 11.10
Scheme representing the additive uncertainty on $G(s)$.

Now let us see how additive and multiplicative uncertainty can be represented according to the scheme in Figure 11.9. First, consider the additive uncertainty as in Figure 11.10. Opening the circuit at the points marked (a) and (b), the scheme can be redrawn as in Figure 11.11, showing that by choosing $T_{zw} = K(I - GK)^{-1}$ the additive uncertainty can be modeled according to the scheme in Figure 11.9.

As for the multiplicative uncertainty of type $\tilde{G}(s) = G(s)(I + \Delta(s))$, represented by Figure 11.12, we can use an equivalent scheme as in Figure 11.13. So, multiplicative uncertainty can be modeled as in Figure 11.9 selecting $T_{zw}(s) = KG(I - KG)^{-1}$.

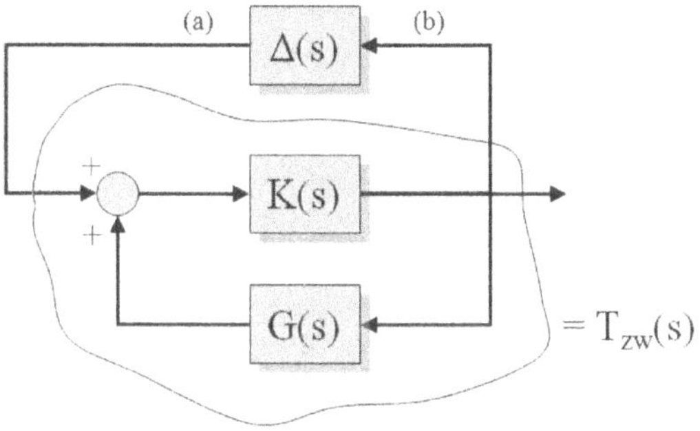

FIGURE 11.11
Formulation of additive uncertainty in terms of feedback between $\Delta(s)$ and $T_{zw}(s)$.

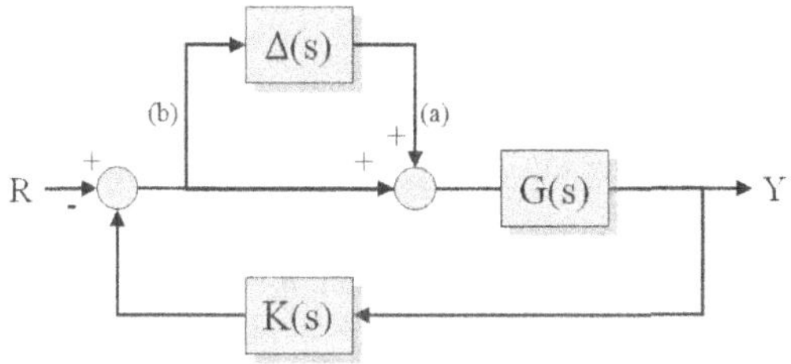

FIGURE 11.12
Scheme representing the multiplicative uncertainty on $G(s)$.

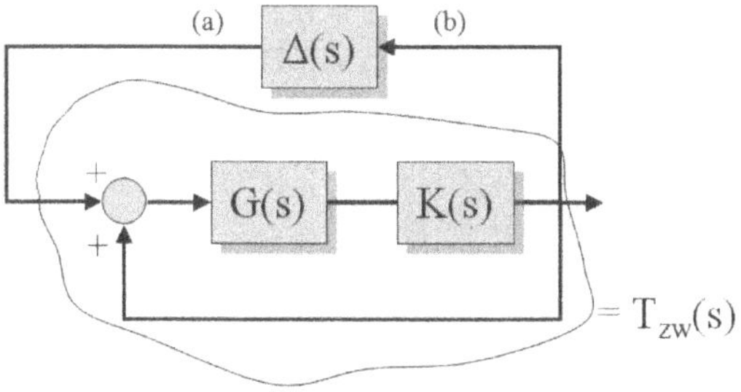

FIGURE 11.13
Formulation of multiplicative uncertainty in terms of feedback between $\Delta(s)$ and $T_{zw}(s)$.

11.4 Exercises

1. Calculate analytically the H_∞ norm for the system with transfer function $G(s) = \frac{2s+1}{s+2}$.
2. Given the continuous-time system with state-space matrices:

$$A = -1; \quad B = \begin{bmatrix} 1 & 2 \end{bmatrix}; \quad C = \begin{bmatrix} 3 \\ 4 \end{bmatrix}$$

   determine analytically the H_∞ norm.
3. Given the continuous-time system with state-space matrices:

$$\mathrm{A} = -3; \quad \mathrm{B} = \begin{bmatrix} 1 & 3 \end{bmatrix}; \quad \mathrm{C} = \begin{bmatrix} 6 \\ 9 \end{bmatrix}$$

   determine analytically the H_∞ norm.
4. Given the system

$$\begin{cases} \dot{x}_1 = x_1 + u + 2w \\ y = x_1 + w \\ z = 2x_1 + 2u \end{cases}$$

   calculate analytically the compensator $C(s)$ with the H_∞ control technique.
5. Calculate the performance of the H_∞ control for the system:

$$\begin{cases} \dot{x} = 2x + 0.1w + u \\ z = 2x + 0.7u \\ y = -x + 2w \end{cases} \tag{11.22}$$

12

Linear Matrix Inequalities for Optimal and Robust Control

CONTENTS

This chapter introduces a technique based on matrix inequalities (the so-called Linear Matrix Inequalities or LMIs) which solves many control problems through a very general formulation. After expressing the main ideas on which the approach relies, the most important control problems are reformulated according to LMI. Whenever this formulation is possible, it paves the way to the use of convex optimization techniques for their solution. In the chapter, the optimal control problems, the H_∞ control and the multi-objective optimization problem are, in particular, focused on. MATLAB includes an efficient toolbox for solving LMI problems, as discussed in the chapter along with several worked examples.

DOI: 10.1201/9781003196921-12

12.1 Definition and Properties of LMI

A LMI is a relation of the following form:

$$\mathrm{A}(\mathbf{x}) = \mathrm{A}_0 + x_1\mathrm{A}_1 + x_2\mathrm{A}_2 + \ldots + x_n\mathrm{A}_n < 0 \tag{12.1}$$

where $\mathbf{x} = \begin{bmatrix} x_1, & x_2, & \ldots, & x_n \end{bmatrix}$ is a vector of unknowns, said decisional variables or optimization variables, $\mathrm{A}_0, \mathrm{A}_1, \mathrm{A}_2, \ldots, \mathrm{A}_n$ are assigned symmetrical matrices, and where the inequality has to be seen in matricial sense ($\mathrm{A}(\mathbf{x})$ is a negative definite matrix).

Note that constraints of type $\mathrm{A}(\mathbf{x}) > 0$ and $\mathrm{A}(\mathbf{x}) < \mathrm{B}(\mathbf{x})$ can be reformulated according to (12.1), considering $-\mathrm{A}(\mathbf{x}) < 0$ and $\mathrm{A}(\mathbf{x}) - \mathrm{B}(\mathbf{x}) < 0$.

Also note that all the points $\mathbf{x}$ that satisfy the LMI (12.1) constitute a convex set. In fact if $\mathrm{A}(\mathbf{y}) < 0$ and $\mathrm{A}(\mathbf{z}) < 0$, then $\mathrm{A}(\frac{\mathbf{y}+\mathbf{z}}{\mathbf{2}}) < 0$.

From the convexity property derives the important consequence that, although the inequality (12.1) has no analytical solution in the general case, it can be solved numerically with the guarantee to find a solution if it exists.

Moreover, since the set of solutions, also called "feasible set," is a convex subset of $\mathbb{R}^n$, finding a solution for the LMI (12.1) is a convex optimization problem. Recall that an optimization problem consists of finding a minimum (or a maximum) in certain regions defined by some constraints on the independent variables.

Furthermore, all the conditions

$$\begin{cases} \mathrm{A}_1(\mathbf{x}) < 0 \\ \mathrm{A}_2(\mathbf{x}) < 0 \\ \quad \ldots \\ \mathrm{A}_k(\mathbf{x}) < 0 \end{cases}$$

are equivalent to one single LMI of type:

$$\mathrm{A}(\mathbf{x}) = \begin{bmatrix} \mathrm{A}_1(\mathbf{x}) & 0 & \ldots & 0 \\ 0 & \mathrm{A}_2(\mathbf{x}) & \ldots & 0 \\ \vdots & & \ddots & \vdots \\ 0 & 0 & \ldots & \mathrm{A}_k(\mathbf{x}) \end{bmatrix} < 0$$

Finally, it is important to notice that nonlinear inequalities of convex type can be transformed in LMI. For example, the following set of inequalities:

$$\begin{cases} \mathrm{R}(\mathbf{x}) > 0 \\ \mathrm{Q}(\mathbf{x}) - \mathrm{S}(\mathbf{x})\mathrm{R}^{-1}(\mathbf{x})\mathrm{S}^T(\mathbf{x}) > 0 \end{cases} \tag{12.2}$$

with symmetric $\mathrm{R}(\mathbf{x})$ and $\mathrm{Q}(\mathbf{x})$ is equivalent to the LMI:

$$\begin{bmatrix} \mathrm{Q}(\mathbf{x}) & \mathrm{S}(\mathbf{x}) \\ \mathrm{S}^T(\mathbf{x}) & \mathrm{R}(\mathbf{x}) \end{bmatrix} > 0 \tag{12.3}$$

In this way, the nonlinear inequality (12.2) can be transformed in a linear inequality (12.3).

In many control problems the LMIs are not in form (12.1), but are structured as follows:

$$\mathbf{L}(X_1, \ldots, X_n) < \mathbf{R}(X_1, \ldots, X_n) \tag{12.4}$$

where **L** and **R** are affine functions of some structured matricial variables $\mathrm{X}_1, \ldots, \mathrm{X}_n$.

For example, consider

$$\mathrm{A}^T\mathrm{X} + \mathrm{X}\mathrm{A} < 0 \tag{12.5}$$

with $\mathrm{A} = \begin{bmatrix} a_{11} & a_{12} \\ a_{21} & a_{22} \end{bmatrix}$. This LMI problem has a particular importance for the study of the stability of linear time-invariant systems. In fact, it is equivalent to the Lyapunov theorem for time-invariant systems. A linear time-invariant system is asymptotically stable if the LMI (12.5) admits a solution $\mathrm{X} > 0$. If we consider $\mathrm{X} = \begin{bmatrix} x_{11} & x_{12} \\ x_{12} & x_{22} \end{bmatrix}$, it is immediate to verify that the inequality $\mathrm{A}^T\mathrm{X} + \mathrm{X}\mathrm{A} < 0$ can be written as $\mathrm{A}_1 x_{11} + \mathrm{A}_2 x_{12} + \mathrm{A}_3 x_{22}$ with A_1, A_2 and A_3 appropriate matrices.

LMIs can be applied to control, identification and design problems. Their importance for these kinds of problems is motivated by several considerations: the design specifications and the design constraints can be expressed in LMI terms; the control problems can be reformulated in terms of convex optimization problems; many of the control problems lack analytical solutions, but they can be treated in LMI terms.

Furthermore, in the period in which LMIs were discovered, the interior point method to solve convex optimization problems was developed. The development of this optimization method was a further motivation to use the LMI.

12.2 LMI Problems

Some convex optimization problems are considered canonical. In particular, many control problems can be formulated in terms of three LMI canonical problems which will be described in detail.

12.2.1 Feasibility Problem

The feasibility problem consists of finding a solution **x** to the LMI problem

$$\mathrm{A}(\mathbf{x}) < 0 \tag{12.6}$$

and it is equivalent to solving the optimization problem:

$$\min t \ \texttt{with constraint} \ \mathrm{A}(\mathbf{x}) < t\mathrm{I} \tag{12.7}$$

with $t < 0$. This problem can be solved with the MATLAB® command `feasp`.

12.2.2 Linear Objective Minimization Problem

Minimizing a convex objective function with a LMI is still a convex problem. In particular, if the objective function is linear it is said to be a *linear objective minimization problem.* This problem is defined by:

$$\min \mathrm{C}^T\mathbf{x} < 0 \ \texttt{with constraint} \ \mathrm{A}(\mathbf{x}) < 0 \tag{12.8}$$

This problem can be solved using the MATLAB command `mincx`.

12.2.3 Generalized Eigenvalue Minimization Problem

The generalized eigenvalue minimization problem consists in minimizing

$$\min \lambda \tag{12.9}$$

with the constraints

$$\begin{cases} \mathrm{A}(\mathbf{x}) < \lambda \mathrm{B}(\mathbf{x}) \\ \mathrm{B}(\mathbf{x}) > 0 \\ \mathrm{C}(\mathbf{x}) < 0 \end{cases} \tag{12.10}$$

This problem can be solved using the MATLAB command `gevp`.

A simplified version of the problem is given by the objective to minimize

$$\min \lambda \tag{12.11}$$

with the constraints

$$\begin{cases} \lambda \mathrm{I} - \mathrm{A}(\mathbf{x}) > 0 \\ \mathrm{B}(\mathbf{x}) > 0 \end{cases} \tag{12.12}$$

Solving a LMI problem means determining first if the problem admits solutions, and so calculating the feasible solution.

The interior point optimization algorithms developed by Nesterov and Nemirovski allowed to solve in an efficient and fast way the previously discussed canonical generical LMI problems.

12.3 Formulation of Control Problems in LMI Terms

Many control problems can be formulated in terms of LMI inequalities. In this section we examine some important examples.

12.3.1 Stability

The system $\dot{\mathbf{x}} = \mathbf{A}\mathbf{x}$ is asymptotically stable if there exists a matrix $\mathrm{P} > 0$ which satisfies $\mathrm{A}^T\mathrm{P} + \mathrm{PA} < 0$. The LMI problem is feasible if the system is asymptotically stable. To study the stability of a linear time-invariant system through LMI, we have to solve a feasibility problem defined by:

$$\begin{aligned} &\mathrm{P} > 0 \\ &\mathrm{A}^T\mathrm{P} + \mathrm{PA} < 0 \end{aligned} \tag{12.13}$$

This LMI problem admits a closed form solution, given by the Lyapunov theorem for linear time-invariant systems.

12.3.2 Closed-loop Stability

The LMI problem (12.13) can be properly adapted to the study of closed-loop stability. In this case the feasibility problem reads

$$\begin{aligned} &\mathrm{P} > 0 \\ &\mathrm{A}_c\mathrm{P} + \mathrm{PA}_c^T < 0 \end{aligned} \tag{12.14}$$

where $\mathrm{A}_c = \mathrm{A} - \mathrm{BK}$ is the closed-loop state matrix. Note that we have considered the equivalent problem of studying the stability of system A_c^T.

Substituting the expression of A_c in equation (12.14), we obtain

$$\begin{aligned} &\mathrm{P} > 0 \\ &(\mathrm{A} - \mathrm{BK})\mathrm{P} + \mathrm{P}(\mathrm{A}^T - \mathrm{K}^T\mathrm{B}^T) < 0 \end{aligned} \tag{12.15}$$

and letting $\mathrm{Q} = \mathrm{KP}$ we obtain

$$\begin{aligned} &\mathrm{P} > 0 \\ &\mathrm{AP} - \mathrm{BQ} + \mathrm{PA}^T - \mathrm{Q}^T\mathrm{B}^T < 0 \end{aligned} \tag{12.16}$$

The LMI problem (12.16) in the two unknowns P and Q allows one to find the control law which stabilizes the system under the assumption that it is controllable. Since P is positive definite, we can calculate the inverse and get K: $\mathrm{K} = \mathrm{QP}^{-1}$.

12.3.3 Simultaneous Stabilizability

The peculiarity of the method described in the previous section can be applied to the *simultaneous stabilizability* problem. This term indicates the possibility to find a control law which stabilizes simultaneously two or more systems.

So, consider m systems (all of order n) $S_1(\mathrm{A}_1, \mathrm{B}_1)$, ..., $S_m(\mathrm{A}_m, \mathrm{B}_m)$. The problem of the existence of a control law $\mathbf{u} = -\mathrm{K}\mathbf{x}$ that stabilizes simultaneously these systems can be viewed as a feasibility problem defined by the following LMIs:

$$\begin{array}{l} \mathrm{P} > 0 \\ \mathrm{A}_1\mathrm{P} - \mathrm{B}_1\mathrm{Q} + \mathrm{P}\mathrm{A}_1^T - \mathrm{Q}^T\mathrm{B}_1^T < 0 \\ \dots \\ \mathrm{A}_m\mathrm{P} - \mathrm{B}_m\mathrm{Q} + \mathrm{P}\mathrm{A}_m^T - \mathrm{Q}^T\mathrm{B}_m^T < 0 \end{array} \tag{12.17}$$

We will see in the following chapter that there exist some analytical conditions for the study of simultaneous stabilizability of two systems, while for most systems the numerical approach here described has to be used. In Section 12.5 an example of simultaneous stabilizability is described.

12.3.4 Positive-real Lemma

The problem of determining if a system is passive, i.e., checking whether $\int_0^T u^T(t)y(t)dt \geq 0$, can be solved through the positive-real lemma discussed in Chapter 9. This problem can also be formulated in LMI terms. Also in this case, this yields a feasibility problem. Particularly, the positive-real lemma can be formulated in terms of two LMIs defined by

$$\begin{array}{l} \mathrm{P} > 0 \\ \begin{bmatrix} \mathrm{A}^T\mathrm{P} + \mathrm{P}\mathrm{A} & \mathrm{P}\mathrm{B} - \mathrm{C}^T \\ \mathrm{B}^T\mathrm{P} - \mathrm{C} & -\mathrm{D}^T - \mathrm{D} \end{bmatrix} \leq 0 \end{array} \tag{12.18}$$

The problem is feasibile if the assumptions of the positive-real lemma are verified.

12.3.5 Bounded-real Lemma

Recall that a system is bounded-real if for any input $u(t)$ the output is such that $\int_0^T y^T(t)y(t)dt \leq \int_0^T u^T(t)u(t)dt$. For linear time-invariant systems this property can be verified through the bounded-real lemma, or in LMI terms verifying that the problem defined by

$$\begin{array}{l} \mathrm{P} > 0 \\ \begin{bmatrix} \mathrm{A}^T\mathrm{P} + \mathrm{P}\mathrm{A} + \mathrm{C}^T\mathrm{C} & \mathrm{P}\mathrm{B} + \mathrm{C}^T\mathrm{D} \\ \mathrm{B}^T\mathrm{P} + \mathrm{D}^T\mathrm{C} & \mathrm{D}^T\mathrm{D} - \mathrm{I} \end{bmatrix} \leq 0 \end{array} \tag{12.19}$$

is feasible.

12.3.6 Calculating the H_∞ Norm Through LMI

The H_∞ norm of a system $G(s) = C(sI - A)^{-1}B$ is α if $\|G(j\omega)\|_\infty \leq \alpha\ \forall\omega$ or equivalently $\|\alpha^{-1}G(j\omega)\|_\infty \leq 1$, i.e., if the system $\tilde{G}(j\omega) = \alpha^{-1}G(j\omega)$ is bounded-real. The approach shown in Chapter 11 is based on the construction of a Hamiltonian matrix $H = \begin{bmatrix} A & \frac{BB^T}{\alpha^2} \\ -C^TC & -A^T \end{bmatrix}$ and on the verification of the minimum value of α for which the matrix has no eigenvalues on the imaginary axis.

Considering that this problem is associated to the Riccati equation $A^TP + PA + C^TC + (\alpha^2)^{-1}PB^TBP = 0$, the problem can be reformulated with a matrix inequality

$$A^TP + PA + C^TC + (\alpha^2)^{-1}PB^TBP < 0$$

where a nonlinear term in P appears. In turn, this nonlinearity inequality can be reformulated as a LMI:

$$\begin{aligned} &P > 0 \\ &\begin{bmatrix} A^TP + PA + C^TC & PB \\ B^TP & 0 \end{bmatrix} \leq \alpha^2 \begin{bmatrix} \varepsilon I & 0 \\ 0 & I \end{bmatrix} \end{aligned} \tag{12.20}$$

The LMI (12.20) is a generalized eigenvalue problem of type $F(x) < \alpha^2 E(x)$. ε is a small quantity introduced to obtain a numerical solution of the problem.

12.4 Solving a LMI Problem

In MATLAB®, it is possible to define any LMI problem and solve it with one of the commands `feasp`, `mincx`, or `gevp`. The definition of the LMI problem does not depend on the type of problem to solve and it is completely general.

The definition of a LMI problem begins with the command `setlmis` and ends with the command `getlmis`. At first it is necessary to define the decisional variables, i.e., the unknowns, of the LMI problem. The command to use is `lmivar` with the following syntax:

```
P=lmivar(type,structure)
```

This command allows one to choose unknown symmetric matrices, rectangular matrices or matrices of other type. Depending on the chosen matrix type, the structure contains different information:

- If `type`=1, the matrix P is square and symmetric. Element (i,1) of `structure` specifies the dimension of the i-block, while element(i,2) specifies the type of block (0 for scalar blocks of type xI, 1 for complete blocks, -1 for zero-blocks);

- If type=2, the matrix P is rectangular of size $m \times n$ as specified in structure=[m,n];
- If type=3, the matrix P is of other type.

MATLAB® Exercise 12.1

Define the following matrices of decisional variables: a 3×3 symmetric matrix X_1; a 2×4 rectangular matrix X_2; and $X_3 = \begin{bmatrix} \Delta & 0 & 0 \\ 0 & \delta_1 & 0 \\ 0 & 0 & \delta_2 I_2 \end{bmatrix}$ with I_2 identity matrix of order two and Δ a 5×5 matrix:

```
X1=lmivar(1,[3 1]);
X2=lmivar(2,[2,4]);
X3=lmivar(1,[5 1; 1 0; 2 0]);
```

A LMI can be specified in MATLAB defining each of its constituent terms with the command `lmiterm`. The LMI is thus specified term by term. The syntax of the command for each term is:

```
lmiterm(termID, A,B,flag)
```

In the `termID` parameter different information are summarized. The first element answers the question: at which LMI belongs the term? `termID(1)` in fact can be $+n$ or $-n$ depending on whether the specified term appears, respectively, in the left- or right-hand of the $n-$th matrix inequality.

The second and the third element answer the question: to which block belongs the specified term? In fact, `termID(2:3)` can be equal to [0 0] for external factors or to $[i, j]$ to indicate the ij block of a generic matrix term that appears in the LMI.

Finally, the fourth element specifies which type of term we are adding to the LMI problem. In fact, `termID(4)` can be 0, X or -X depending on whether the term is constant, of type AXB or $AX'B$.

The other parameters of the command `lmiterm` are more intuitive. `A` and `B` represent the matrices that left- or right- multiply the variable respectively, while the flag is set on 's' to specify with a single command that in the LMI appears not only the given term, but also its symmetric.

In MATLAB it is also possible to use a graphical interface to define a LMI problem. The interface can be launched through the command `lmiedit` and provides the sequence of the commands that define the LMI problem.

MATLAB® Exercise 12.2

As example, we discuss the definition of a LMI problem for the study of the stability of a linear time-invariant system with state matrix $A = \text{diag}(-1, -2, \ldots, -10)$.

```
A=diag([-1:-1:-10]);
setlmis([]);
P=lmivar(1,[10 1]);
lmiterm([-1 1 1 P],1,1);         % LMI #1: P
lmiterm([2 1 1 P],A',1,'s');     % LMI #2: A'*P+P*A
stabproblem=getlmis;
```

Once defined, it is possible to solve the LMI problem with the command `feasp`:

```
[tmin,Psol] = feasp(stabproblem);
```

Running the procedure returns a value $t < 0$. To obtain the solution matrix of the problem we must pass from the decisional matrices to the correspondent matrix with the command `dec2mat`:

```
Pmatrice=dec2mat(stabproblem,Psol,P);
```

MATLAB® Exercise 12.3

Consider the problem of calculating the H_∞ norm for the system with state matrices:

$$\mathrm{A} = \mathrm{diag}(-1, -2, \ldots, -10);\ \mathrm{B} = \begin{bmatrix} 1 \\ 1 \\ \vdots \\ 1 \end{bmatrix};\ \mathrm{C} = \begin{bmatrix} 1 & 1 & \ldots & 1 \end{bmatrix}.$$

We observe that the system is a relaxation system and it is also a strictly proper system. This implies that the H_∞ norm of the system coincides with the value of the frequency response obtained for $\omega \to 0$ so it can be calculated considering $G(0) = \mathrm{C}(-\mathrm{A})^{-1}\mathrm{B}$.
Let us begin defining the system with MATLAB:

```
>> A=diag([-1 -2 -3 -4 -5 -6 -7 -8 -9 -10]);
>> B=ones(10,1);
>> C=ones(1,10);
>> system=ss(A,B,C,0);
```

Then calculate the H_∞ norm exploiting the peculiarities of the system as $\|G(s)\|_\infty = \mathrm{C}(-\mathrm{A})^{-1}\mathrm{B}$:

```
>> normmethod1=C*inv(-A)*B
```

A second method to calculate the H_∞ norm is based on the use of the command `normhinf`:

```
>> normmethod2=normhinf(system)
```

In the LMI toolbox a specific command to calculate the H_∞ norm also exists. In this case the system has to be defined as an object of type `ltisys` and the norm can be calculated as follows:

```
>> g=ltisys(A,B,C,0)
>> [normmethod3, pekf]=norminf(g)
```

Finally, we show the calculation of the H_∞ norm through the definition of a LMI problem which makes use of the equation (12.20):

```
>> setlmis([]);
>> P=lmivar(1,[10 1]);
>> lmiterm([-1 1 1 P],1,1);        % LMI #1: P>0
>> lmiterm([2 1 1 P],A',1,'s');  % LMI #2: A'*P+P*A
>> lmiterm([2 1 1 0],C'*C);       % LMI #2: C'*C
>> lmiterm([2 2 1 P],B',1);       % LMI #2: B'*P
>> lmiterm([-2 1 1 0],0.00001);  % LMI #2: epsilon
>> lmiterm([-2 2 2 0],1);         % LMI #2: 1
>> LMIproblem=getlmis;
>> [a1,po]=gevp(LMIproblem,1);
>> normmethod4=sqrt(a1)
```

Note that the command gevp requires that the LMIs in which the generalized eigenvalue λ to minimize appears are written at the end of the list. In addition we have to specify in how many LMIs λ appears (in this case, it does in a single LMI).

12.5 LMI Problem for Simultaneous Stabilizability

The problem of simultaneous stability is illustrated through an example. Then, an application to the control of nonlinear circuits is discussed.

Consider the system

$$\begin{cases} \dot{x}_1 = x_1^2 + 3x_2 + 2u \\ \dot{x}_2 = x_1 + x_2 \end{cases} \tag{12.21}$$

At $u = 1$ the system admits two equilibrium points $(\bar{x}_1, \bar{x}_2) = (1, -1)$ and $(\tilde{x}_1, \tilde{x}_2) = (2, -2)$. Suppose we want to stabilize the system around both equilibrium points with a unique control law.

The linearized system around the two equilibrium is described by the state matrices:

$$\mathrm{A}_1 = \begin{bmatrix} 2 & 3 \\ 1 & 1 \end{bmatrix}; \mathrm{A}_2 = \begin{bmatrix} 4 & 3 \\ 1 & 1 \end{bmatrix}; \mathrm{B} = \mathrm{B}_1 = \mathrm{B}_2 = \begin{bmatrix} 2 \\ 0 \end{bmatrix}$$

Therefore it is necessary to solve the LMI problem:

$$\begin{array}{l} \mathrm{P} > 0 \\ \mathrm{A}_1\mathrm{P} - \mathrm{BQ} + \mathrm{PA}_1^T - \mathrm{Q}^T\mathrm{B}^T < 0 \\ \mathrm{A}_2\mathrm{P} - \mathrm{BQ} + \mathrm{PA}_2^T - \mathrm{Q}^T\mathrm{B}^T < 0 \end{array} \tag{12.22}$$

Once the matrices are defined in MATLAB®, the LMI problem can be defined through the commands:

```
>> setlmis([]);
>> P=lmivar(1,[2 1]);
>> Q=lmivar(2,[1,2]);
>> lmiterm([-1 1 1 P],1,1);
>> lmiterm([2 1 1 P],A1,1,'s');
>> lmiterm([2 1 1 Q],B,-1,'s');
>> lmiterm([3 1 1 P],A2,1,'s');
>> lmiterm([3 1 1 Q],B,-1,'s');
>> stabilz2=getlmis;
```

The solution can be calculated through the commands:

```
>> [tmin,xfeas] = feasp(stabilz2);
>> Pvalue=dec2mat(stabilz2,xfeas,P);
>> Qvalue=dec2mat(stabilz2,xfeas,Q);
>> k=Qvalue*inv(Pvalue)
```

Note that the obtained value guarantees that the closed-loop eigenvalues have negative real part either if the system works around the equilibrium point $\bar{\mathbf{x}}$ or around $\tilde{\mathbf{x}}$:

```
>> eigenvaluessystem1=eig(A1-B*k)
>> eigenvaluessystem2=eig(A2-B*k)
```

As crosscheck, verify that the controller that, for example, is obtained by assigning the eigenvalues $\lambda_1 = -1$ and $\lambda_2 = -0.5$ in the closed-loop system $A_1 - BK$ does not guarantee the stability of $A_2 - BK$. To do this, it is possible to use the commands:

```
>> openloopeigenvalues=eig(A1)
>> K=acker(A1,B,[-1 -0.5])
>> eig(A1-B*K)
>> eigenvaluessystem1=eig(A1-B*K)
>> eigenvaluessystem2=eig(A2-B*K)
```

MATLAB® Exercise 12.4

In this MATLAB exercise, the LMI approach to simultaneous stability is adopted for the design of an asymptotic observer for a nonlinear circuit.
Let us begin with the linear case, discussing the LMI problem for the observer design for a system described by:

$$\dot{\mathbf{X}} = A\mathbf{X}, \tag{12.23}$$

The dynamical equations of the observer system are:

$$\dot{\hat{\mathbf{X}}} = A\hat{\mathbf{X}} + Ke, \tag{12.24}$$

where K are the observer gains and the error is defined as $e = C\mathbf{X} - C\hat{\mathbf{X}}$ with (A, C) being the state matrices of the system. K has to be chosen in order to ensure the stability of the error system. This can be done following a LMI approach defined starting from the following Lyapunov equation:

$$A_O^T P + PA_O = -\bar{Q}, \tag{12.25}$$

with P and $\bar{Q}$ positive definite matrices and $A_O = A - KC$ the state matrix of the error system.
The corresponding LMI problem is the following:

$$\begin{cases} A_O^T P + PA_O < 0 \\ P > 0 \end{cases} \tag{12.26}$$

where the first constraint is:

$$(A - KC)^T P + P(A - KC) < 0 \tag{12.27}$$

Define now $Q = PK$ and substitute in equation (12.26):

$$\begin{cases} A^T P - C^T Q^T + PA - QC < 0 \\ P > 0 \end{cases} \tag{12.28}$$

Once P and Q solving problem (12.28) are obtained, K is calculated as $K = P^{-1}Q$. The problem is feasible if system (A, C) is observable.
Let us now design the observer for the nonlinear circuit described by the following equations:

$$\begin{cases} \dot{x} = \alpha[y - h(x)] \\ \dot{y} = x - y + z \\ \dot{z} = -\beta y \end{cases} \tag{12.29}$$

with $h(x) = m_1 x + \frac{1}{2}(m_0 - m_1)(|x+1| - |x-1|)$ representing the piece-wise linear nonlinearity. This nonlinear system represents the dimensionless equations governing the so-called Chua's circuit, a nonlinear circuit able to show complex behavior, like chaos. The design of an observer for such kinds of systems can be considered as a solution of the nontrivial problem of chaos synchronization. In the following the parameter values are considered as: $\alpha = 9$, $\beta = 14.286$, $m_0 = -2/7$ and $m_1 = 1/7$.

In this case, the observer has to be effective for each of the possible linear regions in which the observed system and the observer may work. The advantages of the LMI approach for observer design is that other inequalities may be added to the problem so that a set of LMIs has to be solved to find the gains able to simultaneously stabilize more than one system.
Define $\mathbf{e_X} = \mathbf{X} - \hat{\mathbf{X}}$ as the state estimation error. In general, the equation that describes the error system dynamics is:

$$\dot{\mathbf{e}}_\mathbf{X} = \mathrm{A_i}\mathbf{X} - \mathrm{A_j}\hat{\mathbf{X}} - \mathrm{KC}(\mathbf{X} - \hat{\mathbf{X}}) \tag{12.30}$$

where $\mathrm{A_i}$ and $\mathrm{A_j}$ represent respectively the state matrices of the linearized systems in the i-th or j-th region of a generic PWL nonlinearity. If the systems are working in different regions of the PWL nonlinearity, the two matrices $\mathrm{A_i}$ and $\mathrm{A_j}$ are different. Otherwise (i.e., when the observer works in the same region of the observed system), the matrices $\mathrm{A_i}$ and $\mathrm{A_j}$ are equal and the error system dynamic reduces to:

$$\dot{\mathbf{e}}_\mathbf{X} = (\mathrm{A_i} - \mathrm{KC})\mathbf{e_X} \tag{12.31}$$

In this situation the observer can be designed to be stable by solving the following LMI problem:

$$\begin{cases} \mathrm{A_i^T P} - \mathrm{C^T Q^T} + \mathrm{PA_i} - \mathrm{QC} < 0; \quad i = 1, \ldots, q \\ \mathrm{P} > 0 \end{cases} \tag{12.32}$$

where q is the number of regions of the considered PWL nonlinearity. This means that all the LMIs described in each region of the nonlinearity for $\mathrm{A_i} = \mathrm{A_j}$ have to be solved. If the LMI problem is feasible, the gain vector K able to stabilize all the possible error dynamics can be derived.
Although the error system is imposed to be stable only if the observed system and the observer are in the same PWL region (when the two systems are in different regions, the error dynamics is given by equation (12.30)), numerical simulations reveal that these conditions suffice to allow the design of an observer able to reconstruct the dynamics of the observed system.
The considered PWL function has three linear regions, thus the linearized system can be described by the following state matrices:

$$\mathrm{A_1} = \mathrm{A_3} = \begin{bmatrix} -\alpha m_1 & 0 & 0 \\ 1 & -1 & 1 \\ 0 & -\beta & 0 \end{bmatrix}; \mathrm{A_2} = \begin{bmatrix} -\alpha m_0 & 0 & 0 \\ 1 & -1 & 1 \\ 0 & -\beta & 0 \end{bmatrix}$$
$$\mathrm{C} = \mathrm{C_1} = \mathrm{C_2} = \begin{bmatrix} 1 & 0 & 0 \end{bmatrix}$$

Once the matrices are defined in MATLAB®, the LMI problem for the observer design can be set through the commands:

```
>> setlmis([]);
>> P=lmivar(1,[3 1]);
>> Q=lmivar(2,[1,3]);
>> lmiterm([-1 1 1 P],1,1);
>> lmiterm([2 1 1 P],A1',1,'s');
>> lmiterm([2 1 1 Q],C',-1,'s');
>> lmiterm([3 1 1 P],A2',1,'s');
>> lmiterm([3 1 1 Q],C',-1,'s');
>> stabilz2=getlmis;
```

The solution can be calculated through the commands:

```
>> [tmin,xfeas] = feasp(stabilz2);
>> Pvalue=dec2mat(stabilz2,xfeas,P);
>> Qvalue=dec2mat(stabilz2,xfeas,Q);
>> k=inv(Pvalue)*Qvalue
```

Note that the obtained value guarantees that the eigenvalues of the error system have negative real part whether the system works in the first or in the second linear region of the PWL:

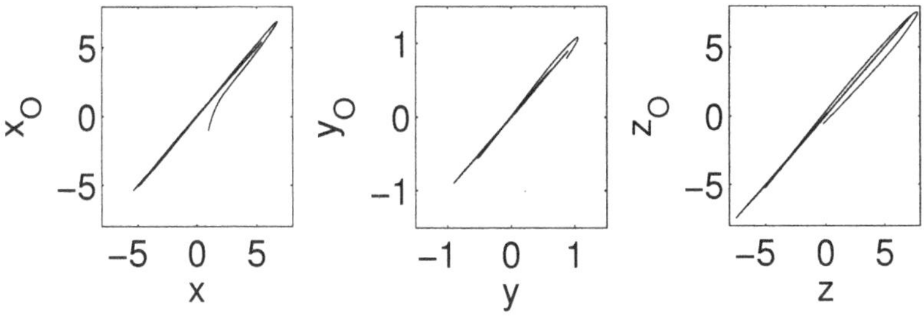

FIGURE 12.1
Representation of the observer state variables as a function of the observed system state variables for Exercise 12.4: all the three state variables are correctly observed.

```
>> eigenvaluessystem1=eig(A1-k*C)
>> eigenvaluessystem2=eig(A2-k*C)
```

The effectiveness of the designed observer is proven in Figure 12.1 where the state variables of the observer, i.e., x_O, y_O and z_O, are reported as a function of the corresponding state variables of the observed system.

12.6 Solving Algebraic Riccati Equations Through LMI

In this paragraph, two examples of control problems formulated in terms of Riccati equations and solved through LMI are dealt with.

MATLAB® Exercise 12.5

Consider first the problem of determining whether a system is bounded-real or not. In this case, assumed that the system is proper, we should establish if there exists or not a solution to the Riccati equation:

$$\mathrm{A}^T\mathrm{P} + \mathrm{PA} + \mathrm{PBB}^T\mathrm{P} + \mathrm{C}^T\mathrm{C} = 0$$

Consider the system defined by

```
>> A=[0 1 0; 0 0 1; -5 -4 -3];
>> B=[0 0 1]';
>> C=[1 0 0];
>> system=ss(A,B,C,0);
```

The system is controllable and observable and has transfer function $G(s) = \frac{1}{s^3+3s^2+4s+5}$. The system is also stable. It is possible to verify immediately that the system is bounded-real from the Bode diagram (in fact the maximum value of $|G(j\omega)|$ is less than 1). Let us now consider how to do it solving a Riccati equation of using the LMI approach.

Define the matrix Q as follows:

```
>> Q=C'*C;
```

the solution of the Riccati equation can be obtained through the command

```
>> Xare=are(A,-B*B',Q)
```

Otherwise, we can solve the LMI problem defined by the minimization of the trace of matrix P subject to a constraint of type

$$\begin{bmatrix} A^T P + PA + C^T C & BP \\ B^T P & -I \end{bmatrix} \leq 0$$

The problem can be defined in MATLAB® with the following commands (note that the problem is of type `mincx`)

```
>> setlmis([]);
>> X=lmivar(1,[3 1]);
>> lmiterm([1 1 1 X],A',1,'s');
>> lmiterm([1 1 1 0],Q);
>> lmiterm([1 2 1 X],B',1);
>> lmiterm([1 2 2 0],-1);
>> lmisys=getlmis;
>> c=mat2dec(lmisys,eye(3))
>> [copt,xopt]=mincx(lmisys,c,[1e-5, 0, 0, 0, 0])
>> Xopt=dec2mat(lmisys,xopt,X)
```

We obtain the same solution found by solving the Riccati equation

$$\text{Xare} = \text{Xopt} = \begin{bmatrix} 1.0528 & 0.5209 & 0.1010 \\ 0.5209 & 0.4682 & 0.1324 \\ 0.1010 & 0.1324 & 0.0445 \end{bmatrix}$$

The existence of a positive definite solution assures that the system is bounded-real. Instead, if we consider $C = \begin{bmatrix} 10 & 0 & 0 \end{bmatrix}$, the system is not bounded-real. In fact, as it is possible to verify in MATLAB®, the Riccati equation does not admit a solution and the LMI problem is not feasible.

MATLAB® Exercise 12.6

Now consider the problem of determining the optimal controller, solving the CARE equation:

$$A^T P + PA - PBB^T P + C^T C = 0$$

To solve the problem using the approach based on the solution of the Riccati equation, we proceed in a completely analogous way to the previous case. Define the system:

```
>> A=[1 -2 1; 3 0 1; 1 -2 -1];
>> B=[1 0 1]';
>> C=[1 1 0];
>> system=ss(A,B,C,0);
```

and solve the Riccati equation with the command `are` or directly with the command `care`:

```
>> Xare=are(A,B*B',C'*C)
```

The control problem can also be solved with the LMI approach. In this case the presence of the minus sign in the nonlinear term requires some adaptations to solve the LMI problem (in fact, this term leads to a non-convex constraint). So we consider $\bar{P} = P^{-1}$, for which the Riccati equation becomes analogous to the case in the MATLAB® Exercise 12.5:

$$A\bar{P} + \bar{P}A^T + \bar{P}C^T C\bar{P} - BB^T = 0$$

Doing so, we get a Riccati equation with a positive nonlinear term. At this point we can consider a new problem of type `mincx`, maximizing, rather than minimizing, the trace of matrix $\bar{P}$ (since it is the inverse of matrix P). We can obtain this changing the sign of the identity matrix with the command `c=mat2dec(lmisys,-eye(3))`, or minimizing an objective function with an opposite sign than the one in the MATLAB® Exercise 12.5. The commands to define and solve the LMI problem are the following:

```
>> setlmis([]);
>> X=lmivar(1,[3 1]);
>> lmiterm([1 1 1 X],A,1,'s');
>> lmiterm([1 1 1 0],-B*B');
>> lmiterm([1 2 1 X],C,1);
>> lmiterm([1 2 2 0],-1);
>> lmisys=getlmis;
>> c=mat2dec(lmisys,-eye(3))
>> [copt,xopt]=mincx(lmisys,c,[1e-5, 0, 0, 0, 0])
>> Xopt=inv(dec2mat(lmisys,xopt,X))
```

Obviously, we obtain the same solution with the two methods:

$$\mathrm{Xare} = \mathrm{Xopt} = \begin{bmatrix} 2.2166 & 0.2635 & 0.6757 \\ 0.2635 & 1.7438 & -0.0275 \\ 0.6757 & -0.0275 & 0.2342 \end{bmatrix}$$

from which we can immediately calculate the value of the gain K of the optimal control law as: $\mathrm{K} = \mathrm{B}^T\mathrm{P} = \begin{bmatrix} 2.89 & 0.24 & 0.91 \end{bmatrix}$.

12.7 Computation of Gramians Through LMI

We discuss here an example of LMI-based calculation of the controllability gramian with two different methods. The procedure can be easily adapted to the observability gramian.

Let us first introduce the LMI problem defined by the maximization of the trace of P subjected to the following constraints:

$$\begin{array}{l} \mathrm{P} > 0 \\ \begin{bmatrix} \mathrm{A}^T\mathrm{P} + \mathrm{PA} & \mathrm{PB} \\ \mathrm{B}^T\mathrm{P} & -\mathrm{I} \end{bmatrix} \leq 0 \end{array} \tag{12.33}$$

The solution P represents the inverse of the controllability gramian of system $S(\mathrm{A}, \mathrm{B})$.

MATLAB® Exercise 12.7

Consider the system with state-space matrices:

$$\mathrm{A}_1 = \begin{bmatrix} -1 & 0 & 0 & 0 & 0 \\ 0 & -2 & 0 & 0 & 0 \\ 0 & 0 & -3 & 0 & 0 \\ 0 & 0 & 0 & -4 & 0 \\ 0 & 0 & 0 & 0 & -5 \end{bmatrix}; \mathrm{B} = \begin{bmatrix} 1 \\ 1 \\ 1 \\ 1 \\ 1 \end{bmatrix} \tag{12.34}$$

Once defined the matrices in MATLAB®, the linear objective problem can be specified as:

```
>> setlmis([]);
>> P=lmivar(1,[5 1]);
```

```
>> lmiterm([-1 1 1 P],1,1);
>> lmiterm([2 1 1 P],A',1,'s');
>> lmiterm([2 2 1 P],B',1);
>> lmiterm([2 2 2 0],1);
>> gramlmi1=getlmis;
```

The linear objective can be defined through the command:

```
>> c=mat2dec(gramlmi1,-eye(5));
```

note that the sign allows to maximize the trace of P. The solution can be calculated through the commands:

```
>> [tmin,xfeas] = mincx(gramlmi1,c,[1e-5 0 0 0 0]);
>> Pvalue=dec2mat(gramlmi1,xfeas,P);
>> Psol=inv(Pvalue)
```

Another approach for the calculation of the controllability gramian is based on the fact that the constraints of the LMI problem of equation (12.33) are equivalent to:

$$\begin{array}{l} \mathrm{Q} > 0 \\ \mathrm{AQ} + \mathrm{QA}^T + \mathrm{BB}^T \leq 0 \end{array} \qquad (12.35)$$

with $\mathrm{Q} = \mathrm{P}^{-1}$. In this case the trace of Q has to be minimized.

MATLAB® Exercise 12.8

Consider again system (12.34) and define the LMI problem:

```
>> setlmis([]);
>> P=lmivar(1,[5 1]);
>> lmiterm([-1 1 1 P],1,1);
>> lmiterm([2 1 1 P],A,1,'s');
>> lmiterm([2 1 1 0],B*B');
>> gramlmi=getlmis;
```

The linear objective is defined through the command:

```
>> c=mat2dec(gramlmi,eye(5));
```

and the solution can be calculated through the commands:

```
>> [tmin,xfeas] = mincx(gramlmi,c,[1e-5 0 0 0 0]);
>> Pvalue=dec2mat(gramlmi,xfeas,P);
```

The solution obtained is comparable with the controllability gramian calculated through the corresponding Lyapunov equation.

12.8 Computation of the Hankel Norm Through LMI

The Hankel norm of a system with stable transfer matrix $\mathrm{G}(s) = \mathrm{C}(s\mathrm{I} - \mathrm{A})^{-1} + \mathrm{D}$ is defined as:

$$\|\mathrm{G}(s)\|_H = \sup_{u \in L_2(\infty,0]} \left(\frac{\int_0^\infty y^T(t)y(t)dt}{\int_{-\infty}^0 u^T(t)u(t)dt} \right)^{\frac{1}{2}} \qquad (12.36)$$

where

$$y(t) = \int_{-\infty}^{0} \mathrm{C}e^{\mathrm{A}(t-\tau)}\mathrm{B}u(\tau)d\tau$$

and

$$L_2(\infty, 0] = \left\{ u : \left(\int_{-\infty}^{0} u^T(t)u(t)dt \right)^{\frac{1}{2}} < \infty \right\}$$

It expresses the quantity of energy that can be transferred through the system from past inputs into future outputs. It can be demonstrated that:

$$\|\mathrm{G}(s)\|_H = \sigma_1 \tag{12.37}$$

where σ_1 is the maximum singular value of the system. For a LTI system the Hankel norm is also the minimum value of γ satisfying the following LMI problem:

$$\begin{cases} \mathrm{P} > 0 \\ \mathrm{Q} > 0 \\ \mathrm{A}^T\mathrm{Q} + \mathrm{QA} + \mathrm{C}^T\mathrm{C} \leq 0 \\ \begin{bmatrix} \mathrm{A}^T\mathrm{P} + \mathrm{PA} & \mathrm{PB} \\ \mathrm{B}^T\mathrm{P} & -\mathrm{I} \end{bmatrix} < 0 \\ \gamma\mathrm{P} - \mathrm{Q} > 0 \end{cases} \tag{12.38}$$

Problem (12.38) is a generalized eigenvalue problem under LMI constraints. It is solved by using the MATLAB command **gevp**, as shown in the following example.

MATLAB® Exercise 12.9

Consider the system with state-space matrices given by:

$$\mathrm{A} = \begin{bmatrix} -1 & 0 & 0 & 0 & 0 \\ 0 & -2 & 0 & 0 & 0 \\ 0 & 0 & -3 & 0 & 0 \\ 0 & 0 & 0 & -4 & 0 \\ 0 & 0 & 0 & 0 & -5 \end{bmatrix}; \mathrm{B} = \mathrm{C}^T = \begin{bmatrix} 1 \\ 1 \\ 1 \\ 1 \\ 1 \end{bmatrix}$$

and calculate the Hankel norm.

Let us first define the matrices:

```
>> A=diag([-1 -2 -3 -4 -5]);
>> B=ones(5,1);
>> C=B';
```

One of the LMI constraints of the problem is not a strict inequality. To deal with such a constraint, one has to introduce a small perturbation matrix:

```
>> m=.00001*eye(5);
```

The LMI problem is now set:

```
>> setlmis([]);
>> P=lmivar(1,[5,1]);
>> Q=lmivar(1,[5,1]);
>> lmiterm([-1 1 1 P],1,1);
>> lmiterm([1 1 1 0],0);
```

```
>> lmiterm([-2 1 1 Q],1,1);
>> lmiterm([2 1 1 0],0);
>> lmiterm([3 1 1 P],1,A,'s');
>> lmiterm([3 2 1 P],B',1);
>> lmiterm([3 2 2 0],-1);
>> lmiterm([4 1 1 Q],1,A,'s');
>> h=C'*C+m;
>> lmiterm([-4 1 1 0],-h);
>> lmiterm([5 1 1 Q],1,1);
>> lmiterm([-5 1 1 P],1,1);
>> lmisys=getlmis;
>> options=[0.00001 0 0 0 0];
>> [Hnorm,PQ]=gevp(lmisys,1,options);
>> P=dec2mat(lmisys,PQ,P);
>> Q=dec2mat(lmisys,PQ,Q);
```

One obtains: $\|G(s)\|_H = 1.1151$.

This result can be compared with the calculation of the maximum singular value of the system as follows:

```
>> Wo2=gram(A',C');
>> Wc2=gram(A,B);
>> S=eig(Wo2*Wc2);
>> s1=max(S)
```

which yields $\sigma_1 = 1.1150$.

12.9 H_∞ Control

In this paragraph, we illustrate the LMI approach to solve the H_∞ control problem through an example. Consider the system:

$$\begin{aligned} \dot{x} &= w + 2u \\ z &= x \\ y &= -x + w \end{aligned} \tag{12.39}$$

Remember that the general equations for a system to which the H_∞ control is applied, are given by:

$$\begin{cases} \dot{\mathbf{x}} = \mathrm{A}\mathbf{x} + \mathrm{B}_1\mathbf{w} + \mathrm{B}_2\mathbf{u} \\ \mathbf{z} = \mathrm{C}_1\mathbf{x} + \mathrm{D}_{11}\mathbf{w} + \mathrm{D}_{12}\mathbf{u} \\ \mathbf{y} = \mathrm{C}_2\mathbf{x} + \mathrm{D}_{21}\mathbf{w} + \mathrm{D}_{22}\mathbf{u} \end{cases} \tag{12.40}$$

System (12.39) is equal to system (12.40) if the following values of the parameters are chosen: $a = 0$; $b_1 = 1$; $b_2 = 2$; $c_1 = 1$; $d_{11} = d_{12} = 0$; $c_2 = -1$; $d_{21} = 1$; $d_{22} = 0$.

Now, let us define in MATLAB the system as a `ltisys` object

```
>> A=0;
>> B1=1;
>> B2=2;
```

```
>> C1=1;
>> D11=0;
>> D12=0;
>> C2=-1;
>> D21=1;
>> D22=0;
>> P=ltisys(A,[B1 B2], [C1; C2],[D11 D12; D21 D22])
```

To solve the problem we can use two methods: the resolution method based on the Riccati equations or the method based on an optimization problem with LMI constraints. In the first case, we need to follow the procedure described in Chapter 11. In the second case we have to consider the LMI minimization problem with linear objective defined by the constraints:

$$\begin{pmatrix} \mathrm{N}_{12} & 0 \\ 0 & \mathrm{I} \end{pmatrix}^T \begin{pmatrix} \mathrm{AR}+\mathrm{RA}^T & \mathrm{RC}_1^T & \mathrm{B}_1 \\ \mathrm{C}_1\mathrm{R} & -\gamma\mathrm{I} & \mathrm{D}_{11} \\ \mathrm{B}_1^T & \mathrm{D}_{11}^T & -\gamma\mathrm{I} \end{pmatrix} \begin{pmatrix} \mathrm{N}_{12} & 0 \\ 0 & \mathrm{I} \end{pmatrix} < 0 \tag{12.41}$$

$$\begin{pmatrix} \mathrm{N}_{21} & 0 \\ 0 & \mathrm{I} \end{pmatrix}^T \begin{pmatrix} \mathrm{A}^T\mathrm{S}+\mathrm{SA} & \mathrm{SB}_1^T & \mathrm{C}_1^T \\ \mathrm{B}_1^T\mathrm{S} & -\gamma\mathrm{I} & \mathrm{D}_{11}^T \\ \mathrm{C}_1 & \mathrm{D}_{11} & -\gamma\mathrm{I} \end{pmatrix} \begin{pmatrix} \mathrm{N}_{21} & 0 \\ 0 & \mathrm{I} \end{pmatrix} < 0 \tag{12.42}$$

$$\begin{pmatrix} \mathrm{R} & \mathrm{I} \\ \mathrm{I} & \mathrm{S} \end{pmatrix} \geq 0 \tag{12.43}$$

where N_{12} and N_{21} are the bases of the null spaces, respectively, of $(\mathrm{B}_2^T, \mathrm{D}_{12}^T)$ and of $(\mathrm{C}_2, \mathrm{D}_{21})$. In MATLAB, we do not need to define the inequalities (12.41)–(12.43) every time, but we can use directly the command `hinflmi`.

To use the method based on the Riccati equations the following command has to be used:

```
>> [gopt,K]=hinfric(P,[1 1])
```

while to use the LMI-based method, the following command has to be used:

```
>> [gopt,K]=hinflmi(P,[1 1])
```

In both cases, the size of the matrix D_{22} (i.e., `[1 1]`) or, in other words, the size of the output vector $\mathbf{y}$ and of the input vector $\mathbf{u}$, has to be specified.

Similar performance is obtained with the two approaches: $\gamma = 1.0088$ with the method based on Riccati equations, and $\gamma = 1.0023$ with the LMI-based method.

To verify the obtained performance, the control scheme in Figure 11.2 has first to be implemented. This can be done by connecting within a feedback loop system $P(s)$ and controller $K(s)$. If the two systems have been defined as ltisys objects, one has to use the command `slft`; while if the two systems have been defined as lti objects, the command `feedback` has to be used.

Consider the first case. The closed-loop system is defined by:

```
>> systemcl=slft(P,K,1,1)
```

sistemacl also is a ltisys object, i.e., it is defined by the realization matrix. Note that in this case, since u is no more an external input of the system, but it derives from the feedback of y, the closed-loop system has a 2×2 state matrix and $\mathrm{D}^{cl} \in \mathbb{R}^{1\times 1}$: sistemacl= $\begin{bmatrix} \mathrm{A}^{cl}_{2\times 2} & \mathrm{B}^{cl}_{1\times 1} \\ \mathrm{C}^{cl}_{1\times 1} & \mathrm{D}^{cl}_{1\times 1} \end{bmatrix}$.

To verify, for instance, that the closed-loop system is stable, one has to find the matrix A^{cl} and then to compute its eigenvalues:

```
>> eig(systemcl(1:2,1:2))
```

this can be also done by using the command:

```
>> spol(sistemacl)
```

Obviously, closed-loop eigenvalues with negative real part ($\lambda_1 = -0.0726 \cdot 10^4$ and $\lambda_2 = -9.8886 \cdot 10^4$) are obtained.

Otherwise, lti objects can be used. For example, the closed-loop system can be rewritten as a lti object (recall that, in this case, the input is w, and the output is z):

```
>>systemcllti1=ss(systemcl(1:2,1:2),systemcl(1:2,3),
        systemcl(3,1:2),systemcl(3,3))
```

Analogously, the closed-loop system can be defined after having defined system $P(s)$ and controller $K(s)$ as lti object, and then use the command feedback:

```
>> Plti=ss(A,[B1 B2],[C1; C2],[D11 D12; D21 D22]);
>> Klti=ss(K(1,1),K(1,2),K(2,1),K(2,2));
>> systemcllti2=feedback(Plti,Klti,2,2,+1)
```

Note that, in this case, the system 'systemcllti' has two inputs and two outputs, since the scheme implemented by the command feedback takes into account both output z and output y and both input w and input u.

By the Bode diagram of the closed-loop system, it can be verified that $\|T_{zw}(s)\|_\infty \leq 1.0023$. The command to use is

```
>> bode(systemcllti1)
```

or

```
>> bode(systemcllti2(1,1))
```

12.10 Multiobjective Control

As discussed above, the objective of the H_∞ control is to find among the controllers that stabilize the closed-loop system the one minimizing the norm $\|T_{zw}(s)\|_\infty$. However, generally, there exist other types of performance that can be optimized. The multiobjective control has been introduced for this reason, allowing to take into account other types of performance, beyond that represented by the H_∞ norm.

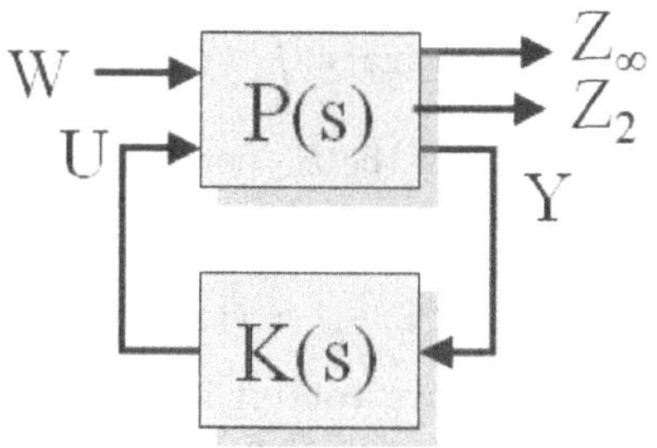

FIGURE 12.2
Scheme of the multiobjective control.

The reference scheme of the multiobjective control is shown in Figure 12.2 where further outputs, i.e., the variables $\mathbf{z}_2$, are highlighted. The state-space equations corresponding to the control scheme of Figure 12.4 are given by:

$$\begin{cases} \dot{\mathbf{x}} = \mathrm{A}\mathbf{x} + \mathrm{B}_1\mathbf{w} + \mathrm{B}_2\mathbf{u} \\ \mathbf{z}_\infty = \mathrm{C}_1\mathbf{x} + \mathrm{D}_{11}\mathbf{w} + \mathrm{D}_{12}\mathbf{u} \\ \mathbf{z}_2 = \mathrm{C}_2\mathbf{x} + \mathrm{D}_{21}\mathbf{w} + \mathrm{D}_{22}\mathbf{u} \\ \mathbf{y} = \mathrm{C}_3\mathbf{x} + \mathrm{D}_{31}\mathbf{w} + \mathrm{D}_{32}\mathbf{u} \end{cases} \tag{12.44}$$

In the multiobjective control what is minimized is the weighted sum of two terms: the first is the H_∞ norm of the transfer matrix from $\mathbf{w}$ to $\mathbf{z}_\infty$; the second is the H_2 norm (defined below) of the transfer matrix from $\mathbf{w}$ to $\mathbf{z}_2$. The idea underlying the multiobjective control is further discussed, after defining the H_2 norm of a system.

The H_2 norm is defined as follows:

$$\begin{aligned} &\|G(s)\|_2 = \left(\frac{1}{2\pi}\int_{-\infty}^{\infty} \mathtt{trace}[G(j\omega)G(j\omega)^*]d\omega\right)^{\frac{1}{2}} = \\ &= \left(\frac{1}{2\pi}\int_{-\infty}^{\infty}\sum_{i=1}^{r}\sigma_i[G(j\omega)]d\omega\right)^{\frac{1}{2}} \end{aligned} \tag{12.45}$$

where r is the rank of $G(j\omega)$.

For systems with stochastic inputs, the physical meaning of the H_2 norm derives from the fact that, when the system is driven by a white noise with zero mean and unit variance, then the H_2 norm is equal to:

$$\|G(s)\|_2 = (E[\mathbf{y}^T(t)\mathbf{y}(t)])^{\frac{1}{2}} \tag{12.46}$$

The H_2 norm thus represents a measure of the energy associated to the output of a system driven by a white noise with zero mean and unit variance.

The H_2 norm can be also viewed as the energy associated to the signal z_2. In fact, if $\mathbf{y}(t)$ is the impulse response of a system, then we have that:

$$\|G(s)\|_2 = (\texttt{trace}[\int_0^\infty \mathbf{y}^T(t)\mathbf{y}(t)])^{\frac{1}{2}} \tag{12.47}$$

As concerns the calculation of the H_2 norm, it can be performed from the controllability and observability gramians:

$$\|G(s)\|_2 = (\texttt{trace}(\mathrm{C}\mathrm{W}_c^2\mathrm{C}^T))^{\frac{1}{2}} = (\texttt{trace}(\mathrm{B}^T\mathrm{W}_o^2\mathrm{B}))^{\frac{1}{2}} \tag{12.48}$$

The aim of the multiobjective control is thus to find a controller that stabilizes the closed-loop system and that, indicated with $\mathrm{T}_{z_\infty,w}$ and $\mathrm{T}_{z_2,w}$ the transfer function matrices from $\mathbf{w}$ to $\mathbf{z}_\infty$ and from $\mathbf{w}$ to $\mathbf{z}_2$, minimizes $\alpha\|\mathrm{T}_{z_\infty,w}\|_\infty^2 + \beta\|\mathrm{T}_{z_2,w}\|_2^2$ with $\alpha \geq 0$ e $\beta \geq 0$.

With MATLAB the multiobjective control can be run with the command `hinfmix`. This also allows one to include a further constraint (also formulated in terms of LMIs) related to the region in which the closed-loop eigenvalues have to lie. Obviously, not all the multiobjective control problems are feasible, since some constraints can be in conflict with each other.

The following examples help to illustrate the multiobjective control.

MATLAB® Exercise 12.10

We go back to the example discussed in Section 12.9, referring to system (12.39), and adding a further specification to the control system: we now also want to minimize the energy associated with the input. To do this, a multiobjective control has to be set by defining a new variable $z_2 = u$. The equations of the system to control become:

$$\begin{aligned} \dot{x} &= w + 2u \\ z_\infty &= x \\ z_2 &= u \\ y &= -x + w \end{aligned} \tag{12.49}$$

System (12.49) is equivalent to system (12.44) with $a = 0$; $b_1 = 1$; $b_2 = 2$; $c_1 = 1$; $d_{11} = d_{12} = 0$; $c_2 = 0$; $d_{21} = 0$; $d_{22} = 1$, $c_3 = -1$; $d_{31} = 1$; $d_{32} = 0$.

Assuming that system P has been defined in MATLAB as a `ltisys` object (as in the examples dealt with previously), the command `hinfmix` can be used. The syntax is:

```
[gopt,h2opt,K,R,S] = hinfmix(P,rv,obj,region)
```

where P is the process to control (a `ltisys` object), `rv` is a vector listing the sizes of z_2, y and u and `obj`= $\begin{bmatrix} VMNormH_\infty & VMNormH_2 & \alpha & \beta \end{bmatrix}$ where $VMNormH_\infty$ and $VMNormH_2$ are the maximum admissible values for $\|\mathrm{T}_{z_\infty,w}\|_\infty$ and $\|\mathrm{T}_{z_2,w}\|_2$, respectively, or in other words the minimum performance that have to be guaranteed. Given this syntax, note that command `hinfmix` can be used in a flexible way. For instance, the H_∞ control can be implemented as follows:

```
>> [gopt,h2opt,K_Hinf]=hinfmix(P,[1 1 1],[0 0 1 0])
```

The performance so obtained is: $\|\mathrm{T}_{z_\infty,w}\|_\infty \leq gopt = 1.0009$. In this case, the control does not take into account in any way the objective defined by the H_2 norm.

To evaluate the performance of the control system, the closed-loop system has first to be calculated:

```
>> systemLTIol=ss(P(1,1),P(1,2:3),P(2:4,1),P(2:4,2:3));
>> compensatorLTI=ss(K_Hinf(1,1),K_Hinf(1,2),
          K_Hinf(2,1),K_Hinf(2,2));
>> systemLTIcl=feedback(systemLTIol,compensatorLTI,2,3,1);
```

The inputs of this system are w and u. The outputs are z_∞, z_2 and $\tilde{y}$. So, to study

the performance of the H_∞ control, the transfer function from input 1 (w) to output 1 (z_∞) has to be investigated.
We can plot the Bode diagram and verify that the norm H_∞ is less than gopt:

```
>> Hs=tf(systemLTIcl);
>> bode(Hs(1,1))
>> normhinf(Hs(1,1))
```

Consider now a H_2/H_∞ problem in which the two norms are weighted in the same way ($\alpha = 1$, $\beta = 1$) and in which no minimum specification is considered:

```
>> [gopt,h2opt,K_H2_Hinf]=hinfmix(P,[1 1 1],[0 0 1 1]);
```

In this case, we obtain $gopt = 1.9104$ and $h2opt = 1.2110$. The performance in terms of H_∞ norm is slightly worse, but now the H_2 norm is quite low. This can be verified by plotting the impulse response of the closed-loop system controlled with the H_∞ technique and the one of the system controlled with the H_2/H_∞ technique.
We first define the new closed-loop system by the commands:

```
>> systemLTIol=ss(P(1,1),P(1,2:3),P(2:4,1),P(2:4,2:3));
>> compensatorLTI=ss(K_H2_Hinf(1,1),K_H2_Hinf(1,2),
        K_H2_Hinf(2,1),K_H2_Hinf(2,2));
>> systemLTIcl2=feedback(systemLTIol,compensatorLTI,2,3,1);
>> Hs2=tf(systemLTIcl2);
```

and then compare the two plots of the impulse response:

```
>> figure(1); impulse(systemLTIcl)
>> figure(2); impulse(systemLTIcl2)
```

In the latter case (H_2/H_∞ control) the variables have smaller amplitudes (and so energy).
The comparison of the Bode diagrams reveals that the H_∞ control leads to a larger bandwidth. Therefore, the system controlled with the H_∞ technique is much faster than that controlled with the H_2/H_∞ technique, but requires more energy.
Note also that the objectives achieved in this way represent the maximum performance that can be obtained with the H_2/H_∞ control. In fact, if we consider for instance:

```
>> [gopt,h2opt,K_H2_Hinf]=hinfmix(P,[1 1 1],[1 1 1 1]);
```

we find that the problem is not feasible.
Finally, we show an example where we also specify the admissible region for the closed-loop eigenvalues. In fact note that the closed-loop eigenvalues of the H_∞ control, obtained by

```
>> pole(Hs)
```

are quite large: $\lambda_1 = -1.4087 \cdot 10^4$, $\lambda_2 = -0.2357 \cdot 10^4$, $\lambda_3 = -1.4087 \cdot 10^4$, $\lambda_2 = -0.2357 \cdot 10^4$.
Let us therefore include a new constraint. We want the closed-loop eigenvalues to be such that $-10 \leq Re\lambda \leq 0$. By using the command

```
>> region=lmireg
```

through a series of multiple choices the given region can be set. Then, by using the command:

```
>>[gopt,h2opt,K_H2_Hinf_R]=hinfmix(P,[1 1 1],[0 0 1 0],region);
```

the H_∞ control with a constraint on the position of the closed-loop eigenvalues is defined.
We now calculate the transfer matrix obtained with this control and indicate it as $Hs3$:

```
>> systemLTIol=ss(P(1,1),P(1,2:3),P(2:4,1),P(2:4,2:3));
>> compensatorLTI=ss(K_H2_Hinf_R(1,1),K_H2_Hinf_R(1,2),
    K_H2_Hinf_R(2,1),K_H2_Hinf_R(2,2));
>> systemLTIcl3=feedback(systemLTIol,compensatorLTI,2,3,1);
>> Hs3=tf(systemLTIcl3);
```

note that now the closed-loop eigenvalues

```
>> pole(Hs3)
```

are much smaller $\lambda_{1,2} = \lambda_{3,4} = -8.9068 + j8.3754$.
By comparing the Bode diagrams and the impulse responses of the controlled systems obtained with the three different techniques we note that, in this latter case, we

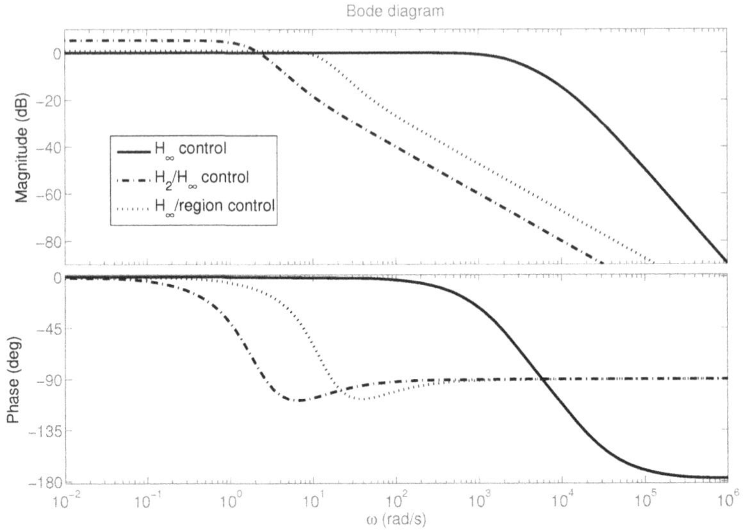

FIGURE 12.3
Performance obtained by the H_∞ control technique, the H_2/H_∞ control technique and the H_∞ control technique with constraint on the position of the closed-loop eigenvalues. The control is applied to system (12.49).

have intermediate performance. The Bode diagrams of the given example are shown in Figure 12.3.

MATLAB® Exercise 12.11

As a second example of multiobjective control, consider the system shown in Figure 12.4. The process to control is first-order with transfer function $G(s) = \frac{1}{s+1}$. The first objective of the control system (beyond closed-loop stability) is to track the input, i.e., to minimize the error $e(t) = r(t) - y(t)$.
First, formulate the control problem in terms of the block scheme of the multiobjective control. To do this, let us write the state-space equations by introducing the variable $\tilde{y} = r - x$ (that is the input of the controller):

$$\begin{aligned} \dot{x} &= -x + u \\ \tilde{y} &= r - x \end{aligned} \tag{12.50}$$

Next, define the variables $\mathbf{z}$: to satisfy the objective of input tracking, fix $z_\infty = e$ and $w = r$. In this way, in fact, we consider the worst case for the error. In fact, for a SISO system the H_∞ norm represents the maximum value of the frequency response, or in other words the maximum amplification for any sinusoidal input.
Let us now consider the objective of the H_2 control. We define $z_2 = y = x$ in such a way to minimize the output energy.
The whole equations of the system to control are thus given by:

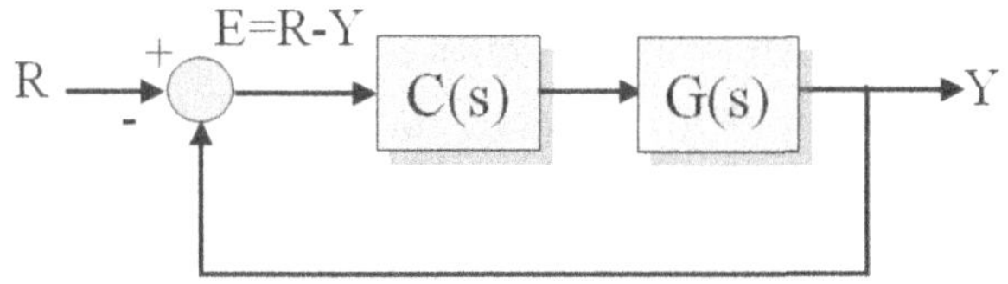

FIGURE 12.4
Example of multiobjective control.

$$\begin{aligned} \dot{x} &= -x + u \\ z_\infty &= w - x \\ z_2 &= x \\ \tilde{y} &= w - x \end{aligned} \tag{12.51}$$

Comparing equations (12.51) with the reference model for multiobjective control as in equations (12.44), we get: $A = -1$; $B_1 = 0$; $B_2 = 1$; $C_1 = -1$; $D_{11} = 1$; $D_{12} = 0$; $C_2 = 1$; $D_{21} = 0$; $D_{22} = 0$; $C_3 = -1$; $D_{31} = 1$; $D_{32} = 0$.
Let us assume that the control objective is to minimize $5\|\mathrm{T}_{z_\infty,w}\|_\infty^2 + 5\|\mathrm{T}_{z_2,w}\|_2^2$ with the further constraint that the closed-loop eigenvalues are such that $-10 \leq Re\lambda < 0$.
In this case we have to use the command:

```
>> [gopt,h2opt,K]=hinfmix(P,[1 1 1],[1 1 5 5],region)
```

The performance obtained in this case is `gopt` $= 1.0000$ and `h2opt` $= 0.6494$, while the controller is given by $K(s) = \left[\begin{array}{c|c} -5.3574 & 1.5066 \\ \hline -0.6449 & 0.3396 \end{array}\right]$.

MATLAB® Exercise 12.12

In this exercise a further example of H_∞ control is given. Let us consider the problem of controlling the vertical position of plasma in the Joint European Torus (JET) (see for more information the paper by Fortuna et *al.* listed in the references of the chapter). The plasma vertical position can be modeled as a cascade of two stages: a first linear system modeling the Poloidal Radial Field Amplifier (PRFA) with transfer function

$$G_1(s) = \frac{100}{1 + sT_a} \tag{12.52}$$

and the plant model

$$G_2(s) = \frac{1}{sT_v}\frac{1}{s - \gamma} \tag{12.53}$$

where $T_v = 10.4ms$ and $\gamma = 80$.
Consider now the following state-space realization:

$$\mathrm{A} = \begin{bmatrix} 0 & 1 \\ 0 & 80 \end{bmatrix}; \mathrm{B} = \begin{bmatrix} 0 \\ 1 \end{bmatrix}; \mathrm{C} = \begin{bmatrix} 9615 & 0 \end{bmatrix}; \mathrm{D} = 0;$$

Consider again the control scheme reported in Figure 12.4 for which the first objective of the control is to track the input, i.e., to minimize the error $e(t) = r(t) - y(t)$.
Rewrite the state space equations by introducing the variable $\tilde{y} = r - y$:

$$\begin{aligned} \dot{\mathrm{X}} &= \mathrm{AX} + \mathrm{B}u \\ \tilde{y} &= w - \mathrm{CX} \end{aligned} \tag{12.54}$$

To satisfy the objective of input tracking, we choose $z_\infty = e$ and $w = r$. The whole equations of the system to control are rewritten as:

$$\begin{aligned} \dot{\mathrm{X}} &= \mathrm{AX} + \mathrm{B}u \\ z_\infty &= w - \mathrm{CX} \\ \tilde{y} &= w - \mathrm{CX} \end{aligned} \tag{12.55}$$

from which it is possible to define $\mathrm{B}_1 = \begin{bmatrix} 0 & 0 \end{bmatrix}^T$, $\mathrm{B}_2 = \begin{bmatrix} 0 & 1 \end{bmatrix}^T$, $\mathrm{C}_1 = \mathrm{C}_2 = -\mathrm{C}$, $D_{11} = D_{21} = 1$, and $D_{12} = D_{22} = 0$.
The realization matrix of the process can be written as:

$$P(s) = \left[\begin{array}{cc|cc} 0 & 1 & 0 & 0 \\ 0 & 80 & 0 & 1 \\ \hline -9615 & 0 & 1 & 0 \\ -9615 & 0 & 1 & 0 \end{array}\right];$$

Applying the procedure to obtain the H_∞ controller we obtain a performance index $\gamma_{opt} = 1.31$ in correspondence of the following transfer function:

$$K(s) = \frac{1.54 \cdot 10^7 s + 153.2}{(s^2 + 4.70 \cdot 10^4 s + 9.28 \cdot 10^8)}$$

However, using the controller $K(s)$ the poles of the closed-loop system are $p_1 = -1.8 \cdot 10^{-5}$, $p_2 = -80$ and $p_{3,4} = -2.34 \cdot 10^4 \pm 1.93 \cdot 10^4 j$.
If we want to design the H_∞ controller choosing a suitable region for the closed-loop system poles, we may use the procedure defined for the multiobjective control imposing that the closed-loop poles have both real and imaginary part between 0 and -20, i.e., the feasible region is a square. In this case the H_∞ controller with a $\gamma_{opt} = 110$ is defined by:

$$K(s) = 0.99 \frac{(s + 0.16)}{(s + 114.2)}$$

leading to the following closed-loop poles: $p_1 = -15.52$, $p_2 = -13.57$ and $p_{3,4} = -9.03 \pm 3.62j$.

The LMI techniques discussed can stimulate the students to formulate other control problems, using this powerful and efficient tool.

The Lagrange's equation provides a method to derive the differential equations governing a system. Moreover, the differential Riccati equation is a design tool to obtain the optimal control law, but other matrix differential equations also arise, as shown in the book of Helmke and Moore, in solving and conceiving more system analysis and design tools. The solution of such problems requires efficient numerical iterative solving procedures.

Since the formulation of many system theory and modern control optimal problems in terms of LMIs leads to convex optimization problems, it allows to use the interior point algorithm which is a universal solver for many problems and requires few iterations to converge to the solution. Therefore, from modelling to design, differential equations are the core of system analysis and control and efficient iterative optimization procedures are the machinery to solve them.

12.11 Exercises

1. Using LMI techniques, determine the H_∞ norm of the system with transfer function $G(s) = \frac{s+1}{(s+2)(s+3)}$.

2. Using LMI techniques, determine if the system with transfer function $G(s) = 10\frac{s+1}{(s^2+5s+10)}$ is bounded-real or not.

3. Given the system:

$$\begin{cases} \dot{\mathbf{x}} = \mathrm{A}\mathbf{x} + \mathrm{B}_1\mathbf{w} + \mathrm{B}_2\mathbf{u} \\ \mathbf{z}_\infty = \mathrm{C}_1\mathbf{x} + \mathrm{D}_{11}\mathbf{w} + \mathrm{D}_{12}\mathbf{u} \\ \mathbf{z}_2 = \mathrm{C}_2\mathbf{x} + \mathrm{D}_{21}\mathbf{w} + \mathrm{D}_{22}\mathbf{u} \\ \mathbf{y} = \mathrm{C}_3\mathbf{x} + \mathrm{D}_{31}\mathbf{w} + \mathrm{D}_{32}\mathbf{u} \end{cases}$$

with $\mathrm{A} = \begin{bmatrix} 0 & 1 \\ 0.2 & -1 \end{bmatrix}$; $\mathrm{B}_1 = \begin{bmatrix} 0 \\ 1 \end{bmatrix}$; $\mathrm{B}_2 = \begin{bmatrix} 0 \\ 1 \end{bmatrix}$; $\mathrm{C}_1 = \begin{bmatrix} 2 & 1 \end{bmatrix}$; $\mathrm{C}_2 = \begin{bmatrix} 0 & 0 \end{bmatrix}$; $\mathrm{C}_3 = \begin{bmatrix} 1 & 0 \end{bmatrix}$; $\mathrm{D}_{11} = \mathrm{D}_{12} = 0$; $\mathrm{D}_{21} = 0$; $\mathrm{D}_{22} = 1$; $\mathrm{D}_{31} = 1$ and $\mathrm{D}_{32} = 0$, design a multiobjective control and verify the obtained performance. Consider then the objective $\|T_{z_\infty w}\|_\infty \leq 0.5$ and $\|T_{z_2 w}\|_2 \leq 2.5$. Is it possible to design a controller satisfying the specifications?

4. Calculate the compensator (regulator and observer) which stabilizes the system

$$\begin{aligned} \dot{\mathbf{x}} &= \mathrm{A}\mathbf{x} + \mathrm{B}\mathbf{u} \\ \mathbf{y} &= \mathrm{C}\mathbf{x} \end{aligned}$$

with $\mathrm{A} = \begin{bmatrix} \lambda_1 & 0 & 0 \\ 0 & \lambda_2 & 0 \\ 0 & 0 & \lambda_3 \end{bmatrix}$; $\mathrm{B} = \begin{bmatrix} b_{1,i} \\ b_2 \\ b_3 \end{bmatrix}$; $\mathrm{C} = \mathrm{B}^T$; $\lambda_1 = -1$; $\lambda_2 = 1$; $\lambda_3 = 5$; $b_{1,1} = 4$; $b_{1,2} = -1$; $b_2 = 1$ and $b_3 = 1$.

5. Using the LMI approach, find the control law that stabilizes simultaneously the two systems:

$$A_1 = \begin{bmatrix} -2 & 0 & 0 \\ 0 & 1 & 0 \\ 0 & 0 & 5 \end{bmatrix}; B_1 = \begin{bmatrix} 0 \\ 1 \\ 1 \end{bmatrix}$$

and

$$A_2 = \begin{bmatrix} -5 & 0 & 0 \\ 0 & -6 & 0 \\ 0 & 0 & 3 \end{bmatrix}; B_2 = \begin{bmatrix} 0 \\ 1 \\ 3 \end{bmatrix}$$

13

The Class of Stabilizing Controllers

CONTENTS

This chapter deals with the problem of determining, rather than a single controller that meets the assigned specifications, a class of controllers that meet them. Using this technique, the class of the controllers can be expressed in function of one or more parameters that allow the designer to specify later additional control objectives. For this reason the problem to find a universal class of controllers which satisfy certain specifications takes the name of parameterization of the controllers. For simplicity, we mainly consider SISO systems, even if the discussed results may be easily generalized to MIMO systems as briefly shown in Section 13.5.

13.1 Parameterization of Stabilizing Controllers for Stable Processes

Referring to Figure 13.1, which shows the classic feedback control scheme, we want to determine the whole class of controllers which internally stabilize the closed-loop system. With internally stable, we mean that at any point of the feedback loop the system must be stable. The objective is that the system defined by the state variables x_1, x_2 and x_3 is stable, i.e., that all the possible transfer functions from any input of the system (r, d and n) to any of the variables x_1, x_2 and x_3 are stable.

DOI: 10.1201/9781003196921-13

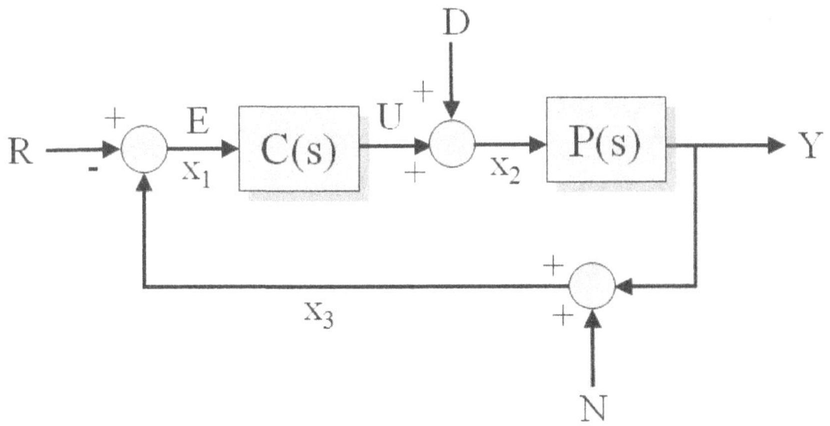

FIGURE 13.1
Feedback control scheme.

In this section we consider the case in which the process to control is asymptotically stable. In the next paragraph we will consider the extension to the case when the process is not stable. Let $\mathbb{H}^\infty$ indicate the set of asymptotically stable processes. The following theorem expresses the class of stabilizing controllers.

Theorem 32 *Given a process $P(s) \in \mathbb{H}^\infty$ the class of the controllers that make the closed-loop system internally stable is given by:*

$$C(s) = \frac{Q(s)}{1 - P(s)Q(s)} \tag{13.1}$$

with $Q(s) \in \mathbb{H}^\infty$.

Applying the parameterization of stabilizing controllers expressed by equation (13.1) we obtain a transfer function given by:

$$F(s) = \frac{C(s)P(s)}{1 + C(s)P(s)} = \frac{\frac{Q(s)}{1-P(s)Q(s)}P(s)}{1 + \frac{Q(s)}{1-P(s)Q(s)}P(s)} = P(s)Q(s)$$

From the expression of the closed-loop transfer function we deduce immediately that, if $P(s)$ and $Q(s)$ are asymptotically stable, also $F(s)$ is. Since the controller (13.1) is an internally stabilizing controller, all nine possible transfer functions between r, d and n and x_1, x_2 and x_3 are asymptotically stable.

We can also get immediately that $Q(s)$ is the transfer function between r and u: $Q(s) = \frac{U(s)}{R(s)}$.

Once the parameterization of stabilizing controllers has been found, it is possible to search the member (or the members) of this class that satisfies further specifications. Consider, for example, the zero error specification of the unit step (closed-loop) response. To impose this specification, we must ensure that the steady-state of the unit step response is equal to one. Applying the final value theorem (that can be applied, as the closed-loop system is asymptotically stable), we obtain:

$$\lim_{t\to+\infty} y(t) = \lim_{s\to 0} sQ(s)P(s)\frac{1}{s} = Q(0)P(0) \tag{13.2}$$

So the condition to impose is

$$Q(0)P(0) = 1$$

Example 13.1

Determine a controller that is internally stabilizing and guarantees zero error for the response to the ramp input for the system $P(s) = \frac{1}{(s+1)(s+5)}$.
Consider $Q(s) = \frac{as+b}{s+1}$. Since $Q(0) = b$ and $P(0) = \frac{1}{5}$, imposing that the unit step error is zero (i.e., that $Q(0) = \frac{1}{P(0)}$), we obtain $b = 5$.
To impose that also the ramp error (indicated with $E_R(t)$) vanishes, consider

$$\lim_{t\to+\infty} E_R(t) = \lim_{s\to 0} s\left(\frac{1}{s^2} - \frac{1}{s^2}F(s)\right) = \lim_{s\to 0}\frac{1}{s}(1 - F(s)) = 0$$

$$\Rightarrow \lim_{s\to 0}\frac{1}{s}\left(1 - \frac{as+5}{s+1}\,\frac{1}{(s+1)(s+5)}\right) = 0$$

$$\Rightarrow \lim_{s\to 0}\frac{s^2+7s+(11-a)}{(s+1)^2(s+5)} = \frac{11-a}{5} = 0 \Rightarrow a = 11$$

So $Q(s) = \frac{11s+5}{s+1}$, from which we get that $C(s) = \frac{Q(s)}{1-Q(s)P(s)} = \frac{11s^3+71s^2+85s+25}{s^2(s+7)}$. As expected, $C(s)$ has two poles at the origin, which are necessary to have a ramp error equal to zero.

13.2 Parameterization of Stabilizing Controllers for Unstable Processes

Now consider the case of an unstable process. In this case, the function $P(s)$ is rewritten as the ratio of two transfer functions that have peculiar properties. In particular, we use a factorization that takes the name of coprime factorization.

Definition 27 *Given an unstable process $P(s)$, we say that $P(s) = \frac{N(s)}{M(s)}$ is coprime factorization, if $N(s)$ and $M(s)$ are stable functions and there exist two transfer functions $X(s)$ and $Y(s)$ so that $\forall s$ the following relation holds*

$$N(s)X(s) + M(s)Y(s) = 1 \tag{13.3}$$

Before enunciating the necessary and sufficient conditions for a factorization to be coprime, we show with an example a factorization that cannot satisfy equation (13.3).

Example 13.2

Consider the process $P(s) = \frac{1}{s-1}$ and the factorization $N(s) = \frac{1}{(s+1)^2}$ and $M(s) = \frac{s-1}{(s+1)^2}$. The two functions $N(s)$ and $M(s)$ are asymptotically stable and $P(s) = \frac{N(s)}{M(s)}$. The condition (13.3) becomes

$$\frac{1}{(s+1)^2}X(s) + \frac{s-1}{(s+1)^2}Y(s) = 1$$

and has to hold $\forall s$. In particular, considering the limit case $s \to \infty$, the condition to satisfy becomes $0 \cdot X + 0 \cdot Y = 1$, which is clearly impossible to solve. So the factorization is not coprime.

In the last example, we can note that $N(s)$ has two zeros at infinity, while $M(s)$ has one zero at infinity and one zero at the unstable pole of $P(s)$. In fact, the examined factorization is not coprime, just because its functions have a common zero (at infinity). Generalizing this result, we can conclude that, whenever there exists a s_0 such that $N(s_0) = M(s_0) = 0$ (i.e., whenever $N(s)$ and $M(s)$ have a common zero), the factorization is not coprime. In fact, this condition is a necessary and sufficient condition, as summarized in the following theorem.

Theorem 33 *Necessary and sufficient condition for a factorization to be coprime is that $N(s)$ and $M(s)$ have no common zero, neither finite nor infinite.*

It follows that a coprime factorization of $P(s) = \frac{1}{s-1}$ is given by: $N(s) = \frac{1}{(s+1)}$ and $M(s) = \frac{s-1}{(s+1)}$.

The coprime factorization permits us to express the class of the stabilizing controllers for unstable processes.

Theorem 34 *Given a process $P(s)$, generally unstable, the class of the controllers that make the closed-loop system internally stable is given by:*

$$C(s) = \frac{X(s) + M(s)Q(s)}{Y(s) - N(s)Q(s)} \tag{13.4}$$

where $Q(s) \in \mathbb{H}^\infty$ and $P(s) = \frac{N(s)}{M(s)}$ with $N(s) \in \mathbb{H}^\infty$ and $M(s) \in \mathbb{H}^\infty$ being a coprime factorization of $P(s)$.

With the parameterization (13.4) we obtain a closed-loop function equal to

$$F(s) = \frac{C(s)P(s)}{1 + C(s)P(s)} = \frac{(X(s) + M(s)Q(s))N(s)}{M(s)Y(s) + X(s)N(s)} = (X(s) + M(s)Q(s))N(s)$$

Also in this case we can immediately say that, if $Q(s) \in \mathbb{H}^{\infty}$, then also $F(s) \in \mathbb{H}^{\infty}$.

The coprime factorization can also be done in the time domain. Consider a system in minimum form $\mathrm{R} = \begin{bmatrix} \mathrm{A} & \mathrm{B} \\ \mathrm{C} & \mathrm{D} \end{bmatrix}$.

The functions $M(s)$ and $N(s)$ are given by

$$M(s) = \left[\begin{array}{c|c} \mathrm{A+BF} & \mathrm{B} \\ \hline \mathrm{F} & \mathrm{I} \end{array}\right] \tag{13.5}$$

$$N(s) = \left[\begin{array}{c|c} \mathrm{A+BF} & \mathrm{B} \\ \hline \mathrm{C+DF} & \mathrm{D} \end{array}\right] \tag{13.6}$$

Note that in both realizations of $N(s)$ and $M(s)$ the state matrix $\mathrm{A}^* = \mathrm{A} + \mathrm{BF}$ is the same. In fact, to factorize $P(s)$, the two functions $N(s)$ and $M(s)$ must have the same poles. Moreover, since $N(s) \in \mathbb{H}^{\infty}$ and $M(s) \in \mathbb{H}^{\infty}$, we need to choose F so that A^* has eigenvalues with negative real part.

Once matrix F has been found solving a problem of eigenvalue placement, $N(s)$ and $M(s)$ can be obtained through the following expressions:

$$M(s) = \mathrm{F}\left(s\mathrm{I} - \mathrm{A} - \mathrm{BF}\right)^{-1}\mathrm{B} + \mathrm{I}$$

$$N(s) = \left(\mathrm{C} + \mathrm{DF}\right)\left(s\mathrm{I} - \mathrm{A} - \mathrm{BF}\right)^{-1}\mathrm{B} + \mathrm{D}$$

To find $X(s)$ and $Y(s)$ we have to solve another problem of eigenvalue placement. In particular, given that

$$X(s) = \left[\begin{array}{c|c} \mathrm{A+HC} & \mathrm{H} \\ \hline \mathrm{F} & 0 \end{array}\right] \tag{13.7}$$

$$Y(s) = \left[\begin{array}{c|c} \mathrm{A+HC} & -\mathrm{B}-\mathrm{HD} \\ \hline \mathrm{F} & \mathrm{I} \end{array}\right] \tag{13.8}$$

we need to find H through eigenvalue placement (chosen arbitrarily). Given the particular structure of $X(s)$ and $Y(s)$ we can prove that they are such that the identity $NX + MY = I$ is satisfied.

Notice that here we have reported the general form as these expressions hold true also in the MIMO case that is briefly discussed at the end of the chapter. Clearly, for SISO systems $\mathrm{I} = 1$.

Example 13.3

Given the system $P(s) = \frac{1}{(s-2)(s-1)}$, design a compensator, such that to stabilize the closed-loop system, assuring zero error of the step response and assuring that the effect of the disturbance $d(t) = A\sin(\omega t)$ with $\omega = 10rad/s$ vanishes for $t \to +\infty$.

At first, since the process $P(s)$ is unstable, we set a coprime representation. For example we can choose $N(s) = \frac{1}{(s+1)^2}$ and $M(s) = \frac{(s-2)(s-1)}{(s+1)^2}$. Imposing that $N(s)X(s) + M(s)Y(s) = 1$, one obtains $X(s) = \frac{19s-11}{s+1}$ and $Y(s) = \frac{s+6}{s+1}$ (functions chosen arbitrary).

To ensure that the error of the step response is null, we must impose that $\lim_{s\to 0} F(s) = 1$ and so that

$$\lim_{s\to 0} N(s)(X(s) + M(s)Q(s)) = 1$$

$$\Rightarrow N(0)X(0) + N(0)M(0)Q(0) = 1$$

$$\Rightarrow Q(0) = 6$$

Finally, consider the specification of the disturbance. From the scheme in Figure 13.1 we can get the transfer function from disturbance to output: $\frac{Y(s)}{D(s)} = \frac{P(s)}{1+C(s)P(s)} = N(s)\left(Y(s) - N(s)Q(s)\right)$. To ensure that the effect of the sinusoidal disturbance is null, we must impose that for $\omega = 10rad/s$ we have $N(j\omega)\left(Y(j\omega) - N(j\omega)Q(j\omega)\right) = 0$ $\Rightarrow Q(j\omega) = -94 + 70j$.
Writing $Q(s)$ as $Q(s) = c_1 + \frac{c_2}{s+1} + \frac{c_3}{(s+1)^2}$, we find $Q(s)$ imposing that:

$$Q(0) = c_1 + c_2 + c_3 = 6$$

and

$$Q(j\omega) = c_1 + \frac{c_2}{1+10j} + \frac{c_3}{(1+10j)^2} = 0$$

Equating the real part and the imaginary part of the first and second member of the last relation, we find that $Q(s) = \frac{-79s^2-881s+6}{(s+1)^2}$.

13.3 Parameterization of Stable Controllers

In this section, we will consider a further very important specification. In the parameterizations seen previously there is no guarantee that the compensator $C(s)$ is stable. Consider now under which conditions it is possible to stabilize a process $P(s)$ through stable compensators. Systems in which it is possible to find a stable and stabilizing controller are called strongly stabilizable systems.

Definition 28 *A system that can be stabilized with a stable compensator is called strongly stabilizable.*

Suppose that the process $P(s)$ has zeros with positive real part, i.e., it is a minimum phase system. For example, consider $P_1(s) = \frac{(s-1)(s-5)}{(s+1)^2}$ and study if it is possible to stabilize $G(s)$ with a compensator of type $C(s) = k$. The root locus, i.e., the locus of the points such that $kP_1(s) + 1 = 0$ for any k, shows the positions of the closed-loop poles at different values of k and allows one to establish whether there exists a value of k for which $C(s) = k$ is a stable and stabilizable compensator. The root locus of the system $P_1(s) = \frac{(s-1)(s-5)}{(s+1)^2}$ is shown in Figure 13.2. For small values of k we find stable and stabilizable compensators.

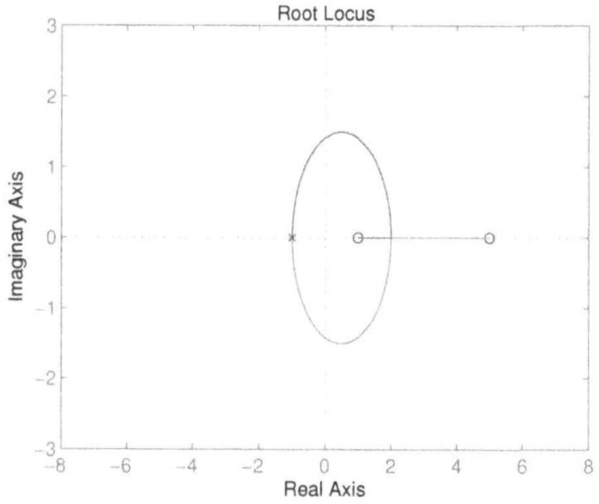

FIGURE 13.2
Root locus of the system $P_1(s) = \frac{(s-1)(s-5)}{(s+1)^2}$.

Consider now the system $P_2(s) = \frac{(s-1)(s-5)}{(s+1)(s-2)}$. Its root locus is shown in Figure 13.3. In this case a controller of type $C(s) = k$ which stabilizes the system does not exist.

The difference between the two systems $P_1(s)$ and $P_2(s)$ is that in the second case between two zeros with positive real part there is an odd number of poles. This condition that we have illustrated for $C(s) = k$ is actually a more general property which guarantees the existence of stable and stabilizing controllers, as expressed in the following theorem.

Theorem 35 *The necessary and sufficient condition for a system $P(s)$ to be strongly stabilizable is that between any pair of zeros with positive real part there is an even number of poles.*

This condition is called parity interlacing property and it has to be verified taking into account also the zeros at infinity.

If a system satisfies the parity interlacing property then it is possible to find stable and stabilizing controllers. Before introducing the parameterization of those controllers we need to define what is meant by unit function.

Definition 29 *A function $Q(s) \in \mathbb{H}^\infty$ is said to be unit if also $\frac{1}{Q(s)} \in \mathbb{H}^\infty$.*

For example, the function $Q(s) = \frac{1}{s+1}$ is not unit, as its inverse $\frac{1}{Q(s)} = s+1$ is not realizable. Instead, $Q(s) = \frac{s+5}{s+2}$ is unit, while $Q(s) = \frac{s-5}{s+2}$ is not unit, because its inverse does not belong to $\mathbb{H}^\infty$.

The parameterization of the class of stable and stabilizing compensators is expressed through the following theorem.

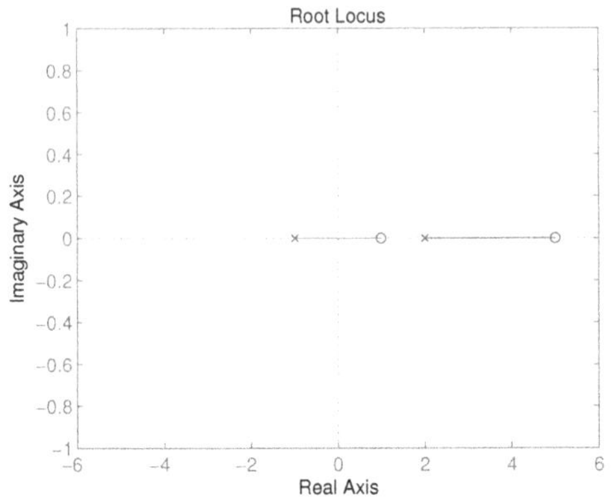

FIGURE 13.3
Root locus of the system $P_2(s) = \frac{(s-1)(s-5)}{(s+1)(s-2)}$.

Theorem 36 *Given a process* $P(s) = \frac{N(s)}{M(s)}$ *with* $N(s)$ *and* $M(s)$ *being a coprime factorization of* $P(s)$*, the class of controllers which are stable and stabilizing the closed-loop system is given by:*

$$C(s) = \frac{U(s) - M(s)}{N(s)} \tag{13.9}$$

where $U(s)$ *is a unit function.*

Example 13.4

Consider for example $P(s) = \frac{s-1}{s(s-0.5)}$. This system verifies the parity interlacing property, as between the zero with positive real part $z = 1$ and the zero at infinity there are no poles. The system $P(s)$ can be stabilized using the compensator (13.9).
Given the coprime factorization $N(s) = \frac{s-1}{(s+1)^2}$ and $M(s) = \frac{s(s-0.5)}{(s+1)^2}$, $C(s)$ is given by:

$$C(s) = \frac{U(s) - \frac{s(s-0.5)}{(s+1)^2}}{\frac{s-1}{(s+1)^2}} = \frac{(s+1)^2U(s) - s(s-0.5)}{s-1} \tag{13.10}$$

In order to obtain a stable $C(s)$, the term $s-1$ at the denominator must be simplified. So, in correspondence of the zero with positive real part ($z = 1$), the numerator of $C(s)$ must be zero. So, we have to impose that $U(1) - M(1) = 0$, i.e., $U(1) = M(1)$.
A second condition to impose derives from the fact that, choosing for simplicity $U(s)$ of first order, the degree of the numerator of $C(s)$ has to be equal to one. In fact the order of $C(s)$ is $n = 1$ and the numerator, to ensure that the system is realizable, must also be a polynomial of degree one. For this, the maximum degree coefficient of the term $(s+1)^2U(s)$ and of the term $s(s-0.5)$ must be equal, so that the numerator of $(s+1)^2U(s) - s(s-0.5)$ is a polynomial of degree two. Remember also that this term has a root $z = 1$ that cancels out the term $s-1$: this makes $C(s)$ realizable.
To impose the condition that the maximum degree coefficients of these two terms are equal, we must set $U(\infty) = M(\infty)$ since $\lim_{s\to\infty} M(s)$ is exactly the value of the maximum degree coefficient.

So we get two interpolation conditions:

$$U(1) = M(1) \tag{13.11}$$

$$U(\infty) = M(\infty) \tag{13.12}$$

Assuming that $U(s) = k\frac{s+\alpha}{s+\beta}$ (unit function if $\alpha > 0$ and $\beta > 0$) and imposing these two conditions we find that

$$k\frac{1+\alpha}{1+\beta} = \frac{1}{8} \tag{13.13}$$

$$k = 1 \tag{13.14}$$

It follows that the parameters of $U(s)$ are related each other according to the equation $\frac{1+\alpha}{1+\beta} = \frac{1}{8}$ which defines a straight line. All the points on this line are suitable values to obtain stable and stabilizing controllers. For example choosing $\alpha = 1$, we have $\beta = 15$. So we have $U(s) = \frac{s+1}{s+15}$ (unit since $U(s) \in \mathbb{H}^\infty$ and $U^{-1}(s) \in \mathbb{H}^\infty$) and $C(s) = -\frac{11.5s+1}{s+15}$.

13.4 Simultaneous Stabilizability of Two Systems

From the techniques introduced in this chapter, the conditions to solve the problem of simultaneous stability of the two systems can be derived. Consider the reference scheme in Figure 13.1 and suppose that one wants to find a compensator that simultaneously stabilizes two systems, $P_1(s)$ and $P_2(s)$. The conditions of simultaneous stabilizability depend on the system characteristics defined by $\Delta(s) = P_1(s) - P_2(s)$. The main result is given by the following theorem.

Theorem 37 *Given two systems $P_1(s)$ and $P_2(s)$ and defined $\Delta(s) = P_1(s) - P_2(s)$, it is possible to find a compensator that stabilizes both systems if $\Delta(s)$ satisfies the parity interlacing property.*

The compensator stabilizing both systems is the stable compensator that stabilizes the system $\Delta(s)$.

This result is only valid for the simultaneous stabilizability of two systems, whereas the technique shown in Chapter 12 can be applied on two or more systems, but it does not allow to establish if the compensator exists.

13.5 Coprime Factorizations for MIMO Systems and Unitary Factorization

Since matrix multiplication is not commutative, for MIMO systems we have to distinguish between left and right coprime factorization. Given the MIMO system $P(s) = N(s)M^{-1}(s)$, the factorization is right coprime if

$$X(s)N(s) + Y(s)M(s) = I$$

In this case, the class of the controllers that make the closed-loop system internally stable (which for SISO systems is given by equation (13.4)) is given by:

$$C(s) = (Y(s) - Q(s)N(s))^{-1}(X(s) + Q(s)M(s)) \tag{13.15}$$

Instead, given $P(s) = \tilde{M}^{-1}(s)\tilde{N}(s)$, the factorization is left coprime if

$$\tilde{N}(s)\tilde{X}(s) + \tilde{M}(s)\tilde{X}(s) = I$$

and the class of the controllers that make the closed-loop system internally stable is given by:

$$C(s) = (\tilde{X}(s) + \tilde{M}(s)Q(s))(\tilde{Y}(s) - \tilde{N}(s)Q(s))^{-1} \tag{13.16}$$

In the coprime factorization obtained through the time domain method we have seen that it is possible to choose in an arbitrary way the eigenvalues of the matrices $\mathrm{A} + \mathrm{BF}$ and $\mathrm{A} + \mathrm{HC}$. When these values are fixed to be equal to the optimal eigenvalues associated to the solution of the CARE equation and the FARE equation, the factorization is said to be unitary coprime factorization.

Given the system $P(s) = N(s)M^{-1}(s)$, indicated with P_2 the CARE solution

$$\mathrm{A}^T\mathrm{P}_2 + \mathrm{P}_2\mathrm{A} - \mathrm{P}_2\mathrm{B}\mathrm{B}^T\mathrm{P}_2 + \mathrm{C}^T\mathrm{C} = 0$$

and with P_1 the FARE solution

$$\mathrm{A}\mathrm{P}_1 + \mathrm{P}_1\mathrm{A}^T - \mathrm{P}_1\mathrm{C}^T\mathrm{C}\mathrm{P}_1 + \mathrm{B}\mathrm{B}^T = 0$$

in the unitary right coprime factorization we fix as eigenvalues of $M(s)$ and $N(s)$ the optimal eigenvalues. So, we choose $\mathrm{F} = -\mathrm{B}^T\mathrm{P}_2$ in the equations (13.5) and (13.6). Concerning $X(s)$ and $Y(s)$, we choose in the equations (13.7) and (13.8) $\mathrm{H} = -\mathrm{P}_1\mathrm{C}^T$. In this way we get:

$$N^T(-s)N(s) + M^T(-s)M(s) = I$$

Unitary factorization is also defined for left factorized MIMO systems $P(s) = \tilde{M}^{-1}(s)\tilde{N}(s)$. The left unitary coprime factorization is defined by

$$\tilde{N}(s)\tilde{N}^T(-s) + \tilde{M}(s)\tilde{M}^T(-s) = I$$

with

$$\tilde{M}(s) = \left[\begin{array}{c|c} \mathrm{A}^* & \mathrm{C} \\ \hline \mathrm{H}^* & \mathrm{I} \end{array}\right] \tag{13.17}$$

and

$$\tilde{N}(s) = \left[\begin{array}{c|c} \mathrm{A}^* & \mathrm{C} \\ \hline \mathrm{B} & 0 \end{array}\right] \tag{13.18}$$

where $\mathrm{A}^* = \mathrm{A} + \mathrm{H}^*\mathrm{C}$ and $\mathrm{H}^* = -\mathrm{P}_1\mathrm{C}^T$.

13.6 Parameterization in Presence of Uncertainty

Consider now the case of a process $P(s)$ with additive uncertainty: $P(s) = P_0(s) + r(s)$. $P_0(s)$ represents the nominal transfer function, while $r(s)$ indicates the additive structural uncertainty. So we have $|P(j\omega) - P_0(j\omega)| < |r(j\omega)|$.

Let us define with $T(s) = \frac{C(s)P_0(s)}{1+C(s)P_0(s)}$ the closed-loop transfer function. With the techniques analyzed above,the class of compensators $C(s)$ that make $T(s)$ asymptotically stable can be parametrized. Now we consider the objective to find the class of compensators that stabilize the closed-loop system in presence of structural uncertainties.

Often the uncertainty is defined considering the following normalization:

$$\frac{|P(j\omega) - P_0(j\omega)|}{|P_0(j\omega)|} < \frac{|r(j\omega)|}{|P_0(j\omega)|}$$

or in the s domain

$$\frac{P(s) - P_0(s)}{P_0(s)} = m(s)$$

with $m(s) = \frac{r(s)}{P_0(s)}$. So we want to find the class of compensators $C(s)$ that stabilizes any process $P(s) = P_0(s)[1 + m(s)]$. To do this we have to impose also that the zeros of $1+C(s)P(s)$ (i.e., the closed-loop poles) are with strictly negative real part.

Since

$$1 + C(s)P(s) = 1 + C(s)P_0(s)(1 + m(s)) =$$

$$= (1 + C(s)P_0(s))[1 + m(s)\frac{C(s)P_0(s)}{1 + C(s)P_0(s)}]$$

$1 + C(s)P(s)$ is given by the product of two terms: $1 + C(s)P_0(s)$ and $1 + m(s)T(s)$. The first term $1 + C(s)P_0(s)$ certainly has roots with negative real part, because it is the denominator of the closed-loop transfer function $T(s)$. The second term $1 + m(s)T(s)$ depends on the closed-loop transfer function and on the uncertainty $m(s)$.

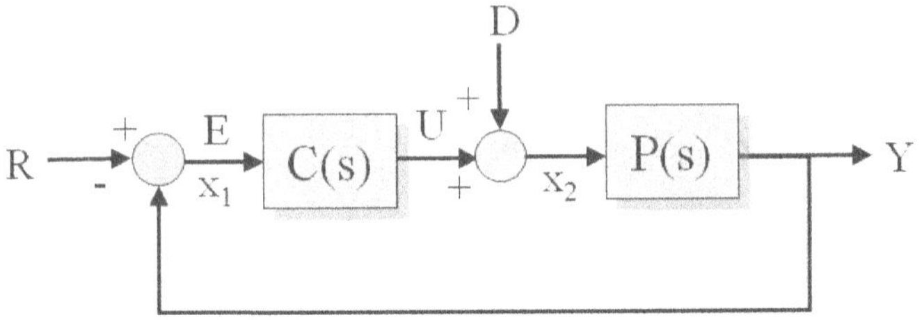

FIGURE 13.4
Feedback control scheme.

If $|m(s)T(s)| < 1$ for s with positive real part, then $1 + m(s)T(s)$ is guaranteed to never vanish for s with positive real part. Therefore, this term does not have roots that would lead the system to instability.

Consider the scheme in Figure 13.4 and consider the transfer functions:

$$\begin{bmatrix} \frac{X_1(s)}{R(s)} & \frac{X_1(s)}{D(s)} \\ \frac{X_2(s)}{R(s)} & \frac{X_2(s)}{D(s)} \end{bmatrix} = \begin{bmatrix} (1+C(s)P(s))^{-1} & -P(s)(1+C(s)P(s))^{-1} \\ C(s)(1+C(s)P(s))^{-1} & (1+C(s)P(s))^{-1} \end{bmatrix}$$

Consider now the class of compensators defined by $C(s) = \frac{Q(s)}{1-P(s)Q(s)}$. We have seen previously that this class of compensators internally stabilizes the system, when $P(s)$ is stable. The use of this class of compensators for unstable systems requires some adaptations that will be described below. At this point, let us suppose we can use $C(s) = \frac{Q(s)}{1-P(s)Q(s)}$ for which $Q(s) = \frac{C(s)}{1+P(s)C(s)}$. This allows the given transfer functions to be rewritten in terms of $P(s)$ and $Q(s)$ (to abbreviate we omit the variable s):

$$\begin{bmatrix} \frac{X_1(s)}{R(s)} & \frac{X_1(s)}{D(s)} \\ \frac{X_2(s)}{R(s)} & \frac{X_2(s)}{D(s)} \end{bmatrix} = \begin{bmatrix} 1-PQ & -P(1-PQ) \\ Q & 1-PQ \end{bmatrix}$$

From this expression, we can get the conditions to impose to $Q(s)$ so that the system is internally stable. The term that represents the transfer function from r to x_2 requires that $Q(s) \in \mathbb{H}^\infty$.

Consider now the term $1 - PQ$:

$$1 - PQ = 1 - \frac{N(s)}{D(s)}\frac{Q_1(s)}{Q_2(s)} = \frac{D(s)Q_2(s) - N(s)Q_1(s)}{D(s)Q_2(s)}$$

The fact that $Q(s) \in \mathbb{H}^\infty$ implies that $Q_2(s)$ has roots with negative real part, but $D(s)$ can have roots with positive real part. These roots must be canceled, imposing some interpolation conditions on the numerator of $1-PQ$. From this it follows that, indicated with $\alpha_1, \alpha_2, \ldots, \alpha_n$ the poles of $P(s)$ with positive real part (i.e., the roots of $D(s)$), we must impose that $P(\alpha_i)Q(\alpha_i) = 1$ $\forall \alpha_i$.

Concerning $P(1-PQ)$, we will clarify below that the necessary adaptation to deal with unstable systems is that this term must be stable.

The last condition to impose derives from some considerations on the process disturbances. We must impose that $|m(s)T(s)| < 1$ $\forall s$ with non-negative real part. But since $T(s) = \frac{C(s)P_0(s)}{1+C(s)P_0(s)}$, $Q(s) = \frac{C(s)}{1+P_0(s)C(s)}$ and $m(s) = \frac{r(s)}{P_0(s)}$, we have $m(s)T(s) = r(s)Q(s)$. If $s = j\omega$ then we have to impose that $|r(j\omega)Q(j\omega)| < 1$ $\forall \omega$. The condition is then that $r(s)Q(s)$ is strictly bounded-real.

In conclusion, the conditions to impose to $Q(s)$ to obtain the class of controllers $C(s) = \frac{Q(s)}{1-P(s)Q(s)}$ that stabilize internally the process $P(s) = P_0(s) + r(s)$ are:

1. $Q(s) \in \mathbb{H}^\infty$;
2. $P(\alpha_i)Q(\alpha_i) = 1$ for any unstable pole α_i of $P(s)$;
3. $|r(j\omega)Q(j\omega)| < 1$ $\forall \omega$.

Note that to apply the condition $P(\alpha_i)Q(\alpha_i) = 1$ since $P(s)$ is unstable requires some adaptation. In particular, the so-called Blaschke product needs to be used.

Consider, for example, an unstable system $P(s) = \frac{1}{(s-1)(s-2)}$ and consider also the system $\tilde{P}(s) = \frac{1}{(s+1)(s+2)}$. The system $\tilde{P}(s)$ can be obtained from the product (called a Blaschke product) of the system $P(s)$ and the all-pass system: $B(s) = \frac{(s-1)(s-2)}{(s+1)(s+2)}$.

The idea is to apply the conditions previously discussed to the system $\tilde{P}(s)$ taking into account the unstable poles of $P(s)$. Applying the conditions 1–3. to the system $\tilde{P}(s)$ we obtain $\tilde{Q}(s)$. From this, we consider $Q(s) = B(s)\tilde{Q}(s)$. In this way $\tilde{Q}(s) \in \mathbb{H}^\infty$ by construction, but also $Q(s) \in \mathbb{H}^\infty$ as it is the product of two stable functions (in fact, also $B(s) \in \mathbb{H}^\infty$).

Note at this point that $\tilde{P}(s)\tilde{Q}(s) = P(s)Q(s)$. So, in this way it is possible to build a transfer function $Q(s)$ which satisfies the conditions 1–3 but starting from $\tilde{P}(s) \in \mathbb{H}^\infty$. At this point, the condition $\tilde{P}(\alpha_i)\tilde{Q}(\alpha_i) = 1$ with α_i indicating the poles of $P(s)$ with positive real part can be applied.

Finally note that, since in this context designing strictly bounded-real functions is more complicated than designing positive-real functions, often we design $Q(s)$ to obtain a positive-real function and then we apply the scattering matrix to obtain a bounded-real function.

13.7 Exercises

1. Given the system $G(s) = \frac{s+5}{s^2+2s+2}$ determine the class of stabilizing compensators. Then determine a closed-loop compensator that assures zero error for the unit step response.
2. Given the system $G(s) = \frac{s+5}{s^2-2s+2}$ determine the class of stabilizing compensators. Then determine a closed-loop compensator that assures zero error for the unit step response.
3. Given the system $G(s) = \frac{1}{s-2}$ determine the class of the stable and stabilizing compensators. Then determine a closed-loop compensator that assures a pole in $p = -5$.
4. Given the system $G(s) = \frac{s-1}{(s-2)(s-5)}$ determine the class of stable and stabilizing compensators.
5. Given the system with transfer function $G(s) = \frac{4}{s-3}$ determine the coprime factorization. Then determine the class of stabilizing compensators with unit step response equal to 1. Finally, calculate the energy associated to the impulse response for the closed-loop system.
6. Determine the right and left unitary coprime factorization of the system $G(s) = \frac{1}{(s-1)^2}$.
7. Calculate the unitary coprime factorization of the system with transfer function $G(s) = \frac{s+1}{s^2+3s+1}$.
8. Calculate a compensator that stabilizes simultaneously the two plants with transfer function $G_1(s) = \frac{1}{s-1}$ and $G_2(s) = \frac{1}{s-2}$.
9. Given the systems

$$\begin{cases} \dot{\mathbf{x}} = \mathrm{A}_i\mathbf{x} + \mathrm{B}_i\mathbf{u} \\ \mathbf{y} = \mathrm{C}_i\mathbf{x} \end{cases}$$

with $\mathrm{A}_i = \begin{bmatrix} \alpha_i & 0 & 0 \\ 0 & -1 & 0 \\ 0 & 0 & -2 \end{bmatrix}$; $\mathrm{B}_i = \mathrm{C}_i^T = \begin{bmatrix} 1 \\ -1 \\ \theta_i \end{bmatrix}$ and with $\alpha_i = i$ and $\theta_i = i+1$ for $i = \{1, 2\}$, determine if it is possible to simultaneously stabilize them and, if so, design the linear state regulator and observer so that the two systems are asymptotically stable. Verify the result.

10. Calculate, if possible, a control law $u = -Kx$ that can simultaneously stabilize the two systems with state-space matrices:

$$A_1 = \begin{bmatrix} -1 & 0 & 0 & 0 & 0 \\ 0 & 2 & 0 & 0 & 0 \\ 0 & 0 & \sqrt{5} & 0 & 0 \\ 0 & 0 & 0 & -3 & 0 \\ 0 & 0 & 0 & 0 & -4 \end{bmatrix}; A_2 = \begin{bmatrix} 1 & 0 & 0 & 0 & 0 \\ 0 & \sqrt{7} & 0 & 0 & 0 \\ 0 & 0 & -4 & 0 & 0 \\ 0 & 0 & 0 & -5 & 0 \\ 0 & 0 & 0 & 0 & -7 \end{bmatrix};$$

$$B_1^T = B_2^T = \begin{bmatrix} 1 & 1 & 1 & 1 & 0 \end{bmatrix}^T.$$

14

Formulation and Solution of Matrix Algebraic Problems through Optimization Problems

CONTENTS

In this chapter, taking into account the possibility to derive, by using a different computational paradigm, the open-loop balanced representation, some matrix computational methods are dealt with. The different computational methods are reformulated in terms of various optimization problems that can be addressed solving nonlinear matrix differential equations. This approach is appealing from a system point of view, in fact the formulation of a suitable optimization problem paves the way to a computational approach that is universal. From a numerical point of view, the solutions can be obtained using simple integration methods, such as the Euler method.

14.1 Solutions of Matrix Algebra Problems Using Dynamical Systems

In the previous chapters, we made use of standard matrix algebra to calculate quantities such as the eigenvalues, the eigenvectors, or the inverse of a matrix. These simple calculations, however, may become highly computational

DOI: 10.1201/9781003196921-14

demanding, when the size of the matrix increases. Similarly, the computation of the solution of Lyapunov equations via the vectorization method discussed in Chapter 2 may also lead to numerical issues. Aim of this chapter is to recast some fundamental operations of matrix algebra in terms of nonlinear dynamical systems, whose steady-state solution provides the result of the given matrix operation. The main idea of this approach is that, rather than computing a matrix operation, a nonlinear dynamical system is let to evolve. This system is such that it converges to the result of the matrix operation that one aims to solve. Hence, this method requires, given the matrix operation to solve, to properly write a nonlinear dynamical system that can converge to the solution of this problem.

Let us begin with four classical problems of matrix algebra and reformulate them in terms of nonlinear dynamical systems. Here, we discuss only a subset of the problems that can be studied via this approach. The technical details and further problems can be retrieved in the literature cited as recommended essential references.

14.1.1 Problem 1: Inverse of a Matrix

Let $\mathrm{A} \in \mathbb{R}^{n\times n}$ be a square matrix. If the inverse of this matrix exists, then, it can be found by considering the following nonlinear dynamical system:

$$\dot{\mathrm{C}}(t) = -\mu \mathrm{A}^T \left[\mathrm{AC}(t) - \mathrm{I}\right] \tag{14.1}$$

where $\mathrm{C}(t) \in \mathbb{R}^{n\times n}$, I is the $n \times n$ identity matrix and μ is a scalar parameter setting the convergence rate to the solution. This nonlinear dynamical system evolves towards a steady-state solution given by:

$$\lim_{t\to+\infty} \mathrm{C}(t) = \bar{\mathrm{C}} \tag{14.2}$$

By definition, the steady-state solution $\bar{\mathrm{C}}$ satisfies $\dot{\mathrm{C}}(t) = 0$ in equation (14.1). Hence, we have that:

$$\left[\mathrm{A}\bar{\mathrm{C}} - \mathrm{I}\right] = 0 \tag{14.3}$$

which gives:

$$\bar{\mathrm{C}} = \mathrm{A}^{-1} \tag{14.4}$$

Notice that system (14.1) has a number of state variables equal to n^2.

Example 14.1

To illustrate the procedure outlined above, let us first discuss a simple example, with A being diagonal:

$$\mathrm{A} = \begin{bmatrix} 3 & 0 \\ 0 & -0.5 \end{bmatrix} \tag{14.5}$$

Let us, hence, consider the dynamical system (14.1) with $n = 2$, set $\mu = 1$, and write the system dynamics of each of the coefficients of the matrix $C(t)$ (the four state variables of the system):

$$\begin{cases} \dot{c}_{11} = -3\,(3c_{11} - 1) \\ \dot{c}_{12} = -9c_{12} \\ \dot{c}_{21} = -0.25c_{21} \\ \dot{c}_{22} = -0.25c_{22} - 0.5 \end{cases} \tag{14.6}$$

As the differential equations of system (14.6) are decoupled, the solution can be analytically found:

$$\begin{cases} \dot{c}_{11}(t) = c_{11}(0)e^{-9t} + \frac{1}{3} \\ \dot{c}_{12}(t) = c_{12}(0)e^{-9t} \\ \dot{c}_{21}(t) = c_{21}(0)e^{-0.25t} \\ \dot{c}_{22}(t) = c_{21}(0)e^{-0.25t} - \frac{1}{2} \end{cases} \tag{14.7}$$

The final step is to calculate $\lim_{t \to +\infty} C(t)$. This yields:

$$\bar{C} = \begin{bmatrix} \frac{1}{3} & 0 \\ 0 & -2 \end{bmatrix} = A^{-1} \tag{14.8}$$

Let us now introduce a general procedure to calculate the evolution of the dynamical system (14.1) by using MATLAB. We use the differential equation solver `ode45` available in MATLAB. The function describing the equations of the system can be written as follows:

```
function dxdt=prob1_MatrixInversion(t,x,A,mu)
   n2=length(x);
   C=reshape(x,sqrt(n2),sqrt(n2));
   dxdt=-mu*A'*(A*C-eye(sqrt(n2)));
   dxdt=dxdt(:);
end
```

The inputs of the function include the matrix A, for which the inverse must be calculated, and the convergence rate μ as well as time and the system state variables. The dynamical system is integrated by using the MATLAB command

```
>> [T,Y]=ode45(@prob1_MatrixInversion,[0:0.01:10],...
                         ...rand(n2,1),[],A);
>> Ainv=reshape(Y(end,:),sqrt(n2),sqrt(n2));
```

where n^2 is the number of entries of matrix A.

MATLAB® Exercise 14.1

Using MATLAB, calculate the inverse of the matrix

$$A = \begin{bmatrix} 0.5377 & 0.3426 & 0.7147 & -1.2075 & 0.2939 & 1.4384 & 0.3192 \\ 1.8339 & 3.5784 & -0.2050 & 0.7172 & -0.7873 & 0.3252 & 0.3129 \\ -2.2588 & 2.7694 & -0.1241 & 1.6302 & 0.8884 & -0.7549 & -0.8649 \\ 0.8622 & -1.3499 & 1.4897 & 0.4889 & -1.1471 & 1.3703 & -0.0301 \\ 0.3188 & 3.0349 & 1.4090 & 1.0347 & -1.0689 & -1.7115 & -0.1649 \\ -1.3077 & 0.7254 & 1.4172 & 0.7269 & -0.8095 & -0.1022 & 0.6277 \\ -0.4336 & -0.0631 & 0.6715 & -0.3034 & -2.9443 & -0.2414 & 1.0933 \end{bmatrix} \tag{14.9}$$

The dynamical system (14.1) can be integrated by using the MATLAB command

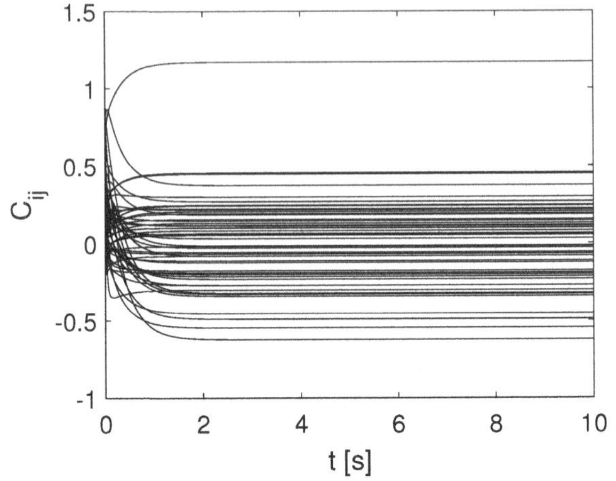

FIGURE 14.1
Trend of the variables $C_{ij}(t)$ for system (14.1) with A as in (14.9).

```
>> [T,Y]=ode45(@prob1_MatrixInversion,[0:0.01:10],rand(numel(A),1),[],A);
```

obtaining the time evolution of the coefficients $C_{ij}(t)$ shown in Figure 14.1. The steady-state solution can be calculated as:

```
>> Cbar=reshape(Y(end,:),length(A),length(A))
```

that is

$$\bar{\mathrm{C}} = \begin{bmatrix} -0.1236 & 0.1453 & -0.2700 & 0.0534 & 0.1135 & -0.0196 & -0.1893 \\ 0.1910 & 0.1001 & 0.1300 & -0.0812 & 0.0525 & -0.0669 & 0.0625 \\ 0.2161 & -0.1920 & -0.1124 & 0.1165 & 0.2996 & 0.2221 & -0.1763 \\ -0.5461 & 0.2693 & -0.0821 & 0.1839 & -0.2133 & 0.4540 & -0.2704 \\ -0.0166 & 0.0311 & -0.2312 & -0.2004 & -0.0292 & 0.4473 & -0.4537 \\ 0.2049 & 0.1540 & 0.2423 & 0.2454 & -0.3013 & -0.0570 & 0.0818 \\ -0.3218 & 0.3738 & -0.6225 & -0.4894 & -0.3405 & 1.1699 & -0.3272 \end{bmatrix} \quad (14.10)$$

The inverse of the matrix A is thus given by: $\mathrm{A}^{-1} = \bar{\mathrm{C}}$.

14.1.2 Problem 2: Eigenvalues of a Matrix

Let H be a $n \times n$ square matrix and consider the following nonlinear dynamical system:

$$\dot{\mathrm{H}} = \mathrm{H}\,(\mathrm{HN} - \mathrm{NH}) - (\mathrm{HN} - \mathrm{NH})\,\mathrm{H} \quad (14.11)$$

where

$$\mathrm{N} = \begin{bmatrix} \mu_1 & 0 & \cdots & 0 \\ 0 & \mu_2 & \cdots & 0 \\ \vdots & \vdots & \ddots & \vdots \\ 0 & 0 & \cdots & \mu_n \end{bmatrix} \quad (14.12)$$

with $\mu_1 > \mu_2 > \ldots > \mu_n > 0$ positive real numbers. It can be demonstrated that the dynamical system in equation (14.11) converges to a diagonal matrix $\bar{\mathrm{H}}$ whose diagonal entries are the eigenvalues of the matrix H(0).

Therefore, to calculate the eigenvalues of a matrix A, one can integrate system (14.11), starting from initial condition given by H(0) = A. To this aim, the following MATLAB procedure can be employed. Once again, the differential equation solver `ode45` can be used, after writing the system equations as follows:

```
function dxdt=prob2_MatrixEigenvalues(t,x)
   n2=length(x);
   H=reshape(x,sqrt(n2),sqrt(n2));
   N=diag([sqrt(n2):-1:1]);
   dxdt=H*(H*N-N*H)-(H*N-N*H)*H;
   dxdt=dxdt(:);
end
```

The dynamical system is then integrated by using the MATLAB command

```
>> [T,Y]=ode45(@prob2_MatrixEigenvalues,[0:0.01:10],A);
>> E=reshape(Y(end,:),length(A),length(A));
```

where the initial conditions have been set equal to the matrix A, whose eigenvalues have to be computed.

MATLAB® Exercise 14.2

Calculate the eigenvalues of the matrix

$$\mathrm{A} = \begin{bmatrix} 0.6930 & 0.7707 & 0.1762 \\ 0.2608 & 0.0757 & 0.5628 \\ 0.1127 & 0.5856 & 0.1239 \end{bmatrix} \tag{14.13}$$

The dynamical system (14.11) can be integrated by using the MATLAB command

```
>> [T,Y]=ode45(@prob2_MatrixEigenvalues,[0:0.01:10],A);
```

where the initial conditions are set equal to the matrix A. The time evolution of the state variables $H_{ij}(t)$ obtained in this way are shown in Figure 14.2. They converge to the steady-state matrix

```
>> Hb=reshape(Y(end,:),length(A),length(A));
```

that reads

$$\bar{\mathrm{H}} = \begin{bmatrix} 1.115 & 0 & 0 \\ 0 & 0.2976 & 0 \\ 0 & 0 & -0.52 \end{bmatrix} \tag{14.14}$$

The diagonal entries of this matrix are equal to the eigenvalues of A as it can be checked by using the MATLAB command

```
>> E=eig(A)
```

Note that the eigenvalues must be real quantities, otherwise the system (14.11) diverges or does not converge to a steady-state solution.

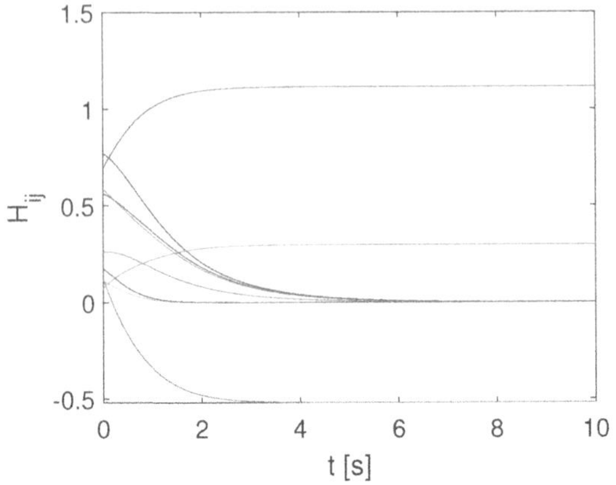

FIGURE 14.2
Trend of the variables $H_{ij}(t)$ for system (14.11) with A as in (14.13).

14.1.3 Problem 3: Eigenvectors of a Symmetric Positive Definite Matrix

Let us consider the following nonlinear dynamical system:

$$\dot{\Theta} = \mathrm{A}\Theta\mathrm{N} - \Theta\mathrm{N}\Theta^T\mathrm{A}\Theta \tag{14.15}$$

where A is symmetric positive definite, N is a diagonal matrix as in equation (14.12) and $\Theta(0)$ an orthonormal matrix that could be set equal to the identity matrix. In this case, it can be shown that the columns of the matrix $\Theta_\infty = \lim_{t\to+\infty} \Theta(t)$ are the eigenvectors of A (which are real quantities, as the matrix is symmetric positive definite and thus has real eigenvalues and eigenvectors).

To obtain Θ_∞, the dynamical system (14.15) is integrated by using MATLAB. The first step is to write the system equations as follows:

```
function dxdt=prob3_MEigenv(t,x,A)
   n2=length(x);
   Theta=reshape(x,sqrt(n2),sqrt(n2));
   N=diag([sqrt(n2):-1:1]);
   dxdt=A*Theta*N-Theta*N*Theta'*A*Theta;
   dxdt=dxdt(:);
end
```

The dynamical system is solved by using the MATLAB command

```
>> [T,Y]=ode45(@prob3_MEigenv,[0:0.01:10],eye(length(A)),[],A);
>> Q=reshape(Y(end,:),length(A),length(A));
```

where the initial conditions are set to the identity matrix of the suitable dimensions and A is the symmetric positive definite matrix whose eigenvectors have to be calculated.

MATLAB® Exercise 14.3

Calculate the eigenvectors of the matrix

$$A = \begin{bmatrix} 7.0975 & -8.4650 & 5.8477 & -6.3906 & -0.9032 \\ -8.4650 & 25.2876 & -19.9587 & 2.3167 & 15.7190 \\ 5.8477 & -19.9587 & 16.8034 & 1.0059 & -14.2585 \\ -6.3906 & 2.3167 & 1.0059 & 24.2867 & -12.0941 \\ -0.9032 & 15.7190 & -14.2585 & -12.0941 & 29.2483 \end{bmatrix} \tag{14.16}$$

The dynamical system (14.15) is integrated by using the MATLAB commands

```
>> I=eye(5);
>> [T,Y]=ode45(@prob3_MEigenv,[0:0.01:10],I(:),[],A);
>>Q=reshape(Y(end,:),5,5)
```

that produces the steady-state solution

$$Q = \begin{bmatrix} 0.1413 & 0.3313 & 0.1294 & 0.9199 & -0.0855 \\ -0.5879 & -0.3389 & -0.2894 & 0.1930 & -0.6469 \\ 0.4969 & 0.1778 & 0.2990 & -0.2525 & -0.7539 \\ 0.1560 & -0.7877 & 0.5604 & 0.1880 & 0.0763 \\ -0.6026 & 0.3511 & 0.7043 & -0.1321 & 0.0092 \end{bmatrix} \tag{14.17}$$

whose columns are the eigenvectors of A as it can be verified by using the MATLAB command

```
>> [V,E]=eig(A)
```

or calculating the diagonal matrix

```
D=Q'*A*Q
```

which yields

$$D = \begin{bmatrix} 59.6802 & 0 & 0 & 0 & 0 \\ 0 & 33.1344 & 0 & 0 & 0 \\ 0 & 0 & 6.9494 & 0 & 0 \\ 0 & 0 & 0 & 2.5407 & 0 \\ 0 & 0 & 0 & 0 & 0.4125 \end{bmatrix} \tag{14.18}$$

14.1.4 Problem 4: Observability and Controllability Gramian

Let us consider the gramian equations (5.1) and (5.2) and compute their solutions via the approach based on the integration of dynamical systems. Let us begin with the observability gramian. It can be found as the matrix X that minimizes the following performance index

$$J = \frac{1}{2}||\mathrm{E}||_2^2$$

where $E = \mathrm{A}^T\mathrm{X} + \mathrm{XA} + \mathrm{C}^T\mathrm{C}$. This problem can be reformulated in terms of the following dynamical system:

$$\dot{\mathrm{X}} = \mathrm{A}^T\mathrm{X} + \mathrm{XA} + \mathrm{C}^T\mathrm{C} \tag{14.19}$$

whose steady-state solution is the observability gramian. In fact, considering $\dot{\mathrm{X}} = 0$ leads to the gramian equation.

Similarly, for the controllability gramian the following system has to be considered:

$$\dot{\mathrm{X}} = \mathrm{AX} + \mathrm{XA}^T + \mathrm{BB}^T \tag{14.20}$$

Notice that the algebraic problems we are discussing in this chapter are related to optimization problems which, in turn, are solved by a matrix that can be also obtained as the steady-state solution towards which a dynamical system converges. The choice of the initial conditions from which the nonlinear dynamical system starts is often a crucial step to obtain the desired solution.

The MATLAB functions to use for the computation of the observability and controllability gramians are

```
function dxdt=ObsvGram(t,x,A,C)
   n2=length(x);
   X=reshape(x,sqrt(n2),sqrt(n2));
   dxdt=A'*X+X*A+C'*C;
   dxdt=dxdt(:);
```

and

```
function dxdt=CtrbGram(t,x,A,B)
   n2=length(x);
   X=reshape(x,sqrt(n2),sqrt(n2));
   dxdt=A*X+X*A'+B*B';
   dxdt=dxdt(:);
```

The two dynamical systems are then integrated by using the MATLAB commands

```
>> [T,Y]=ode45(@CtrbGram,[0:0.01:10],eye(length(A)),[],A,B);
>> Wc=reshape(Y(end,:),length(A),length(A));
>> [T,Y]=ode45(@ObsvGram,[0:0.01:10],eye(length(A)),[],A,C);
>> Wo=reshape(Y(end,:),length(A),length(A));
```

14.2 Computation of the Open-loop Balanced Representation via the Dynamical System Approach

The problems presented in the previous section can be jointly used to obtain more complex operations. In particular, in this section, we illustrate how to employ them to derive the open-loop balanced representation of a linear time-invariant system.

Let us define the matrix

$$\mathrm{P}_\infty = \mathrm{W}_c^{(-\frac{1}{2})} \left(\mathrm{W}_c^{\frac{1}{2}} \mathrm{W}_o \mathrm{W}_c^{\frac{1}{2}} \right)^{\frac{1}{2}} \mathrm{W}_c^{(-\frac{1}{2})} \tag{14.21}$$

where W_o and W_c are the observability and controllability gramians calculated for a given state-space representation (A, B, C, D). It can be proved that with the transformation matrix $\tilde{\mathrm{T}} = \mathrm{P}_\infty^{-\frac{1}{2}}$ one obtains a state space representation $\left(\tilde{\mathrm{A}}, \tilde{\mathrm{B}}, \tilde{\mathrm{C}}, \mathrm{D}\right)$ for which it holds that

$$\tilde{\mathrm{W}}_c = \tilde{\mathrm{W}}_o \tag{14.22}$$

that is, the two gramians are equal. However, in general they are not diagonal. The matrix P_∞ can be found as the steady-state solution of the nonlinear dynamical system

$$\dot{\mathrm{P}} = \mathrm{P}^{-1}\mathrm{W}_o\mathrm{P}^{-1} - \mathrm{W}_c \tag{14.23}$$

assuming to select the initial conditions as $\mathrm{P}(0) = \mathrm{P}_0$, where P_0 is a positive semi-definite matrix.

The MATLAB function defining the equations for system (14.23) can be written as

```
function dxdt=PinfSystem(t,x,Wo,Wc)
   n2=length(x);
   P=reshape(x,sqrt(n2),sqrt(n2));
   [T,Y]=ode45(@prob1_MatrixInversion,[0:0.1:10],rand(n2,1),[],P);
   Pinv=reshape(Y(end,:),sqrt(n2),sqrt(n2));
   dxdt=Pinv*Wo*Pinv-Wc;
   dxdt=dxdt(:);
```

Notice that this function makes use of the routine of problem 1 to obtain the inverse matrix. The system can be integrated with the commands

```
>> In2r=reshape(eye(sqrt(n2)),n2,1);
>> [T,Y]=ode45(@PinfSystem,[0:0.1:100],In2r,[],WoD,WcD);
>> Pinf=reshape(Y(end,:),sqrt(n2),sqrt(n2))
```

Notice that condition (14.22) does not guarantee open-loop balancing, for which it required that the two gramians are equal and diagonal:

$$\bar{\mathrm{W}}_c = \bar{\mathrm{W}}_o = \begin{bmatrix} \sigma_1 & 0 & \cdots & 0 \\ 0 & \sigma_2 & \cdots & 0 \\ \vdots & \vdots & \ddots & \vdots \\ 0 & 0 & \cdots & \sigma_n \end{bmatrix} \tag{14.24}$$

where $\sigma_1, \sigma_2, \ldots, \sigma_n$ are the Hankel singular values.

To address this issue, let us define

$$\Lambda = \begin{bmatrix} \sigma_1 & 0 & \cdots & 0 \\ 0 & \sigma_2 & \cdots & 0 \\ \vdots & \vdots & \ddots & \vdots \\ 0 & 0 & \cdots & \sigma_n \end{bmatrix} \tag{14.25}$$

Now, as $\tilde{\mathrm{W}}_c = \tilde{\mathrm{T}}\mathrm{W}_c\tilde{\mathrm{T}}^T$ and $\tilde{\mathrm{W}}_c = \mathrm{Q}\Lambda\mathrm{Q}^T$, we have that $\mathrm{W}_c =$

$\tilde{\mathrm{T}}^{-1}\mathrm{Q}\Lambda\mathrm{Q}^T\tilde{\mathrm{T}}^{-1}$, and so $\Lambda = \mathrm{Q}^T\tilde{\mathrm{T}}\mathrm{W}_c\tilde{\mathrm{T}}^T\mathrm{Q}$. Therefore, to calculate the open-loop balanced realization $\left(\bar{\mathrm{A}}, \bar{\mathrm{B}}, \bar{\mathrm{C}}, \mathrm{D}\right)$ the following transformation has to be considered:

$$\bar{\mathrm{T}} = \tilde{\mathrm{T}}^T Q \tag{14.26}$$

Accordingly, one can define an algorithm to obtain the open-loop balanced representation starting from the state-space representation (A, B, C) entirely based on the dynamical system approach:

1. Calculate the gramians W_c and W_o using problem 4;
2. Calculate the transformation matrix $\tilde{\mathrm{T}}$ combining problem 1 to calculate the inverse of P and problem in equation (14.23) to obtain the matrix P_∞;
3. Calculate the eigenvector matrix Q for the matrix $\tilde{\mathrm{W}}_c = \tilde{\mathrm{T}}\mathrm{W}_c\tilde{\mathrm{T}}^T$ by using problem 3;
4. Calculate the inverse of $\bar{\mathrm{T}} = \tilde{\mathrm{T}}^T\mathrm{Q}$ using again problem 1;
5. Check the singular values and the open-loop balancing by calculating the eigenvalues of the matrix $\Lambda = \mathrm{Q}^T\tilde{\mathrm{T}}\mathrm{W}_c\tilde{\mathrm{T}}^T\mathrm{Q}$ by using problem 2.

MATLAB® Exercise 14.4

Let us consider again the system in Exercise 5.2, characterized by the following state-space representation:

$$\mathrm{A} = \begin{bmatrix} -1 & 0 & 0 \\ 1/2 & -1 & 0 \\ 1/2 & 0 & 1 \end{bmatrix}; \quad \mathrm{B} = \begin{bmatrix} 1 & 0 \\ 0 & -1 \\ 0 & 1 \end{bmatrix}; \quad \mathrm{C}\begin{bmatrix} 0 & 0 & 1 \\ 1 & 1 & 0 \end{bmatrix} \tag{14.27}$$

Let us now apply the procedure outlined above.
The first step is the computation of the gramians for the given state-space representation as

```
>> [T,Y]=ode45(@ObsvGram,[0:0.01:10],reshape(eye(3),9,1),[],A,C);
>> Wo=reshape(Y(end,:),3,3);
```

and

```
>> [T,Y]=ode45(@CtrbGram,[0:0.01:10],reshape(eye(3),9,1),[],A,B);
>> Wc=reshape(Y(end,:),3,3);
```

Now, the matrix P_∞ can be calculated as

```
>> [T,Y]=ode45(@PinfSystem,[0:0.01:10],reshape(eye(3),9,1),[],Wo,Wc);
>> Pinf=reshape(Y(end,:),3,3);
```

Next, we calculate the transformation matrix $\tilde{\mathrm{T}}$:

```
>> Tt=Pinf^(-0.5);
```

and its inverse

```
>> [T,Y]=ode45(@prob1_MatrixInversion,[0:0.01:10],rand(numel(Tt),1),[],Tt);
>> TtInv=reshape(Y(end,:),3,3);
```

With these matrices, the state-space representation $\left(\tilde{\mathrm{A}}, \tilde{\mathrm{B}}, \tilde{\mathrm{C}}, \mathrm{D}\right)$ where the two gramians are equal can be computed:

```
>> At=TtInv*A*Tt;
>> Bt=TtInv*B
>> Ct=C*Tt
```

```
>> [T,Y]=ode45(@CtrbGram,[0:0.01:10],reshape(eye(3),9,1),[],At,Bt);
>> Wct=reshape(Y(end,:),3,3)
>> [T,Y]=ode45(@ObsvGram,[0:0.01:10],reshape(eye(3),9,1),[],At,Ct);
>> Wot=reshape(Y(end,:),3,3)
```

The next step is to obtain the eigenvector matrix Q as

```
>> [T,Y]=ode45(@prob3_MEigenv,[0:0.01:50],reshape(eye(3),9,1),[],Wct);
>> Q=reshape(Y(end,:),3,3)
```

and therefore

```
>> Tbal=Tt'*Q
```

is the transformation matrix which allows to derive an open-loop balanced representation as:

```
>> Abal=inv(Tbal)*A*Tbal
>> Bbal=inv(Tbal)*B
>> Cbal=C*Tbal
```

This ultimately yields the open-loop balanced representation

$$
\begin{array}{ll}
\mathrm{A}_b = \begin{bmatrix} -0.8005 & -0.1686 & -0.2657 \\ -0.2866 & -0.7578 & 0.3816 \\ 0.3317 & -0.2803 & -1.4417 \end{bmatrix}; & \mathrm{B}_b = \begin{bmatrix} 0.9327 & -0.7801 \\ -0.3406 & -0.8712 \\ -0.3442 & -0.0330 \end{bmatrix}; \\
\mathrm{C}_b = \begin{bmatrix} -0.2508 & -0.9325 & 0.2431 \\ 1.1896 & 0.0733 & 0.2458 \end{bmatrix} &
\end{array}
\tag{14.28}
$$

14.3 Concluding Remarks

In this chapter, algorithms to solve classic matrix algebra problems based on the integration of continuous-time dynamical systems have been presented. By the joint application of some of these problems, we focused on an algorithm to determine the open-loop balanced realization of a linear time-invariant system.

As the introduced dynamical systems are nonlinear, their evolution is calculated by means of numerical integration, hence they are recast as discrete-time dynamical models. This naturally leads to the question whether it is possible to conceive a way to solve the dynamical systems directly in the continuous-time, analog, domain. There exists a dichotomy between electronic analog circuits and continuous-time dynamical systems. Reformulating matrix algebra problems in terms of dynamical systems paves the way to the design of analog computing devices able to solve them in real-time.

14.4 Exercises

1. Using the procedure introduced in this chapter, calculate the inverse of the matrix $A = \begin{bmatrix} 1 & 0 & 0 \\ 0 & 2 & 0 \\ 0 & 0 & a \end{bmatrix}$ with $a = 0.1$, $a = 0.01$ and $a = 0.001$. Evaluate the time needed to reach the steady-state in the three cases.

2. Calculate the eigenvalues of the matrix A in equation (14.13) perturbing the matrix N with a gaussian noise. Evaluate the effect of the perturbation on the steady-state matrix.

3. Calculate the eigenvectors of the matrix A in equation (14.16) perturbing the matrix N with a gaussian noise. Evaluate the effect of the perturbation on the steady-state matrix.

4. Determine the controllability and observability gramians for the continuous-time LTI system with state-space realization

$$A = \begin{bmatrix} 0 & 1 \\ -3 & -2 \end{bmatrix}; \quad B = \begin{bmatrix} 0 \\ 1 \end{bmatrix}; \quad C \begin{bmatrix} -1 & 1 \end{bmatrix} \tag{14.29}$$

5. Calculate the singular values of the system with transfer function $G(s) = \frac{s^2-5s+4}{s^2+5s+4}$.

6. Calculate an open-loop balanced realization for the system with state-space matrices:

$$A = \begin{bmatrix} -0.5 & -1 & 0 & 0 \\ 1 & -0.5 & 0 & 0 \\ 0 & 0 & -3 & 0 \\ 0 & 0 & 0 & -4 \end{bmatrix}; \quad B = \begin{bmatrix} 1 \\ -1 \\ -1 \\ 1 \end{bmatrix}; \tag{14.30}$$
$$C = \begin{bmatrix} 0 & 1 & -1 & 1 \end{bmatrix}$$

15

Time-delay Systems

CONTENTS

Aim of this chapter is to provide a few main results related to time-delay systems. The concepts explored in this chapter have a fundamental importance in robust control theory due to the implications of time-delay in practical implementations of control systems. Indeed, time-delay systems are accurate models of real systems. Control engineers tend to neglect the presence of time-delay in feedback loops assuming that the control input is presented in the proper time to observe the desired response. However, many systems are based on the existence of a time-delay, such as radars and sonars whose working principle relies on the time-delay between the emitted wave and the reception of the echo. Moreover, the presence of a human operator in a control loop is modeled as a time-delay, an aspect of crucial importance in the theory of human-machine interaction.

15.1 Modeling Systems with Time-delays

In industrial automation time-delays are a common feature, always appearing in plants and systems involved in automatic factories. In the evaluation of just-in-time production, the queue delay systems are the key points of the optimization. Time-delays are present in the temperature control system for our showers and, if we think to the internal combustion engines, also the production of torque is delayed with respect to the required one. An interesting series of examples of systems with time-delays include different classes of systems:

DOI: 10.1201/9781003196921-15

- Fluid flow models for a congested router in TCP/AQM controlled network;
- Car following systems;
- Rotating cutting and milling machines;
- Heating systems.

In the last decade, particular interest has been also devoted to internet congestion with many contributions devoted to understand the role of time-delay in this field. In particular, the topic of robust control of time-delay systems is dealt with in several books, listed as recommended essential references for this chapter, where important applications in underwater control systems, in biosystems analysis and control, and in the area of mathematical modeling have been also discussed.

In this chapter we will discuss only some fundamental results about this topic. The chapter is organized as follows: in the second part the main problems related to time-delay systems are reported with practical examples in order to introduce in a concrete way the topic. In the part three a classification of the various categories of time-delay systems will be presented. Our discussion is referred to closed-loop schemes of delay systems corresponding to the feedback design, and some particular results regarding the possibility of stabilizing a class of delay systems in the state-space domain are presented. The referenced books on delay systems include exhaustive theoretical results, whereas in this chapter we give a brief introduction focusing more on reference examples rather than theoretical results.

15.2 Basic Principles of Time-delay Systems

Let us consider the continuous-time linear system S_1, where the output $y(t)$ is related to the input $u(t)$ by the relationship $y(t) = u(t - \tau)$. This system represents an ideal time-delay. In fact, the output is a replica of the input after a delay τ. The transfer function of this system is given by

$$G(s) = e^{-s\tau} \tag{15.1}$$

which is a direct application of one of the properties of the Laplace transform.

The characteristic of $G(s)$ is that it is an all-pass stable system, i.e., $|G(j\omega)| = 1$, $\forall\omega$. The system is BIBO stable, the unique singularity is for $s \to -\infty$ and the unique zero is at $s \to \infty$. The system belongs to the class of infinite dimensional linear systems.

Let us now consider the closed-loop system reported in Figure 15.1. The transfer function is

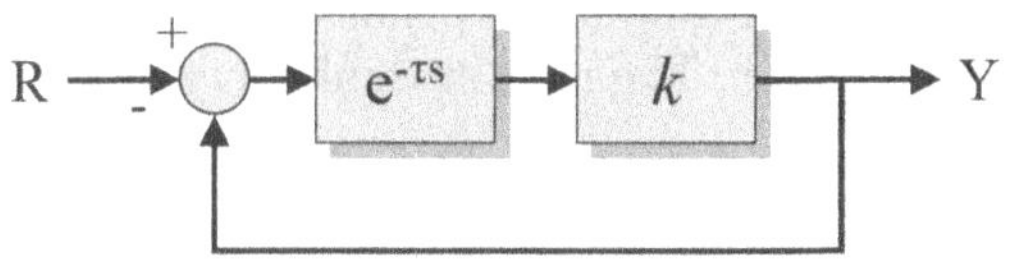

FIGURE 15.1
A closed-loop system with a time-delay block.

$$F(s) = \frac{ke^{-s\tau}}{1 + ke^{-s\tau}} \tag{15.2}$$

In order to check its stability the poles of the systems must be computed by solving

$$1 + ke^{-s\tau} = 0 \tag{15.3}$$

Equation (15.3) is a transcendental equation that can be rewritten as

$$ke^{-\sigma\tau}e^{-j\omega\tau} = -1 \tag{15.4}$$

that is a complex equation corresponding to the following conditions on the real and imaginary part:

$$\begin{aligned} ke^{-\sigma\tau}\cos\omega\tau &= -1 \\ ke^{-\sigma\tau}\sin\omega\tau &= 0 \end{aligned} \tag{15.5}$$

The second condition yields $\sin\omega\tau = 0$, that is solved by $\omega\tau = i\pi$, with $i = 1, 2, \ldots, n$. The first condition implies that the index i must be odd. In this case, the solution is given by $\sigma = \frac{\ln k}{\tau}$. The system is therefore stable if $0 < k < 1$.

The same result can be obtained considering the small-gain theorem. In fact, since it is an unitary control feedback scheme, then $|ke^{-j\omega\tau}|$ must be less than 1, $\forall\omega\tau$ that leads to the condition $0 < k < 1$.

Another possibility to derive the same result is to consider the Nyquist plot and determine the number of encirclements of the critical point $(-1, 0)$. In fact, in this case, we find that for each $k > 1$ the critical point is encircled, and thus the system is unstable.

From this simple example, we have seen that the poles of time-delay system are an infinite number. In order to have information about the stability of a time-delay system, several approaches can be used:

1. The analytical one, that means to find the infinite roots of the characteristic equation;
2. The classical approach in the frequency domain by using the Nyquist criterion and the Bode diagrams.

The stability of delayed systems depends not only on the static gain k, but also on the time-delay τ. The next example illustrates another important case study.

Example 15.1

Let us consider the classical control scheme shown in Figure 15.2. It can be physically interpreted as a delay speed control system for a motion controlled system. The controlled system may represent a car, an airplane, a bicycle and so on, whereas the control action is performed with some delay, due to the physiological characteristics of the human response.

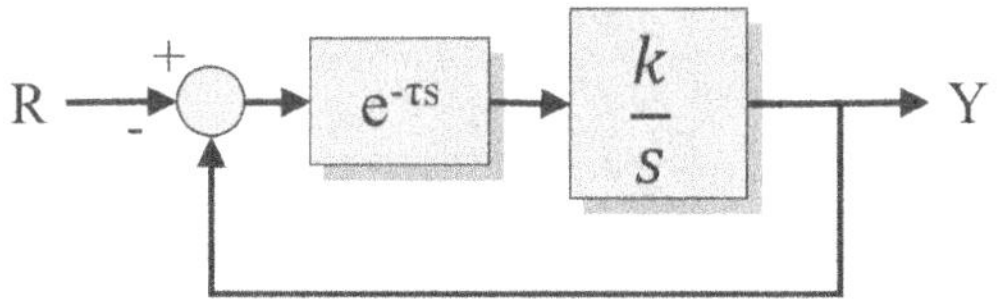

FIGURE 15.2
The closed-loop system used in Example 15.1.

To obtain the stability condition, the phase margin of the open-loop system must be considered. Stability requires that it is positive. In order to compute it, the so-called crossover frequency can be calculated as

$$\left| \frac{k}{j\omega_c} e^{-j\omega_c \tau} \right| = 1 \tag{15.6}$$

Therefore, we have $\omega_c = k$. The phase margin is given as follows:

$$m_\phi = \pi - \psi_{\omega_c} = \pi - \frac{\pi}{2} - \omega_c \tau = \frac{\pi}{2} - \omega_c \tau = \frac{\pi}{2} - k\tau \tag{15.7}$$

Therefore, the stability condition yields $k\tau < \frac{\pi}{2}$.
It is evident that high values of the time-delay τ require to decrease the gain, at the expenses of the precision of the control. On the contrary, for small values of the time-delay τ, an higher gain can be used, with better performance.

The examples reported above show the effect of the time-delay in simple case studies. More in general, systems may include more than a single delay. In similar cases, in order to get some insights on the system behavior, it is often convenient to introduce an approximation of the various time-delay elements. It therefore useful to get a rational function approximation of the general delay element $e^{-s\tau}$ that reflects its important properties. The Padé approximation is the approximation more commonly used in many applications. It is derived matching the first coefficient of the Taylor expansion of $e^{-s\tau}$ with that of a rational transfer function of order n that must have all-pass characteristics. Therefore, if the approximation function is indicated as $R_n(s) = \frac{N_n(s)}{D_n(s)}$, we have

$$D_n(s) = N_n(-s) \tag{15.8}$$

The Padé approximation is already implemented in MATLAB. The following commands can be used to obtain the third-order Padé approximation of $e^{-s\tau}$ with $\tau = 1$:

```
>> tau=1
>>[N,D]=pade(tau,3)
```

15.3 Stability of Time-delay Systems

The state-space representation of a linear time-invariant (LTI) system with time-delays is the following:

$$\dot{\mathbf{x}}(t) = \mathrm{A}_0\mathbf{x}(t) + \sum_{i=1}^{q} \mathrm{A}_i\mathbf{x}(t - \tau_i) \tag{15.9}$$

where $\mathrm{A}_k \in \mathbb{R}^{n\times n}$ with $k = 0, \ldots, q$ are constant state matrices and $\tau_i \geq 0$ for $i = 1, \ldots, q$ are time-delays.

If $\tau_i = i\tau$, then the delays are said to be commensurate and the system to be a commensurate delay system:

$$\dot{\mathbf{x}}(t) = \mathrm{A}_0\mathbf{x}(t) + \sum_{i=1}^{q} \mathrm{A}_i\mathbf{x}(t - i\tau) \tag{15.10}$$

Otherwise, the system is said to be an incommensurate delay system.

The stability of system (15.9) is fully determined by its characteristic quasi-polynomial $p(s, \tau_i)$, which is given by:

$$p(s, \tau_i) = \det\left(s\mathrm{I} - \mathrm{A}_0 - \sum_{i=1}^{q} \mathrm{A}_i e^{-s\tau_i}\right) \tag{15.11}$$

or equivalently

$$p(s, \tau_i) = p_0(s) + \sum_{i=1}^{q} p_i(s)e^{-s\tau_i} \tag{15.12}$$

where $p_i(s)$ with $i = 0, \ldots, q$ are polynomials related to the state matrices A_i. In particular, they can be expressed as $p_0(s) = s^n + \sum_{j=0}^{n-1} p_{0,j}s^j$ and $p_i(s) = \sum_{j=0}^{n} p_{k,j}s^j$ for $i = 1, \ldots, q$.

For commensurate delay system, the characteristic quasi-polynomial may be written as:

$$p(s,\tau) = \det\left(sI - A_0 - \sum_{i=1}^{q} A_i e^{-si\tau}\right) \tag{15.13}$$

or equivalently as

$$p(s,\tau) = \sum_{i=0}^{q} p_i(s)e^{-si\tau} \tag{15.14}$$

where $p_k(s)$ have the same meaning than in the case of systems with incommensurate time delays.

Note that the quasi-polynomial admits infinite roots. If all the roots are in the left-half plane, the time-delay system is asymptotically stable.

For simplicity, in the rest of the chapter we will mainly focus on commensurate delay systems.

Let us now define formally the stability property for a time delay system from the properties of its characteristic quasi-polynomial and, in particular, introduce the notion of delay-independent stability.

Definition 30 *The characteristic quasi-polynomial (15.13) is said to be stable if all the roots of the characteristic quasi-polynomial $p(s,\tau)$ lie in the open left half plane. It is said to be delay-independent stable if this condition is valid for all $\tau \geq 0$.*

System (15.9) is stable if and only if its characteristic quasipolynomial is stable. In addition, it is stable independent of delay if its characteristic quasi-polynomial is such.

15.4 Stability of Time-delay Systems with $q = 1$

Let us consider the class of SISO time-delay systems with the following transfer function

$$\tilde{G}(s) = \frac{N_1(s)}{N_2(s) + N_3(s)e^{-s\tau}} \tag{15.15}$$

where $N_i(s)$ are polynomials of order n_i. In particular, we assume that the orders of the polynomials are such that $n_1 \leq n_2$ and $n_3 \leq n_2$.

The denominator can be considered the characteristic quasi-polynomial of a time-delay system with $q = 1$.

$\tilde{G}(s)$ can be rewritten as follows

$$\tilde{G}(s) = \frac{F_1(s)}{1 + F_2(s)e^{-s\tau}} \tag{15.16}$$

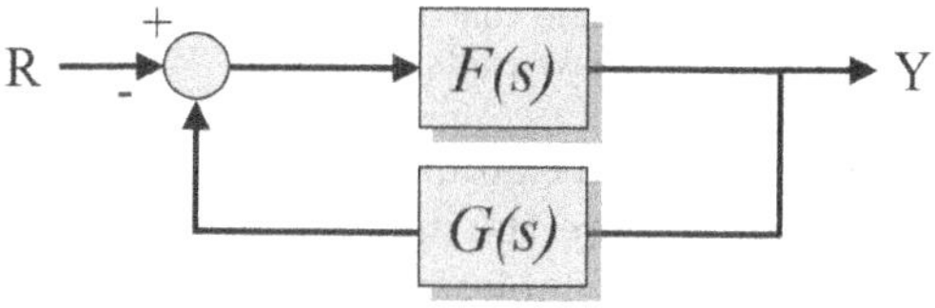

FIGURE 15.3
Feedback control scheme. The control scheme is a negative feedback loop with $F(s)$ in the direct chain and $G(s)$ in the indirect one. The reference signal is labeled as R and the controlled output is labeled as Y.

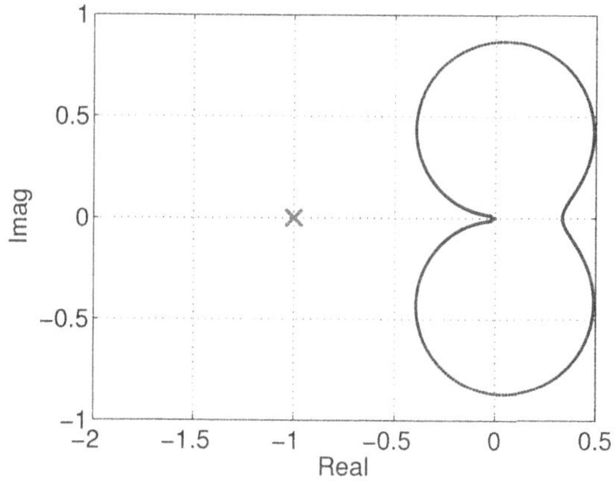

FIGURE 15.4
Nyquist plot of $G(s)$ as in equation (15.18).

being $F_1(s) = \frac{N_1(s)}{N_2(s)}$ and $F_2(s) = \frac{N_3(s)}{N_2(s)}$ rational transfer functions. This system corresponds to the classical feedback control scheme (Figure 15.3) with $F(s) = F_1(s)$ and $G(s) = \frac{F_2(s)}{F_1(s)}e^{-s\tau}$. This formalization allows us to use the classical results of control theory to analyze this class of time-delay systems, by inspecting the closed-loop properties of the transfer function $F(s)G(s)$.

Example 15.2

Let us consider the following transfer function

$$\tilde{G}(s) = \frac{1}{(s+1)^2 + (s+2)e^{-s}} \tag{15.17}$$

It can be rewritten as

$$\tilde{G}(s) = \frac{\frac{1}{(s+1)^2}}{1 + \frac{(s+2)}{(s+1)^2}e^{-s}} \tag{15.18}$$

The Nyquist plot, shown in Figure 15.4, does not encircle the critical point and therefore the quasi-polynomial admits only roots that are in the left-half plane.
To confirm this results, we calculate the phase margin from the Bode diagram, obtaining

$m_\phi = 46^\circ$, which allows us to conclude that the system $\tilde{G}(s)$ is asymptotically stable and the quasi-polynomial does not have roots for $\sigma > 0$.

Now, we present some classical examples that are reported in the literature which confirm how critical can be the analysis of the stability of time-delay systems. The analysis of these examples will be made by using also MATLAB.

Example 15.3

Let us consider the following transfer function

$$\tilde{G}(s) = \frac{1}{s + 1 + se^{-s}} \tag{15.19}$$

Here, $p(s) = s+1+se^{-s}$. Letting $p(s) = 0$ yields $\frac{s+1}{s} = -e^{-s}$. Considering $s = \sigma + j\omega$, the equation does not admits solution for $\sigma > 0$. Moreover, for $\sigma = 0$ it follows:

$$-\frac{j\omega + 1}{j\omega} = e^{-j\omega} \tag{15.20}$$

whose modulus yields

$$\frac{\sqrt{\omega^2 + 1}}{\omega} = 1 \tag{15.21}$$

that admits infinite solutions for $\omega \to \infty$. This leads to have solutions on the imaginary axis with multiplicity greater than 1, so that the system is unstable.
This can also be shown by computing the H_∞ norm of the system $\tilde{G}(s)$ by considering the denominator $p(j\omega) = (j\omega + 1) + j\omega(\cos(\omega) - j\sin(\omega))$ and using the following MATLAB commands

```
for o=0.001:0.001:20
R=o*sin(o)+1;
I=o*cos(o)+o;
M=R^2+I^2;
plot(o,M,'.r');
hold on
end
```

The minimum of the modulus tends to zero, as shown in Fig. 15.5, therefore the H_∞ norm tends to infinity, thus $\tilde{G}(s)$ does not belong to the $\mathbb{H}^\infty$ class.
Moreover, if we consider a Padé approximation of the time-delay e^{-s} of order $n_P = 50$, the impulse response calculated with MATLAB oscillates, as shown in Figure 15.6, thus showing that the system is not asymptotically stable.
If we use the Nyquist plot of $\frac{s}{s+1}e^{-s}$, we cannot draw conclusions on the stability since the critical point is crossed by the plot.

Example 15.4

Let us consider the system with transfer function

$$\tilde{G}(s) = \frac{1}{(s+1)\left[(s+1) + se^{-s}\right]} \tag{15.22}$$

Even if the poles of $\tilde{G}(s)$ are those of the previous example with a further pole in $s = -1$, this example is quite different as the two systems have a different behavior for $\omega \to \infty$. The transfer function in (15.22), in fact, belongs to the $\mathbb{H}^\infty$ class as shown by using the following MATLAB commands which calculate the magnitude of the polynomial $p(j\omega) = (j\omega + 1)(j\omega + 1 + j\omega(\cos(\omega) - j\sin(\omega)))$, that is the denominator of $\tilde{G}(s)$.

```
for o=0.001:0.001:20
   R=o*sin(o)+1-o^2*cos(o)-o^2;
   I=o*cos(o)+2*o+o^2*sin(o);
   M=R^2+I^2;
```

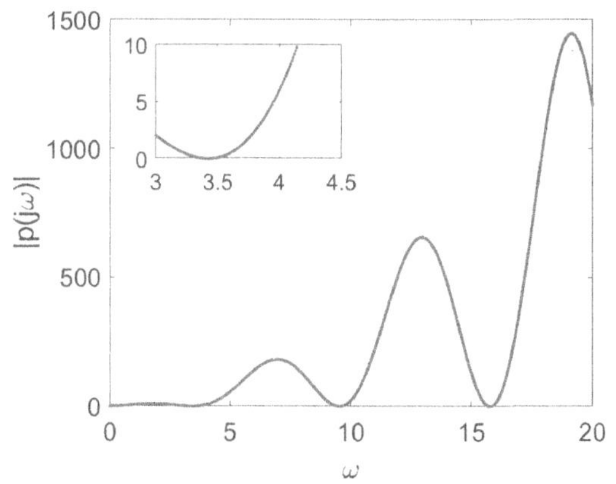

FIGURE 15.5
Magnitude of the denominator $p(j\omega)$ of the transfer function $\tilde{G}(s)$ as in (15.18). The zoom in the inset allows to verify that the magnitude reaches the minimum value of 0, thus leading to an infinite H_∞ norm.

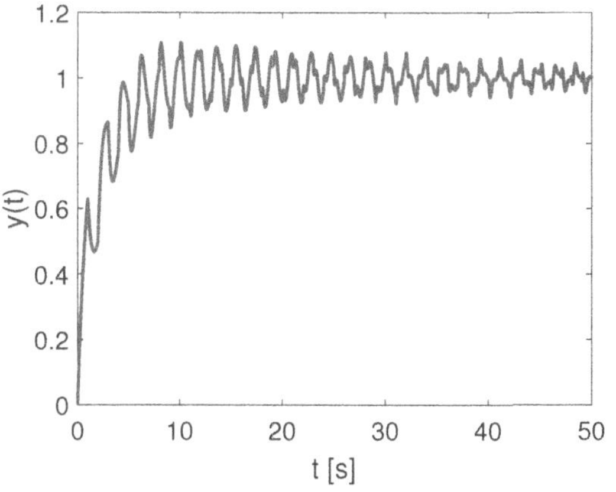

FIGURE 15.6
Unit step response of $\tilde{G}(s)$ as in (15.18) with a 50-th order Padé approximation of the time-delay e^{-s}.

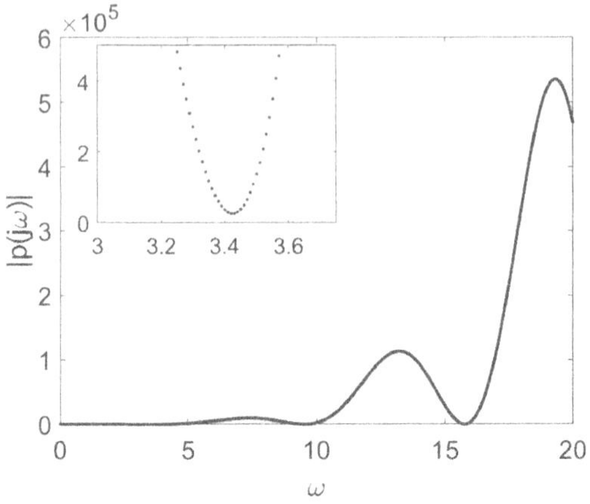

FIGURE 15.7
Magnitude of the denominator $p(j\omega)$ of transfer function $\tilde{G}(s)$ as in (15.22). The zoom in the inset allows to verify that the magnitude reaches the minimum value of 0.25, thus leading to $|\tilde{G}(s)|_\infty \approx 4$.

```
   plot(o,M,'.b');
   hold on
end
```

We consider now the Padé approximation of the time-delay e^{-s} of order $n_P = 50$ and calculate the impulse response with MATLAB as reported in Fig. 15.8, thus showing that the system is stable. However, this example is very sensitive to the parameters and the stability is fragile.

Example 15.5

Let us consider the following transfer function

$$\tilde{G}(s) = \frac{1}{(\frac{s}{a}+1)\left[(s+1)+se^{-s}\right]} \tag{15.23}$$

This example generalizes the previous one with the introduction of the parameter a, which indicates the location of a further pole. The fragility of the stability depends on a since, even if $\tilde{G}(s)$ belongs to the $\mathbb{H}^\infty$ class, the parameter a modulates the distance of the roots from the imaginary axis. The following MATLAB code allows to calculate the minimum magnitude of the polynomial $p(j\omega, a) = (\frac{s}{a}+1)\left[(s+1)+s(\cos(\omega)-j\sin(\omega))\right]$ for different values of the parameter a.

```
for k=1:20
   a=.1*k
   for i=1:20000
      o=0.001*i;
      R=-o^2/a-o^2/a*cos(o)+1+o*sin(o);
      I=o/a+o^2/a*sin(o)+o+o*cos(o);
      p=R^2+I^2;
      H(i)=p;
   end
   M=min(H);
   V(k)=M;
```

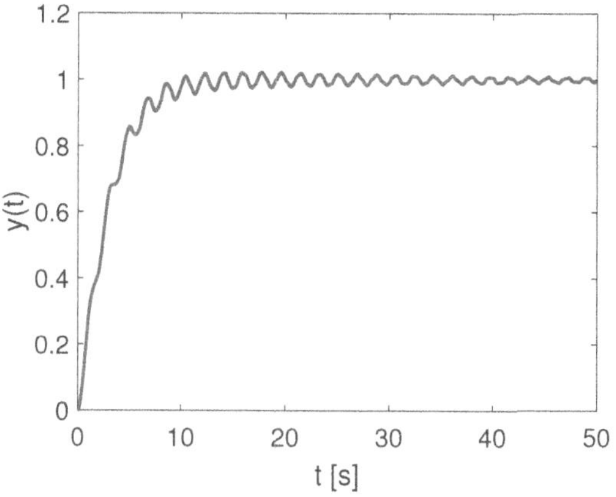

FIGURE 15.8
Unit step response of $\tilde{G}(s)$ as in (15.22) with a 50-th order Padé approximation of the time-delay e^{-s}.

```
end
plot(0.1:0.1:2,V)
```

The curve reported in Figure 15.9 shows that for increasing values of α the minimum tends towards zero, thus leading to an infinite H_∞ norm.

15.5 Direct Method

This section is devoted to illustrate the *direct method* that provides a criterion to evaluate the stability of quasi-polynomials of the form:

$$p(s,\tau) = p_0(s) + p_1(s)e^{-s\tau} \tag{15.24}$$

under the hypothesis that the system without delay is stable, that is, the roots of $p(s,0)$ are all in the left-half plane. The limit value of the delay, τ^*, for which the delay system is stable, is found by considering when the characteristic equation $p(s,\tau) = 0$ has some solution on the imaginary axis. Thanks to the complex conjugate symmetry of the complex roots, this root is also a solution of the equation $p(-s,\tau) = 0$ for the same value of τ. Using this result, one can look for simultaneous solutions of $p(s,\tau) = 0$ and $p(-s,\tau) = 0$ for $s = j\omega$:

$$\begin{aligned} p_0(j\omega) + p_1(j\omega)e^{-j\omega\tau} = 0 \\ p_0(-j\omega) + p_1(-j\omega)e^{j\omega\tau} = 0 \end{aligned} \tag{15.25}$$

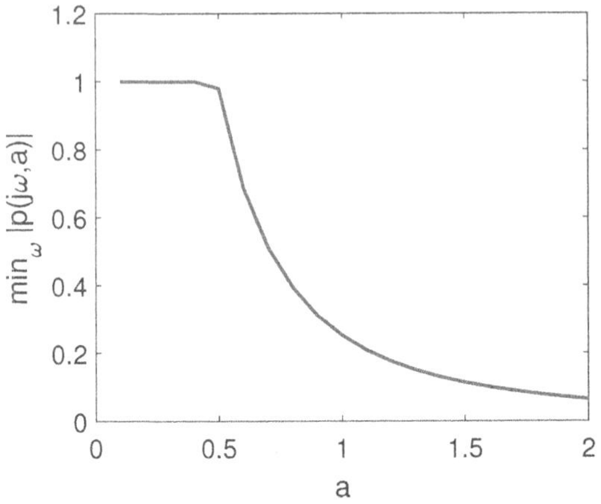

FIGURE 15.9
Minimum over ω of the magnitude of the polynomial $p(j\omega)$ for different values of the parameter a.

that prompts for the derivation of the exponential term as follows:

$$e^{j\omega\tau} = -\frac{p_0(-j\omega)}{p_1(-j\omega)} \tag{15.26}$$

Substituting this expression in (15.25), we get

$$p_0(j\omega)p_0(-j\omega) - p_1(j\omega)p_1(-j\omega) = 0 \tag{15.27}$$

or equivalently

$$|p_1(j\omega)|^2 - |p_0(j\omega)|^2 = 0 \tag{15.28}$$

When equation (15.27) (or equation (15.28)) admits a solution, say $\bar{\omega}$, then, the limit delay for stability can be derived from (15.26) with $\omega = \bar{\omega}$. Otherwise, if equation (15.27) (or equation (15.28)) does not have any solution, then the system with characteristic equation $p(s,\tau) = 0$ is stable for any value of τ.

Example 15.6

Consider the following time-delay system

$$\dot{x}(t) = -ax(t) - bx(t-\tau) \tag{15.29}$$

with $a > 0$ and $b > 0$. If we take the Laplace transform of the previous equation, we find

$$(s + a + be^{-s\tau})X(s) = 0 \tag{15.30}$$

Therefore, system (15.33) has a characteristic quasi-polynomial $p(s,\tau)$ of the form (15.24) with $p_0(s) = s + a$ and $p_1(s) = b$. Let us then study for which values of the parameters a and b the system is delay-independent stable.

For system (15.33), equation (15.27) reads:

$$(j\omega + a)(-j\omega + a) - b^2 = 0 \tag{15.31}$$

and hence

$$a^2 + \omega^2 - b^2 = 0 \tag{15.32}$$

that has no solution if $a > b > 0$.
Consider now the transfer function $F(s) = \frac{p_1(s)}{p_0(s)} = \frac{b}{s+a}$ and notice that, if $a > b > 0$, then $F(s)$ is strictly bounded-real, $|F(j\omega)| = |\frac{b}{j\omega+a}| < 1 \ \forall\omega$. As we will discuss below, this result holds in general.

Consider the control scheme in Figure 15.10. If $F(s)$ is set as $F(s) = \frac{p_1(s)}{p_0(s)}$, then the characteristic equation of this feedback configuration is $p(s,\tau) = 0$ with $p(s,\tau)$ as in (15.24).

In order to assess when the feedback system in Figure 15.10 is time-independent stable, the small gain theorem can be applied. Consequently, if $F(s)$ is strictly bounded-real, then we can conclude that $p(s,\tau) = p_0(s) + p_1(s)e^{-s\tau}$ is time-independent stable.

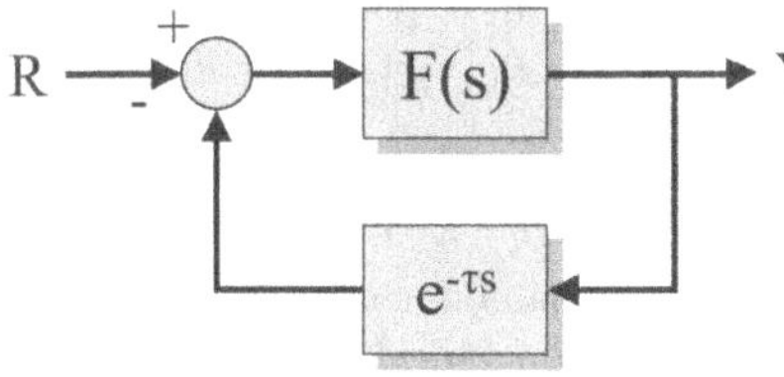

FIGURE 15.10
Feedback scheme having the characteristic quasi-polynomial (15.24) when $F(s) = \frac{p_1(s)}{p_0(s)}$.

Example 15.7

Let us consider the following time-delay system:

$$\dot{\mathbf{x}}(t) = \begin{bmatrix} 0 & 1 \\ -1 & -k \end{bmatrix} \mathbf{x}(t) + \begin{bmatrix} 0 & 0 \\ 0 & -1 \end{bmatrix} \mathbf{x}(t-\tau) \tag{15.33}$$

Taking the Laplace transform of the previous equation, we get:

$$s\mathbf{X}(s) = \begin{bmatrix} 0 & 1 \\ -1 & -k \end{bmatrix} \mathbf{X}(s) + \begin{bmatrix} 0 & 0 \\ 0 & -1 \end{bmatrix} e^{-s\tau}\mathbf{X}(s) \tag{15.34}$$

and so:

$$\left\{ sI - \begin{bmatrix} 0 & 1 \\ -1 & -k \end{bmatrix} - \begin{bmatrix} 0 & 0 \\ 0 & -1 \end{bmatrix} e^{-s\tau} \right\} \mathbf{X}(s) = 0 \tag{15.35}$$

The characteristic quasi-polynomial can be computed as follows:

$$p(s,\tau) = \det \begin{bmatrix} s & -1 \\ 1 & s + k + e^{-s\tau} \end{bmatrix} \tag{15.36}$$

and hence

$$p(s, \tau) = s^2 + sk + se^{-s\tau} + 1 \tag{15.37}$$

The quasi-polynomial (15.37) is in the form (15.24) with $p_0(s) = s^2 + sk + 1$ and $p_1(s) = s$. Notice that $p(s, 0)$ is Hurwitz if $k > 0$.
Therefore, to derive the condition on k yielding a delay-independent stable system, we have to consider

$$F(s) = \frac{s}{s^2 + sk + 1} \tag{15.38}$$

and study when it is strictly bounded-real. To this aim, let us consider a state-space representation of $F(s)$ as follows:

$$\mathrm{A} = \begin{bmatrix} 0 & 1 \\ -1 & -k \end{bmatrix}; \mathrm{B} = \begin{bmatrix} 0 \\ 1 \end{bmatrix}; \mathrm{C} = \begin{bmatrix} 0 & 1 \end{bmatrix}; \mathrm{D} = 0 \tag{15.39}$$

and calculate the Hamiltonian matrix associated to the bounded-real lemma:

$$\mathrm{H} = \begin{bmatrix} \mathrm{A} & \mathrm{BB}^T \\ -\mathrm{C}^T\mathrm{C} & -\mathrm{A}^T \end{bmatrix} \tag{15.40}$$

We obtain:

$$\mathrm{H} = \begin{bmatrix} 0 & 1 & 0 & 0 \\ -1 & -k & 0 & 1 \\ 0 & 0 & 0 & 1 \\ 0 & -1 & -1 & k \end{bmatrix} \tag{15.41}$$

The characteristic polynomial of H is:

$$\det(\lambda \mathrm{I} - \mathrm{H}) = \lambda^4 + (3 - k)^2\lambda^2 + 1 \tag{15.42}$$

Studying the roots of this polynomial, one finds that, for $k > 1$, H has no eigenvalues on the imaginary axis. Consequently, $F(s)$ is bounded-real for $k > 1$ and system (15.33) is delay-independent stable.

Example 15.8

Consider now the time-delay system:

$$\dot{\mathbf{x}}(t) = \mathrm{A}_0\mathbf{x}(t) + \mathrm{A}_1\mathbf{x}(t - \tau) \tag{15.43}$$

with

$$\mathrm{A}_0 = \begin{bmatrix} -2 & 0 \\ 0.5 & -2 \end{bmatrix}; \mathrm{A}_1 = \begin{bmatrix} 0 & 0.5 \\ 0 & 0 \end{bmatrix} \tag{15.44}$$

Let us compute the characteristic quasi-polynomial:

$$p(s, \tau) = \det(s\mathrm{I} - \mathrm{A}_0 - \mathrm{A}_1 e^{-s\tau}) \tag{15.45}$$

We get:

$$p_0(s) = s^2 + 4s + 4 \tag{15.46}$$

and

$$p_1(s) = -\frac{1}{4} \tag{15.47}$$

The transfer function $F(s) = -\frac{\frac{1}{4}}{s^2+4s+4}$ is strictly bounded-real and, therefore, the system is delay-independent stable.
Another criterion to assess delay-independent stability is now presented. According to

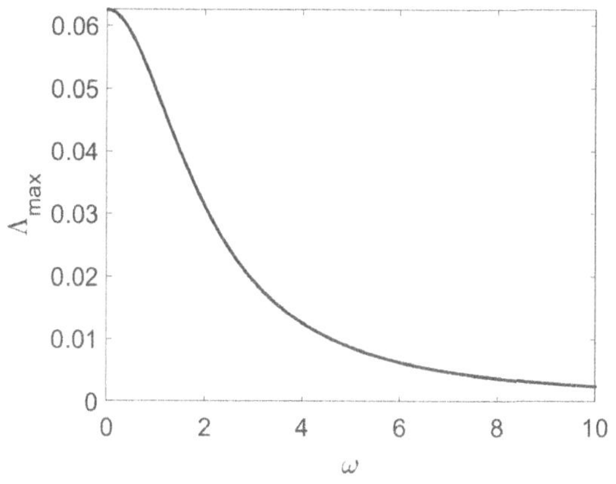

FIGURE 15.11
Trend of the maximum eigenvalue Λ_{max} of $(j\omega\mathrm{I} - \mathrm{A}_0)^{-1}\mathrm{A}_1$. Since it is strictly less than 1 for any ω, the system (15.43) is delay-independent stable.

this criterion, one has to check if the maximum eigenvalue Λ_{max} of $(j\omega\mathrm{I} - \mathrm{A}_0)^{-1}\mathrm{A}_1$ is strictly less than one, for any ω, provided that A_0 and $\mathrm{A}_0 + \mathrm{A}_1$ are Hurwitz.
This criterion can be applied to system (15.43) with the following MATLAB code:

```
for o=0:0.01:10
   AA=inv(j*o*eye(2)-A0)*A1;
   L=max(abs(eig(AA)));
   plot(o,L,'b.');
   hold on
end
```

The trend of $\Lambda_{max}(\omega)$ obtained in this case is reported in Fig. 15.11.

Example 15.9

Let us consider the following time-delay system:

$$\dot{\mathbf{x}}(t) = \begin{bmatrix} 0 & 1 \\ -1 & -k \end{bmatrix} \mathbf{x}(t) + \begin{bmatrix} 0 & 0 \\ 0 & -1 \end{bmatrix} \mathbf{x}(t-\tau) + \mathbf{u}(t) \tag{15.48}$$

with $k = 0.1$. As shown in Example 15.7, for this value of k, the system is not delay-independent stable if $\mathbf{u}(t) = 0$. Is it possible to find a control action $\mathbf{u}(t) = -L\mathbf{x}(t)$, with L a scalar quantity, such that the controlled system is delay-independent stable? To address this problem, let us consider the controlled system

$$\dot{\mathbf{x}}(t) = \begin{bmatrix} 0 & 1 \\ -1 & -k \end{bmatrix} \mathbf{x}(t) + \begin{bmatrix} 0 & 0 \\ 0 & -1 \end{bmatrix} \mathbf{x}(t-\tau) - \begin{bmatrix} L & 0 \\ 0 & L \end{bmatrix} \mathbf{x}(t) \tag{15.49}$$

and calculate its characteristic quasi-polynomial:

$$p(s,\tau) = \det(s\mathrm{I} - \mathrm{A}_0 - \mathrm{A}_1 e^{-s\tau} + L\mathrm{I}) \tag{15.50}$$

We obtain:

$$p_0(s) = s^2 + s\left(\frac{1}{10} + 2L\right) + \left(1 + L^2 + \frac{L}{10}\right) \tag{15.51}$$

and

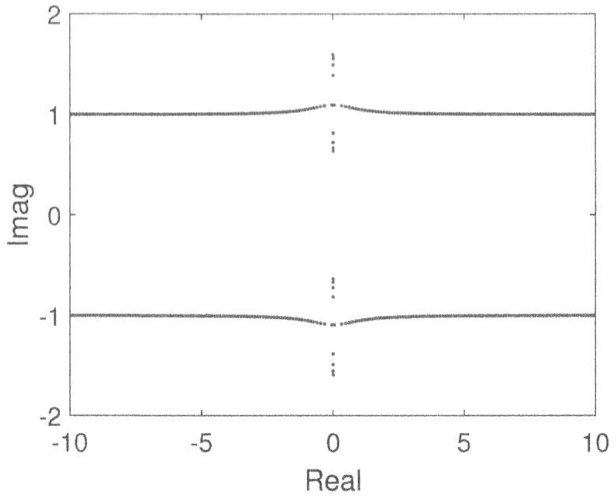

FIGURE 15.12
Closed-loop eigenvalues for the controlled system (15.49) at different values of the controller gain L.

$$p_1(s) = s + L \tag{15.52}$$

Solving the control problem is, therefore, equivalent to find L such that the function $F(s) = \frac{p_1(s)}{p_0(s)}$ is bounded-real. To this aim, we can use the following MATLAB commands:

```
for l=10:-.1:0.1
    A=[0 1;(-1-l^2-l/10) (-1/10 -2*l)];
    B=[0;1];
    C=[l 1];
    H=[A B*B';-C'*C -A'];
    cl_eig=eig(H)
    plot(real(cl_eig),imag(cl_eig),'b.')
    hold on
end
```

Since the eigenvalues of the Hamiltonian matrix associated to the bounded-real lemma lie on the imaginary axis for $L \leq 0.4$, then, as shown in Figure 15.12, the system is delay-independent stable for $L \geq 0.4$.

Let us fix for example $L = 0.5$ and check that the controlled system is delay-independent stable with the direct method. For this value of L, we have:

$$p_0(s) = s^2 + \frac{11}{10}s + 1.3 \tag{15.53}$$

and

$$p_1(s) = s + 0.5 \tag{15.54}$$

Application of the direct method requires that equation (15.28) has no solution. For this system, equation (15.28) becomes:

$$(1.3 - \omega^2)^2 + \frac{121}{100}\omega^2 - \left(\frac{1}{4} + \omega^2\right) = 0 \tag{15.55}$$

and thus

$$\omega^4 - \frac{239}{100}\omega^2 + \frac{36}{25} = 0 \tag{15.56}$$

that has no real solution.

It is important to remark that, in the previous example, the problem of delay-independent stability is solved with a control law with a single parameter.

The result is valid for systems with commensurate delays, as stated in the following general theorem.

Theorem 38 *Let us consider a commensurate time-delay system as in equation (15.10) with* A_0 *and* $A_0 + \sum_{i=1}^{q} A_i$ *stable matrices. Then, there exists a linear state-space regulator* $\mathbf{u}(t) = -L\mathbf{x}(t)$ *with scalar gain* L *such that the controlled system*

$$\dot{x}(t) = \bar{A}_0 x(t) + \sum_{i=1}^{q} A_i x(t - i\tau) \tag{15.57}$$

where $\bar{A}_0 = A_0 - LI$ *is delay-independent stable.*

15.6 Exercises

1. Propose an approximation for $G(s) = e^{-s\tau}$ with $\tau = 1$, $\tau = 0.5$ and $\tau = 0.1$ by using the Padé method and numerically compute the H_∞ norm of the error between $G(s)$ and its approximation.
2. Determine, if possible, the compensator $C(s)$ that stabilizes the closed-loop system of Figure 15.13 with $G(s) = \frac{k}{s}$, $k = 1$ and $\tau = 2$.

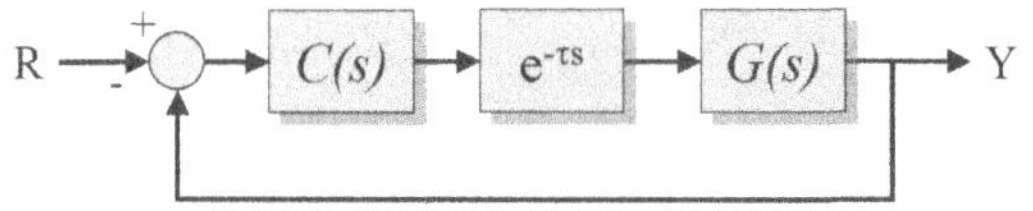

FIGURE 15.13
Block scheme for Exercise 2.

3. Consider the block scheme in Figure 15.14 with $G(s) = \frac{1}{s+\alpha}$. Determine the value of α such that the system is asymptotically stable.
4. Consider the system in Figure 15.15 with $G(s) = \frac{1}{(s+1)^3}$ and determine for which values of τ the system is asymptotically stable.

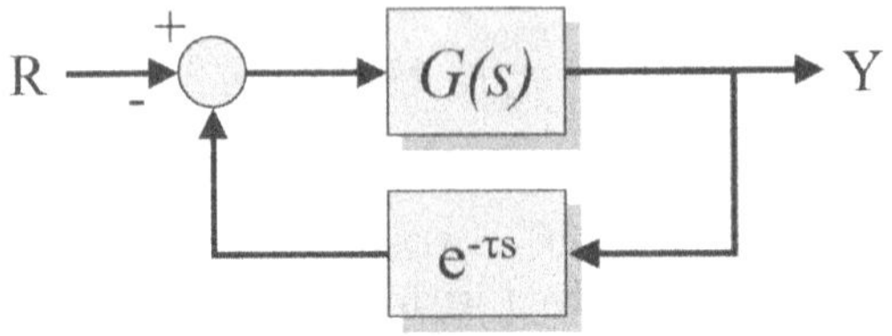

FIGURE 15.14
Block scheme for Exercise 3.

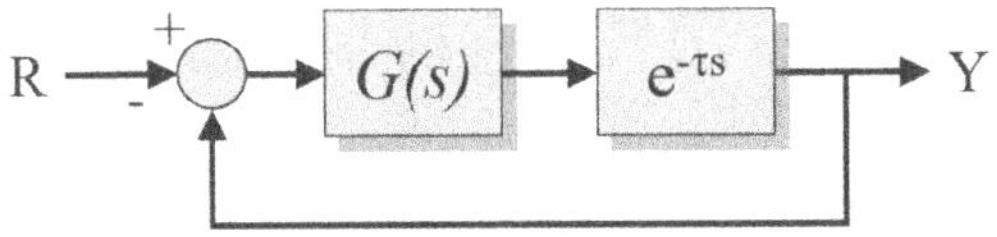

FIGURE 15.15
Block scheme for Exercises 4 and 5.

5. With reference to system in Figure 15.15 with $G(s) = \frac{k}{(s+1)^3}$ and $\tau = 0.1$, determine for which values of k the feedback system is asymptotically stable.

Recommended Essential References

We report here a list of recommended essential references for each chapter of the book. The reader can find an extensive discussion of fundamental results that in our book have been only briefly summarized as well as new results and techniques to deepen the knowledge on the given topic.

Chapter 1

1. G. Franklin, J. D. Powell, A. Emami-Naeini, *Feedback Control of Dynamic Systems*, 5th Edition, Prentice Hall, Englewood Cliffs, NJ, 2004.
2. J. C. Doyle, B. A. Francis, A. R. Tannenbaum, *Feedback Control Theory*, Macmillan Publishing Company, New York, 1992.
3. W. S. Levine, *The Control Handbook*, CRC Press and IEEE Press, Boca Raton, 1996.
4. S. P. Bhattacharyya, H. Chapellat, L. H. Keel, *Robust Control. The Parametric Approach*, Prentice Hall PTR, 2000.
5. K. Zhou, J. C. Doyle *Essentials of Robust Control*, Prentice Hall PTR, 1999.

Chapter 2

1. T. Kailath, *Linear Systems*, Prentice Hall, Englewood Cliffs, NJ, 1979.
2. D. G. Schultz, J. L. Melsa, *State Functions and Linear Control Systems*, McGraw-Hill, New York, 1967.
3. V. L. Kharitonov, "Asymptotic stability of an equilibrium position of a family of systems of linear differential equations," *Differential'nye Uraveniya*, 1978, 14, 83–1485.

4. R. C. Dorf, R. H. Bishop, *Modern Control Systems*, 12th Edition, Prentice Hall, Englewood Cliffs, NJ, 2010.
5. W. S. Levine, *The Control Handbook*, CRC Press and IEEE Press, Boca Raton, 1996.

Chapter 3

1. H. Kwakernaak, R. Sivan, *Linear Optimal Control Systems*, John Wiley & Sons, 1972.
2. W. J. Rugh, *Linear System Theory*, Prentice Hall, Englewood Cliffs, NJ, 1966.
3. T. Kailath, *Linear Systems*, Prentice Hall, Englewood Cliffs, NJ, 1979.
4. C. Piccardi, S. Rinaldi, *I sistemi lineari: teoria, modelli, applicazioni*, Città Studi, Biella, 1998.
5. N. Wiener, *Nonlinear Problems in Random Theory*, The Massachusetts Institute of Technology, 1958.

Chapter 4

1. G. H. Golub, C. F. Van Loan, *Matrix Computations*, The Johns Hopkins University Press, Baltimore, MD, 1996.
2. T. Kailath, *Linear Systems*, Prentice Hall, Englewood Cliffs, NJ, 1979.
3. W. S. Levine, *The Control Handbook*, CRC Press and IEEE Press, Boca Raton, 1996.

Chapter 5

1. K. Zhou K., J. C. Doyle, K. Glover, *Robust and Optimal Control*, Prentice Hall, Upper Saddle River, 1996.

2. F. W. Fairman, *Linear Control Theory: The State Space Approach*, John Wiley & Sons, West Sussex, England, 1998.
3. R. A. Roberts, C. T. Mullis, *Digital Signal Processing*, Addison-Wesley Publishing Company, 1987.
4. C. K. Chui, G. Chen, *Discrete H_∞ Optimization*, 2nd Edition, Springer, London, 1997.

Chapter 6

1. B. Moore, "Principal component analysis in linear systems: Controllability, observability, and model reduction," *IEEE Trans. Automatic Control*, 1981, 26(1), 17–32.
2. G. Obinata, B. D. O. Anderson, *Model Reduction for Control System Design*, Springer-Verlag, London, 2001.
3. M. Green, D. J. N. Limebeer, *Linear Robust Control*, Prentice Hall, Englewood Cliffs, NJ, 1995.
4. F. Fagnani, J. Willems, "Representations of symmetric linear dynamical systems," *SIAM Journal on Control and Optimization archive*, 31(5), 1993.
5. L. Fortuna, A. Gallo, C. Guglielmino, G. Nunnari, "On the solution of a nonlinear matrix equation for MIMO realizations," *Systems and Control Letters*, 1988, 11, 79–82.
6. L. Fortuna, G. Nunnari, A. Gallo, *Model Order Reduction Techniques With Applications in Electrical Engineering*, Springer-Verlag, London, 1992.

Chapter 7

1. B. D. O. Anderson, J.B. Moore, *Optimal Control. Linear Quadratic Methods*, Prentice-Hall, Englewoods Cliff, NJ, 1989.
2. H. Kwakernaak, R. Sivan, *Linear Optimal Control Systems*, John Wiley & Sons, 1972.
3. T. Kailath, *Linear Systems*, Prentice Hall, Englewood Cliffs, NJ, 1979.

4. G. C. Goodwin, S. F. Graebe, M. E. Salgado, *Control Systems Design*, Valparaiso, 2000.
5. A. Sinha, *Linear Systems: Optimal and Robust Control*, CRC Press, Boca Raton, 2007.
6. D. Kleinman, "On an iterative technique for Riccati equation computations," *IEEE Trans. Automatic Control*, 1968, 13(1), 114–115.
7. K. J. Astrom, R. M. Murray, *Feedback Systems. An Introduction for Scientists and Engineers*, Princeton University Press, Princeton, NJ, 2008.
8. W. S. Levine, *The Control Handbook*, CRC Press and IEEE Press, Boca Raton, 1996.

Chapter 8

1. E. A. Jonckheere, L. M. Silverman, "A new set of invariant for linear systems. Applications to reduced order compensator design," *IEEE Trans. Automatic Control*, 1981, 26(1), 17–32.
2. L. Fortuna, G. Nunnari, A. Gallo, *Model Order Reduction Techniques With Applications in Electrical Engineering*, Springer-Verlag, London, 1992.
3. L. Fortuna, G. Muscato, G. Nunnari, "Closed-loop balanced realizations of LTI discrete-time systems," *Journal of the Franklin Institute*, 1993, 330(4), 695–705.
4. K. J. Astrom, R. M. Murray, *Feedback Systems. An Introduction for Scientists and Engineers*, Princeton University Press, Princeton, NJ, 2008.

Chapter 9

1. B. D. O. Anderson, S. Vongpanitlerd, *Network Analysis and Synthesis: A Modern Systems Theory Approach*, Dover Publications Inc., New York, 2006.
2. S. Boyd, L. El Ghaoui, E. Feron, V. Balakrishnan, *Linear Matrix Inequalities in System and Control Theory*, SIAM Studies in Applied Mathematics, Philadelphia, PA, 1994.

3. I. R. Petersen, A. Lanzon (2010). "Feedback control of negative-imaginary systems". *IEEE Control Systems Magazine*, 30(5), 54–72.
4. A. Ferrante, A. Lanzon, L. Ntogramatzidis (2017). "Discrete-time negative imaginary systems". *Automatica*, 79, 1–10.

Chapter 10

1. A. Buscarino, L. Fortuna, M. Frasca, M. G. Xibilia (2012). "An analytical approach to one-parameter MIMO systems passivity enforcement". *International Journal of Control*, 85(9), 1235–1247.
2. A. Buscarino, L. Fortuna, M. Frasca (2016). "Forward action to make a system negative imaginary". *Systems & Control Letters*, 94, 57–62.
3. M. Bucolo, A. Buscarino, L. Fortuna, M. Frasca (2019). "Forward action to make time-delay systems positive-real or negative-imaginary", *Systems & Control Letters*, 131, 104495.

Chapter 11

1. R. T. Stefani, B. Shahian, C. J. Savant, G. H. Hostetter, *Design of Feedback Control Systems*, Oxford Series in Electrical and Computer Engineering, Oxford University Press, Oxford, 2001.
2. P. Dorato, R. K. Yedavalli, *Recent Advances in Robust Control*, IEEE Press, New York, 1990.
3. S. Boyd, V. Balakrishnan, P. Kabamba, "A Bisection Method for Computing the H_∞ Norm of a Transfer Matrix and Related Problems," *Math. Control Signals Systems*, 1989, 2, 207–219.
4. P. Dorato, L. Fortuna, G. Muscato, *Robust Control for Unstructured Perturbations-An Introduction*, Lecture Notes in Control and Information Sciences, Springer-Verlag, London, 1992.
5. R. Chiang, M. G. Safonov, G. Balas, A. Packard, *Robust Control Toolbox*, 3rd Edition, The Mathworks Inc., Natick, MA, 2007.
6. B. A. Francis, *A course in H_∞ control theory*, Lecture Notes in Control and Information Sciences, Springer-Verlag, Berlin, 1987.

7. J. C. Doyle, K. Glover, P. P. Khargonekar, B. A. Francis, "State-space solutions to standard H_2 and H_∞ control problems", *IEEE Trans. Automatic Control*, 1989, 34, 831–847.
8. W. S. Levine, *The Control Handbook*, CRC Press and IEEE Press, Boca Raton, 1996.

Chapter 12

1. S. Boyd, L. El Ghaoui, E. Feron, V. Balakrishnan, *Linear Matrix Inequalities in System and Control Theory*, SIAM Studies in Applied Mathematics, Philadelphia, PA, 1994.
2. R. E. Skelton, T. Iwasaki, K. M. Grigoriadis, *A Unified Algebraic Approach to Linear Control Design*, Taylor & Francis, 1998.
3. P. Gahinet, A. Nemirovski, A. J. Laub, M. Chilali, *LMI Control Toolbox. For Use with MATLAB®*, The Mathworks Inc., Natick, MA, 1995.
4. Y. Nesterov and A. Nemirovski, *Interior Point Polynomial Algorithms in Convex Programming*, Studies in Applied Mathematics, SIAM, 1993.
5. U. Helmke, J. B. Moore, *Optimization and Dynamical Systems*, Springer-Verlag, London, 1996.
6. L. Fortuna, M. Frasca, M. G. Xibilia, *Chua's Circuit Implementations: Yesterday, Today and Tomorrow*, World Scientific, Singapore, 2009.
7. W. S. Levine, *The Control Handbook*, CRC Press and IEEE Press, Boca Raton, 1996.
8. L. Fortuna, A. Gallo, G. Nunnari, "A self-tuning adaptive control implementation by using a transputer-based parallel architecture", *Trans. Inst. MC*, 1990, 12, 156–164.

Chapter 13

1. D. C. Youla, J. J. Bongiorno, C. N. Lu, *Single-loop Feedback Stabilization of Linear Multivariable Dynamical Plants*, American Math Society, Colloquium Publications, vol. XX, Providence, RI, 1956.

2. D. C. Youla, H. A. Jabr, J. J. Bongiorno, "Modern Wiener-Hopf design of optimal controllers. Part I: the single-input-output case," *IEEE Trans. Automatic Control*, 1974, 10, 159–173.
3. P. Dorato, L. Fortuna, G. Muscato, *Robust Control for Unstructured Perturbations-An Introduction*, Lecture Notes in Control and Information Sciences, Springer-Verlag, London, 1992.
4. J. C. Doyle, B. A. Francis, A. R. Tannenbaum, *Feedback Control Theory*, Macmillan Publishing Company, New York, 1992.
5. B. D. O. Anderson, J.B. Moore, *Optimal Control. Linear Quadratic Methods*, Prentice-Hall, Englewood Cliffs, NJ, 1989.
6. W. S. Levine, *The Control Handbook*, CRC Press and IEEE Press, Boca Raton, 1996.

Chapter 14

1. U. Helmke, and J. B. Moore. *Optimization and dynamical systems.* Springer Science & Business Media, 2012.
2. A. Cichocki, R. Unbehauen, and R. W. Swiniarski. *Neural networks for optimization and signal processing.* Vol. 253. New York: Wiley, 1993.

Chapter 15

1. C. V. Hollot, V. Misra, D. Towsley, W. Gong, (2002). "Analysis and design of controllers for AQM routers supporting TCP flows". *IEEE Transactions on Automatic Control*, 47(6), 945–959.
2. K. Gu, J. Chen, and V. L. Kharitonov, (2003). *Stability of time-delay systems*, Springer Science & Business Media.
3. V. Kharitonov, (2012). *Time-delay systems: Lyapunov functionals and matrices*, Springer Science & Business Media.
4. J. E. Marshall, H. Gorecki, A. Korytowski, and K. Walton, (1992). *Time-Delay Systems: Stability and Performance Criteria with Applications.* London: Ellis Horwood.
5. L. Mirkin, *Time Delays in Control Systems*, available online at url "http://leo.technion.ac.il/Courses/TDCE/TDCSnotes.pdf"

6. L. Belhamel, A. Buscarino, L. Fortuna, M.G. Xibilia, (2020). "Delay Independent Stability Control for Commensurate Multiple Time-Delay Systems", *IEEE Control Systems Letters*, 5(4), 1249–1254.

Appendix A. Norms

In this Appendix the main norms used in the text or, in any case, significant for the control theory, are summarized.

Norm of a Vector

The norm of a vector $\mathbf{x} \in \mathbb{C}^n$ is a function $f : \mathbb{C}^n \to \mathbb{R}$ satisfying the following properties:

- $f(\mathbf{x}) \geq 0$;
- $f(\mathbf{x}) = 0 \leftrightarrow \mathbf{x} = 0$;
- $f(\mathbf{x} + \mathbf{y}) \leq f(\mathbf{x}) + f(\mathbf{y})$ with $\mathbf{x}$ and $\mathbf{y} \in \mathbb{C}^n$;
- $f(\alpha \mathbf{x}) = |\alpha| f(\mathbf{x})$ with $\alpha \in \mathbb{R}$.

Let $\mathbf{x} = \begin{bmatrix} x_1 \\ x_2 \\ \vdots \\ x_n \end{bmatrix}$, the 1-norm is:

$$\|\mathbf{x}\|_1 = \sum_{i=1}^{n} |x_i| \tag{A.1}$$

The 2-norm is instead defined by:

$$\|\mathbf{x}\|_2 = \left(\sum_{i=1}^{n} |x_i|^2 \right)^{\frac{1}{2}} \tag{A.2}$$

The 1-norm and the 2-norm are special cases of a more general family of norms, the p-norm of a vector:

$$\|\mathbf{x}\|_p = \left(\sum_{i=1}^{n} |x_i|^p \right)^{\frac{1}{p}} \tag{A.3}$$

Another commonly used norm of the p-norm family is the ∞-norm defined as:

$$||\mathbf{x}||_\infty = \max_i |x_i| \tag{A.4}$$

Norm of a Matrix

By considering the role of matrices as linear operators, the norm of a matrix can be defined extending the norm of a vector to matrices. The norm is said to be an *induced norm*, since it depends on the choice of the vector norm. The p-norm (induced by the vector p-norm) is defined as follows:

$$||\mathrm{A}||_p = \sup_{\mathbf{x}\neq 0} \frac{||\mathrm{A}\mathbf{x}||_p}{||\mathbf{x}||_p} = \max_{||\mathbf{x}||=1} ||\mathrm{A}\mathbf{x}||_p \tag{A.5}$$

The most commonly used norms are the 1-norm, the 2-norm and the ∞-norm, defined as follows:

$$||\mathrm{A}||_1 = \max_j \sum_{i=1}^{n} |a_{ij}| \tag{A.6}$$

$$||\mathrm{A}||_2 = \sigma_1(\mathrm{A}) = \max_i \lambda_i(\mathrm{A}^*\mathrm{A}) \tag{A.7}$$

$$||\mathrm{A}||_\infty = \max_i \sum_{j=1}^{n} |a_{ij}| \tag{A.8}$$

Additional norms (which are not induced norms) can be defined. We report only the Frobenius norm, which is defined as follows:

$$||\mathrm{A}||_F = \left(\sum_{j=1}^{n} \sum_{i=1}^{n} |a_{ij}|^2 \right)^{\frac{1}{2}} \tag{A.9}$$

Norm of a Scalar Signal

Given a signal $g(t) \in \mathbb{R}$ with $t \in \mathbb{R}$, the 1-norm, the 2-norm and the ∞-norm are defined as follows:

$$||g(t)||_1 = \int_{-\infty}^{+\infty} |g(t)| dt \tag{A.10}$$

$$\|g(t)\|_2 = \left(\int_{-\infty}^{+\infty} |g(t)|^2 dt \right)^{\frac{1}{2}} \tag{A.11}$$

$$\|g(t)\|_\infty = \sup_{t \in \mathbb{R}} |g(t)| \tag{A.12}$$

Norm of a Vector Signal

For a vector-valued signal $\mathbf{g}(t) = \begin{bmatrix} g_1(t) \\ g_2(t) \\ \vdots \\ g_n(t) \end{bmatrix}$, the 1-norm, the 2-norm and the ∞-norm are given by:

$$\|\mathbf{g}(t)\|_1 = \int_{-\infty}^{+\infty} \|\mathbf{g}(t)\|_2 dt \tag{A.13}$$

$$\|\mathbf{g}(t)\|_2 = \left(\int_{-\infty}^{+\infty} \|\mathbf{g}(t)\|_2^2 dt \right)^{\frac{1}{2}} \tag{A.14}$$

$$\|\mathbf{g}(t)\|_\infty = \sup_{t \in \mathbb{R}} \|\mathbf{g}(t)\|_2 \tag{A.15}$$

Norm of a Transfer Function Matrix

For transfer function matrices, usually only two norms are commonly used: the 2-norm and the ∞-norm. Given a transfer function matrix $\mathrm{G}(s)$, and indicated as $\mathrm{g}(t)$ the impulse response, i.e., $\mathrm{g}(t) = \mathcal{L}^{-1}(G(s))$, the 2-norm and the ∞-norm can be defined either in the frequency domain or in the time domain. In the frequency domain the norms are defined as:

$$\|\mathrm{G}(s)\|_2 = \left(\frac{1}{2\pi} \int_{-\infty}^{+\infty} \|\mathrm{G}(j\omega)\|_F^2 d\omega \right)^{\frac{1}{2}} \tag{A.16}$$

$$\|\mathrm{G}(s)\|_\infty = \sup_{\omega \in \mathbb{R}} \sigma_1(\mathrm{G}(j\omega)) \tag{A.17}$$

while in the time domain the two norms are defined by:

$$\|\mathrm{G}(s)\|_2 = \left(\frac{1}{2\pi}\int_{-\infty}^{+\infty} \|\mathrm{g}(t)\|_F^2 dt\right)^{\frac{1}{2}} \tag{A.18}$$

$$\|\mathrm{G}(s)\|_\infty = \sup_{u\neq 0} \frac{\|(\mathrm{G}u)(t)\|_2}{\|u(t)\|_2} \tag{A.19}$$

Appendix B. Algebraic Riccati Equations

Algebraic Riccati equations appear in many problems discussed in this text. In this section we summarize the main algebraic Riccati equations for continuous-time systems.

Algebraic Riccati Equations for Optimal Control and Optimal Observer

$$\mathrm{PA} + \mathrm{A}^T\mathrm{P} - \mathrm{PBR}^{-1}\mathrm{B}^T\mathrm{P} + \mathrm{Q} = 0 \tag{B.1}$$

$$\mathrm{A}\Pi + \Pi\mathrm{A}^T - \Pi\mathrm{C}^T\mathrm{M}_v^{-1}\mathrm{C}\Pi + \mathrm{M}_d = 0 \tag{B.2}$$

CARE and FARE for Closed Loop Balancing

$$\mathrm{A}^T\mathrm{P} + \mathrm{PA} - \mathrm{PBB}^T\mathrm{P} + \mathrm{C}^T\mathrm{C} = 0 \tag{B.3}$$

$$\mathrm{A}\Pi + \Pi\mathrm{A}^T - \Pi\mathrm{C}^T\mathrm{C}\Pi + \mathrm{BB}^T = 0 \tag{B.4}$$

CARE and FARE for Closed Loop Balancing for Not-strictly Proper Systems

$$\mathrm{A}^T\mathrm{P} + \mathrm{PA} - (\mathrm{PB} + \mathrm{C}^T\mathrm{D})(\mathrm{I} + \mathrm{D}^T\mathrm{D})^{-1}(\mathrm{B}^T\mathrm{P} + \mathrm{D}^T\mathrm{C}) + \mathrm{C}^T\mathrm{C} = 0 \tag{B.5}$$

$$\mathrm{A}\Pi + \Pi\mathrm{A}^T - (\Pi\mathrm{C}^T + \mathrm{BD}^T)(\mathrm{I} + \mathrm{DD}^T)^{-1}(\mathrm{C}\Pi + \mathrm{DB}^T) + \mathrm{BB}^T = 0 \tag{B.6}$$

Algebraic Riccati Equations for Positive-real Systems

$$\mathrm{PA} + \mathrm{A}^T\mathrm{P} = -(-\mathrm{PB} + \mathrm{C}^T)(\mathrm{D} + \mathrm{D}^T)^{-1}(-\mathrm{PB} + \mathrm{C}^T)^T \tag{B.7}$$

$$\Pi\mathrm{A}^T + \mathrm{A}\Pi = -(-\Pi\mathrm{C}^T + \mathrm{B})(\mathrm{D} + \mathrm{D}^T)^{-1}(-\Pi\mathrm{C}^T + \mathrm{B})^T \tag{B.8}$$

Algebraic Riccati Equations for Bounded-real Systems

$$\begin{aligned}&\mathrm{P}(\mathrm{A} + \mathrm{B}(\mathrm{I} - \mathrm{D}^T\mathrm{D})^{-1}\mathrm{D}^T\mathrm{C}) + (\mathrm{A}^T + \mathrm{C}^T\mathrm{D}(\mathrm{I} - \mathrm{D}^T\mathrm{D})^{-1}\mathrm{B}^T)\mathrm{P}+\\&+\mathrm{PB}(\mathrm{I} - \mathrm{D}^T\mathrm{D})^{-1}\mathrm{B}^T\mathrm{P} + \mathrm{C}^T\mathrm{D}(\mathrm{I} - \mathrm{D}^T\mathrm{D})^{-1}\mathrm{D}^T\mathrm{C} + \mathrm{C}^T\mathrm{C} = 0\end{aligned} \tag{B.9}$$

$$\begin{aligned}&\mathrm{P}(\mathrm{A}^T + \mathrm{C}^T(\mathrm{I} - \mathrm{DD}^T)^{-1}\mathrm{DB}^T) + (\mathrm{A} + \mathrm{BD}^T(\mathrm{I} - \mathrm{DD}^T)^{-1}\mathrm{C})\mathrm{P}+\\&+\mathrm{PC}^T(\mathrm{I} - \mathrm{DD}^T)^{-1}\mathrm{CP} + \mathrm{BD}^T(\mathrm{I} - \mathrm{DD}^T)^{-1}\mathrm{DB}^T + \mathrm{BB}^T = 0\end{aligned} \tag{B.10}$$

Algebraic Riccati Equations for Strictly Proper Bounded Real Systems

$$\mathrm{PA} + \mathrm{A}^T\mathrm{P} + \mathrm{PBB}^T\mathrm{P} + \mathrm{C}^T\mathrm{C} = 0 \tag{B.11}$$

$$\mathrm{PA}^T + \mathrm{AP} + \mathrm{PC}^T\mathrm{CP} + \mathrm{BB}^T = 0 \tag{B.12}$$

Algebraic Riccati Equations for H_∞ Control

$$\begin{aligned}&(\mathrm{A} - \mathrm{B}_2\tilde{\mathrm{D}}_{12}\mathrm{D}_{12}^T\mathrm{C}_1)\mathrm{X}_\infty + \mathrm{X}_\infty(\mathrm{A} - \mathrm{B}_2\tilde{\mathrm{D}}_{12}\mathrm{D}_{12}^T\mathrm{C}_1)^T+\\&+\mathrm{X}_\infty(\gamma^{-2}\mathrm{B}_1\mathrm{B}_1^T - \mathrm{B}_2\tilde{\mathrm{D}}_{12}\mathrm{B}_2^T)\mathrm{X}_\infty + \tilde{\mathrm{C}}_1^T\tilde{\mathrm{C}}_1 = 0\end{aligned} \tag{B.13}$$

$$\begin{aligned}&(\mathrm{A} - \mathrm{B}_2\tilde{\mathrm{D}}_{12}\mathrm{D}_{12}^T\mathrm{C}_1)^T\mathrm{X}_\infty + \mathrm{X}_\infty(\mathrm{A} - \mathrm{B}_2\tilde{\mathrm{D}}_{12}\mathrm{D}_{12}^T\mathrm{C}_1)+\\&+\mathrm{X}_\infty(\gamma^{-2}\mathrm{C}_1^T\mathrm{C}_1 - \mathrm{C}_2^T\tilde{\mathrm{D}}_{21}\mathrm{C}_2)\mathrm{X}_\infty + \tilde{\mathrm{B}}_1\tilde{\mathrm{B}}_1^T = 0\end{aligned} \tag{B.14}$$

with $\tilde{\mathrm{C}}_1 = (\mathrm{I} - \mathrm{D}_{12}\tilde{\mathrm{D}}_{12}\mathrm{D}_{12}^T)\mathrm{C}_1$, $\tilde{\mathrm{B}}_1 = \mathrm{B}_1(\mathrm{I} - \mathrm{D}_{21}^T\tilde{\mathrm{D}}_{21}\mathrm{D}_{21})$, $\tilde{\mathrm{D}}_{12} = (\mathrm{D}_{12}^T\mathrm{D}_{12})^{-1}$ and $\tilde{\mathrm{D}}_{21} = (\mathrm{D}_{21}\mathrm{D}_{21}^T)^{-1}$.

Cross Riccati Equations for Optimal Control and Optimal Observer (for Symmetrical Systems)

For symmetrical systems, cross-Riccati equations can be defined. As an example, we report those related to optimal control and optimal observer:

$$\mathrm{AP}^* + \mathrm{P}^*\mathrm{A} - \mathrm{P}^*\mathrm{BCP}^* + \mathrm{BC} = 0 \tag{B.15}$$

Appendix C. Invariance Under Frequency Transformations

In this appendix, we briefly summarize some important properties of systems subject to a frequency transformation that is loss-less. In particular, we show that for these systems the singular values, the characteristic values and the shape of the Nyquist plot remain the same of the original (i.e., not transformed) system. The proofs and technical details can be found in the references reported at the end of this appendix.

Frequency Transformations

Given a linear time-invariant system, either SISO or MIMO, with transfer function matrix $\mathrm{G}(s)$, a frequency transformation replaces s with a SISO transfer function $F(s)$. The frequency transformation is indicated as $s \leftarrow F(s)$ and the transformed system as $\tilde{\mathrm{G}}(s) = \mathrm{G}(F(s))$.

Frequency transformations are important in filter design: for instance, in the SISO case, if $G(s)$ is a prototype low-pass filter and $F(s)$ a loss-less system, then a multi-bandpass/bandstop high-order filter $G(F(s))$ can be obtained.

Invariance of Hankel Singular Values

The first result concerns the controllability and observability gramians of the system $\mathrm{G}(F(s))$. Let us indicate them as $\mathcal{W}_c$ and $\mathcal{W}_o$. Let W_c and W_o be the gramians of system $\mathrm{G}(s)$, and, finally, let $\mathrm{P_I}$ the positive definite matrix solution of equations (9.13) and (9.14) for the loss-less system $F(s)$. If $\mathrm{G}(s)$ is stable and in minimal form and $F(s)$ is a loss-less strictly proper transfer function, $\mathcal{W}_c$ and $\mathcal{W}_o$ are related to W_c and W_o by the following relationships:

$$\begin{aligned} \mathcal{W}_c &= \mathrm{W}_c \otimes \mathrm{P_I}^{-1} \\ \mathcal{W}_o &= \mathrm{W}_o \otimes \mathrm{P_I} \end{aligned} \tag{B.1}$$

From these relationships, it is immediate to derive that the singular values of G($F(s)$), here indicated as $\sigma_1^{GF}, \sigma_2^{GF}, \ldots, \sigma_{n_G n_F}^{GF}$ are equal to those of G(s) (indicated as $\sigma_1^G, \sigma_2^G, \ldots, \sigma_{n_G}^G$):

$$\sigma_i^{GF} = \sigma_j^G \tag{B.2}$$

with $j = 1, \ldots, n_G$ and $(j-1)n_F + 1 \leq i \leq jn_F$.

Invariance of Characteristic Values

The second result concerns the solution of the CARE and FARE for the system G($F(s)$). Let us indicate with $\mathcal{P}$ and $\mathbf{\Pi}$ the positive definite solutions of the CARE and FARE for the trasformed system G($F(s)$). In addition, let P_G and Π_G be the positive definite solutions of the CARE and FARE for system G(s). Then we have that, if G(s) is a controllable and observable system of order n_G and $F(s)$ a loss-less strictly proper transfer function of order n_F, for the system obtained by applying the substitution $s \leftarrow F(s)$, i.e., G($F(s)$), which is of order $n_G n_F$, the following relationships hold:

$$\begin{array}{l} \mathcal{P} = \mathrm{P}_G \otimes \mathrm{P}_\mathrm{I} \\ \mathbf{\Pi} = \Pi_G \otimes {\mathrm{P}_\mathrm{I}}^{-1} \end{array} \tag{B.3}$$

The direct consequence of this result is that, indicated with $\mu_1^G \geq \mu_2^G \geq \ldots \geq \mu_{n_G}^G$ the characteristic values of G(s) and with $\mu_1^{GF} \geq \mu_2^{GF} \geq \ldots \geq \mu_{n_G n_F}^{GF}$ those of G($F(s)$), then:

$$\mu_i^{GF} = \mu_j^G \tag{B.4}$$

with $j = 1, \ldots, n_G$ and $(j-1)n_F + 1 \leq i \leq jn_F$.

Invariance of the Shape of the Nyquist Plot

Also the shape of Nyquist plot is invariant to frequency transformation. The key property here is that the frequency transformation $s \leftarrow F(s)$ is an odd function and entirely spans the Nyquist path, i.e., for these functions it occurs that $F(j\omega) = j\bar{F}(\omega)$. This is a class more general than loss-less positive-real functions, but includes them. For simplicity here let us consider loss-less transformations and a SISO system $G(s)$.

Let $\mathcal{C}_s$ be the Nyquist path for a linear time-invariant continuous-time SISO system $G(s)$ in the complex plane and consider the Nyquist plot of $G(s)$, i.e., the closed oriented curve $\mathcal{C}_G$ obtained evaluating $G(s)$ for every $s \in \mathcal{C}_s$. The curve $\mathcal{C}_G$ acan be viewed as the image of a mapping $\gamma : (-\infty, +\infty) \rightarrow \mathbb{C}$

through $G(j\omega)$. This prompts for the definition of the *shape of the Nyquist plot* as the codomain Γ_G of γ. Clearly, the shape can be also viewed as the locus of points in the complex plane visited by $\mathcal{C}_G$.

For systems under frequency transformations we have the following result.

Given a SISO system $G(s)$ and a loss-less frequency transformation $s \leftarrow F(s)$, then the closed oriented curve $\mathcal{C}_{\tilde{G}}$ in the complex plane, representing the Nyquist plot of $\tilde{G} = G(F(s))$, has the same shape of the closed oriented curve $\mathcal{C}_G$, representing the Nyquist plot of $G(s)$. In addition, let m_G be the number of clockwise turnings of $\mathcal{C}_G$, then the curve $\mathcal{C}_{\tilde{G}}$ performs a number of clockwise turnings $m_{\tilde{G}}$ given by $m_{\tilde{G}} = -I_c m_G$, where I_c is the Cauchy index of $\overline{F}(\omega)$.

This result is very important to study the closed-loop stability of the transformed system by considering the characteristics of the original system $G(s)$ and of the frequency transformation $F(s)$.

Also notice that the result, here presented for a SISO system, also holds in the more general case of MIMO systems.

Positive-real Systems Under Loss-less Transformations

Other interesting results of loss-less frequency transformations apply when the original system G(s) is positive-real. In this case, it can be demonstrated that the transformed system G$(F(s))$, under a loss-less frequency transformation, is also positive-real.

In addition, one can consider the so-called stochastic balancing, which is obtained starting from the Riccati equation associated to the positive-real lemma:

$$\mathrm{PA} + \mathrm{A}^T\mathrm{P} + (-\mathrm{PB} + \mathrm{C}^T)(\mathrm{D} + \mathrm{D}^T)^{-1}(-\mathrm{PB} + \mathrm{C}^T)^T = 0 \qquad \text{(B.5)}$$

and its dual:

$$\Pi\mathrm{A}^T + \mathrm{A}\Pi + (-\Pi\mathrm{C}^T + \mathrm{B})(\mathrm{D} + \mathrm{D}^T)^{-1}(-\Pi\mathrm{C}^T + \mathrm{B})^T = 0 \qquad \text{(B.6)}$$

In the stochastic balanced form the two solutions are equal and diagonal. The elements of the diagonal are system invariants and relationships similar to (B.4) can be written for them (in fact, similarly to the characteristic values, they also derive from solutions of Riccati equations). The consequence of this is that, for positive-real transformed systems under a loss-less frequency transformation one can build a reduced order model starting from the stochastic balanced realization of the original system, truncating it and then, as a final step, using the frequency transformation.

References

S. Koshita, Y. Mizukami, T. Konno, M. Abe, M. Kawamata (2008). "Analysis of second-order modes of linear continuous-time systems under positive real transformations", *IEICE Trans. Fundam.* E91-A(2), 575–583.

A. Buscarino, L. Fortuna, M. Frasca, M. G. Xibilia (2016). "Invariance of characteristic values and L_∞ norm under lossless positive real transformations", *Journal of the Franklin Institute* 353, 2057–2073.

A. Buscarino, L. Fortuna, M. Frasca, M. G. Xibilia (2017). "Positive-real systems under lossless transformations: Invariants and reduced order models". *Journal of the Franklin Institute*, 354(11), 4273–4288.

A. Buscarino, L. Fortuna, M. Frasca (2019). "Nyquist plots under frequency transformations", *Systems & Control Letters*, 125, 16–21, doi.org/10.1016/j.sysconle.2019.01.004.

M. Bucolo, A. Buscarino, L. Fortuna, M. Frasca (2021). "Nyquist plots for MIMO systems under frequency transformations", *IEEE Control Letters*, 10.1109/LCSYS.2021.3053660.

Index

Printed in the United States
by Baker & Taylor Publisher Services